D0848966

europe

PLACE NAMES OF THE WORLD

HISTORICAL CONTEXT, MEANINGS AND CHANGES

JOHN EVERETT-HEATH

This work is dedicated to my wife, Judith, who has accompanied me round the world, making the most of the thick and the thin.

I dedicate this first volume to my elder daughter, Catharine.

europe

PLACE NAMES OF THE WORLD

HISTORICAL CONTEXT, MEANINGS AND CHANGES

JOHN EVERETT-HEATH

D
905
.E94
2000

© John Everett-Heath, 2000.

All rights reserved.
No part of this publication may be reproduced, stored in or introduced into a retrieval system, or transmitted, in any form or by any means (electronic, mechanical, photocopying, recording or otherwise) without prior written permission of the publisher unless in accordance with the provisions of the Copyright Designs and Patents Act 1988, or under the terms of any licence permitting limited copying issued by the Copyright Licensing Agency, 90 Tottenham Court Road, London W1P 9HE, UK.
Further, any person, company or firm found to be reproducing or assisting others in producing mailing lists, or machine-readable versions derived from this publication will be prosecuted for copyright infringement.

Published in the United Kingdom by

 MACMILLAN PRESS LTD, 2000
Brunel Road, Houndmills, Basingstoke, Hants RG21 6XS
London and Oxford
Companies and representatives throughout the world

British Library Cataloguing in Publication Data:
Everett-Heath, John
    Place names of the world Europe: historical context, meanings and changes
    1. Names, Geographical – Europe
    2. Names, Geographical – Europe – History
    I. Title
    914'.0014

Macmillan ISBN: 0-333-77759-X

Distributed by Macmillan Distribution Ltd
Brunel Road, Houndmills, Basingstoke, Hants RG21 6XS

Published in the United States and Canada by
ST. MARTIN'S PRESS, 2000
175 Fifth Avenue, New York, NY 10010

St. Martin's Press ISBN: 1-56159-300-1

A catalog record for this book is available from the Library of Congress

Typeset by The Florence Group, Stoodleigh, Devon

Printed and bound in Great Britain by Antony Rowe Ltd, Chippenham, Wiltshire

43376650

CONTENTS

CONTENTS

LIST OF MAPS

ACKNOWLEDGEMENTS

This volume would not have been possible without the help of many people who have given generously of their time and knowledge. It is a pleasure to acknowledge, in particular, my grateful thanks to Felicity Cave, Dragan Djukic, Ana Carla Duta, Olga Grahor, František Krs, Andrea Madden, Elizabeth and Hans Magnusson, Tony Makepeace-Warne, Boris Malakhov, Vasile Palcău, Claudia Pugh-Thomas, Vyacheslav Pronin, Antonio de Rienzo and Denis Shulakov who have provided information and advice, and read sections of the text. I should also like to thank the staff of the Ministry of Defence library in Whitehall for their general assistance and in finding answers to specific questions. My son, Tom, has been particularly helpful, even while editing his own book on Central Asia and holding down a full-time job.

Stephen Ramsey should be congratulated for turning my very rough drafts into such excellent maps and my editor, Diana Levy, who performed wonders in preparing the book for publication. I owe a special debt of thanks to Andrea Hartill, my publisher, for her constant encouragement and advice throughout.

Lastly, I must thank my wife who has not just tolerated the diversion of my attention, but done everything possible to ease the gestation of this work.

INTRODUCTION

Place names are a window onto the history and characteristics of a country. They reflect the migrations of peoples, their religious and cultural traditions, local languages, conquests, fortifications long since disappeared, topography and even industrial development. Thus history and geography are inextricably linked. The *status quo* never lasts long. States are brought into existence while others disappear; some are absorbed by their neighbours or amalgamate with them. Boundaries change and names change. In medieval Europe territory was divided into fluid spheres of influence rather than defined by formal frontiers. It is different now; in 1907, Lord Curzon, a Viceroy of India and British Foreign Secretary, said: "Frontiers are indeed the razor's edge on which hang suspended the modern issues of war and peace, of life and death to nations."

The ways in which a place gets its name are many and varied, but there is no inhabited place, however obscure, without a name. Many have a meaning, but it may not be known now or it may be disputed. Different theories are put forward and often there is little likelihood of the matter being resolved.

A place may take its name from the person who discovered it, for example, *Tasmania*, or from an individual to be honoured, such as *Gagarin*, *Kimberley* and *Washington*. A country may be named after the people who inhabited it (not necessarily its first inhabitants), such as *France* or *Slovakia*. So-called folk names are quite common. *Amiens* is named after the *Ambiani* tribe, *Alsace* (Elsazzun) means 'those living on the outside' (that is, west of the Rhine), and *Bohemia* is named after the *Boii*. Many towns have been named after saints, for example *St Albans*, particularly if they were founded on the saint's feast day.

A town or even a country may owe its name to a local natural resource, silver in the case of *Srebrenica* in Bosnia, or a natural feature such as the Turkish town of *Trabzon* which takes its name from the Greek word *trapeza*, table, a reference to the flat-topped mountain close by. Quite often places are named after the river on which they lie and are therefore based on the local and contemporary word for water. *Rijeka* simply means river. Some names are descriptive like *Chad*, meaning lake, or *Japan*, land of the rising sun; others are often somewhat unimaginative: *Naples* (Neapolis) and *Novgorod* are nothing more than 'New Town' while *Nowa Huta* in Poland simply means 'New Steelworks', a town built to house the workers at the nearby steel mills.

Countries that have gained independence from a colonial power have often retained names originally transferred from the home country by their former rulers. This is especially prevalent in the United States and the former Commonwealth dominions of Australia, Canada, New Zealand and South Africa. In some cases, however, names were changed: *Salisbury* in *Southern Rhodesia* became *Harare* in *Zimbabwe*. New names can be contrived: *Pakistan* comes from the initials of *P*unjab, *A*fghanistan and *K*ashmir with the Persian word *stan*, land, added.

For new governments coming to power and wishing to make an impact, or when they acquire new territory by conquest or peace treaty, the temptation to change names can be difficult to resist. The desire to exorcise the past can be very strong and changing names can be a method of underpinning the new 'legitimacy'. The clearest example of this was the Soviet Union's wish, almost mania, to eradicate the imperial era while honouring the heroes of the Bolshevik Revolution, only to be followed by the Russian impulse to rid itself of these trappings of its communist past after the dissolution of the Soviet Union in 1991.

Language is very often the key to finding the origin of a name. Some require little more than a straight-forward translation. *Brod* in Croat can mean a ford so *Slavonski Brod* means 'a crossing place (over the River Sava) into Slavonia'. This is a name by association and is a widespread practice: there is no shortage of names that include bridge or ford in many languages.

Historical developments and individual events sometimes provide an explanation for the origin of a name. *Ukraine* derives its name from the Russian *okrainy*, meaning the border lands or marches, denoting the time when they were the frontier lands of medieval Russia and the Polish-Lithuanian Commonwealth. *Christmas Island* received its name when it was discovered on Christmas Day 1643.

Where foreign invaders have displaced their predecessors they have adopted existing names, modified others to suit their own language and devised new ones. Original names may be corrupted because newcomers could not pronounce them correctly. In Europe names given by the Greeks and Romans abound, and can often be seen in modified form: *Mérida* in Spain was developed as a place for Roman veterans (*emereti*) to retire. Dedicated to the Emperor Augustus, it was called *Augusta Emerita*, becoming in due course *Mérida*. England, for example, is rich in place names which have Roman, Anglo-Saxon, Scandinavian and French associations, all a legacy of invasion. It is not unusual for medieval names to be shortened: the Domesday Book's *Glaestingeberia* has become *Glastonbury*. The Roman *Colonia Claudia Ara Agrippinensis* eventually became known simply as the Colony or *Cologne*. Names with Celtic roots can be found throughout Europe and others with Germanic and Slavonic roots are more widespread than might be expected; around the Mediterranean are to be found names with Arabic roots.

Place name changes make for confusion, particularly if the change is not widely publicized. The names of Napoleon's battles are well known, but today these places have different names: *Austerlitz* is *Slavkov*, *Eylau* is *Bagrationovsk* and *Friedland* is *Pravdinsk*. In a similar fashion, treaties are usually named after the place in which they are signed and some of these places have different names: for example, *Pressburg* is now *Bratislava*, the capital of Slovakia.

It is easy to make incorrect assumptions. *Cambridge* may be thought to indicate a bridge over the River Cam but it does not. The Old English name was *Grantanbrycg*, bridge over the Grante. Furthermore, names may be derived from myth and legend, but not always convincingly. A well-known example is *Rome*, said to be derived from Romulus, the legendary founder with Remus of the city. Both, however, were mythical figures. The well-known Greek explanation of the origin of the name Rome, while attractive, is nothing more than an invention.

Place names may be derived from natural resources, but they also give their names to manufactured goods. The Czech plastic explosive known as Semtex, so coveted by terrorists, derives its name from the village where it is made, *Semtin*. Croatia, in Croat *Hrvatska*, has given us cravat, a scarf worn by Croat mercenaries in French service at Versailles. Some items of food are also qualified by the name of a place: Parma ham and Parmesan cheese, Budweiser beer, and champagne, the sparkling wine named after the region from which it comes and which itself is derived from the Latin *campania*, land of plains.

This volume is devoted to Europe, although what comprises this continent has never been universally agreed. There is no obvious barrier between it and Asia and therefore, for the sake of convenience only, limits have been imposed. There is a signpost on the south side of the Bosporus in Istanbul proclaiming the start of Asia. Therefore, this volume excludes Turkey despite the fact that there is a significant part of that country north of the Bosporus in Europe. To the east, the Ural Mountains might be considered a barrier of sorts, but there are Russians on both sides and there is no merit in splitting a country in two; thus the whole of Russia is included here. The Caucasus Mountains in the south provide a more substantial barrier, although by no means a distinct one. Nevertheless, for the purposes of this volume, they are taken to be the south-eastern limits of Europe and Armenia, Azerbaijan and Georgia are excluded.

As a result, this volume is confined to 45 states in Europe. Because of their small size Andorra, Gibraltar, Liechtenstein, Malta, Monaco, San Marino and Vatican City have received abbreviated treatment. The work endeavours to cover those regions (republics, provinces, districts, counties), islands, cities and towns of particular importance or interest; villages and ruins, mountains and rivers are not included. In many cases, regions take the name of their capital city and only one entry appears. To be included the origin of a place name must be documented or it must have been changed. Most places falling into these categories choose themselves, but the selection of some, unavoidably, has been subjective. A dearth of information has excluded some deserving towns of special interest.

There is a further minor complication, in that a local name may be quite different to the name known to those who speak English. English speakers know *Montenegro*, but to Montenegrins their republic is *Crna Gora*. This book, therefore, includes both the so-called 'English' name, the local name where different, and former names; in some cases, a former name is still used in neighbouring countries or, indeed, a variant. Where current and former names are very similar in sound or spelling, for example *Trenčin* in Czech/*Trentschinn* in German, they do not qualify as a change; alternate names in other languages, for example *Londres* for London, are not given. Excluded also are names whose meaning is obvious.

Personal names have generally been anglicized – for example, Alexander for the Russian Aleksandr, Christian for the Scandinavian Kristian and Charles for Karl. Where names are more widely known by their local name, for example, King Juan Carlos of Spain or Mikhail Gorbachev, then they have been used.

To place the names in some sort of context a country 'profile' precedes each list. It is not a history, but attempts to bring forward those events which have had an impact on the formation of a country and its geographical extent, foreign relations, and those influences that may have played a part in shaping place names. Internal affairs are not mentioned unless relevant. There is an old saying: 'A country is a group of people united by a common misconception about their ancestry and a common dislike of their neighbours'. Certainly, historical events are interpreted in different ways, usually from a nationalist point of view. To the Russians, for example, the Second World War is of less consequence than their Great Patriotic War against Nazi Germany which only began in 1941.

There is today a discernible trend towards people identifying, not with their state, but with their nation – their cultural homeland with its distinctive ethnic origins and language. Few are the states whose borders coincide with ethnic boundaries to form a genuine nation-state, a state whose population comprises a single ethnic culture. Not only are there many states which include a number of nations (the United Kingdom, for example, with the English, Irish, Welsh and Scots), but there are some nations which spread over more than one state (25 million Russians live outside Russia, mainly in the republics of the former Soviet Union which are now sovereign states in their own right). No more does language define a nation: the Swiss speak French, German, Italian and Romansch, while both Austrians and Germans speak German. A state can have more than one official language. Austro-Hungary had eight: German, Hungarian, Croat, Czech, Italian, Polish, Romanian and Slovene.

If ethnic minorities become assertive in their demands for 'human rights' they may be seen as potential security threats; counter-action may then increase the risk of fragmentation. Since 1991 Czechoslovakia, the Soviet Union and Yugoslavia have all disintegrated, and separatist movements are strong in other countries. Sadly, it is often the case that when intermixed ethnic groups are separated into their own nations tranquillity is not achieved but rather an intensification of the hatred and loathing between them. States are generally adamant that they will not allow minorities to break away; hence the violence in Chechnya (Russia), Kosovo (Yugoslavia) and Transdniester (Moldova) within Europe alone.

After the unification of Italy and Germany in the second half of the 19th century, Europe began to break up. Compared to the 38 major countries covered in this book, Europe consisted of only 21 sovereign states in 1900 and 15 in 1800. In 1900, of the 21 all but France and Switzerland were monarchies. By the end of the century only eight monarchies remained. This century has witnessed the disintegration of both the Russian and Soviet empires into eight separate states in Europe today and the Austro-Hungarian empire into seven. From the reduction and collapse of the Ottoman Empire in the Balkans during the 19th and 20th centuries six new states emerged. One, Montenegro, has 'disappeared', having been subsumed into Yugoslavia.

The fall of the Berlin Wall in November 1989 and the subsequent collapse of Communism had profound effects throughout the world. The end of the Cold War and the disappearance of superpower rivalry have not led to entirely positive developments, the eruption of ethnic conflicts and the growth of international terrorism among them. In many ways the world is a more unstable and dangerous place.

More change in Europe is inevitable, with both centripetal and centrifugal forces strong. Greater integration is sought through the European Union; some people, perhaps disillusioned with the concept of the nation-state and the destructive power of nationalism, even seek a United States of Europe. On the other hand, the war in Kosovo in 1999 demonstrated a divisive tendency which could be echoed in other parts of Europe in this new century. However, there is no place in this book to discuss current affairs or to make predictions for the future.

The statistics box at the end of each country profile does not claim to be exhaustive. There are any number of yearbooks that give full statistics. The population figures, estimated for 2000, are taken from the Statesman's Yearbook 2000. The date given for independence is the most recent, some countries having lost and regained their independence; the Baltic states, however, insist that their current independence dates from the end of the First World War and not from 1991.

English-speaking Name	Local Name	Former Names	Notes
Europe			Europe is said to be named after Europa, a Phoenician princess of Greek mythology who was carried off by Zeus, the supreme ruler of the Greek gods. It is also said to mean 'Mainland', that is the vast territory to the north of ancient Greek horizons. It might be derived from the Phoenician *ereb*, west or land of the setting sun.
Andorra			Principality of Andorra (Principat d'Andorra). Independent since 803 when Charlemagne drove the Muslims out of the area. Because of a quarrel about ownership between the Spanish bishops of Urgel and the French counts of Urgel, it was agreed (1278) that Andorra should be governed jointly. French rights passed to the head of state (1572). Name derived from the local word *andurrial*, shrub-covered land.
Andorra-la-Vella			Means 'Andorra the Old'. Capital.
Balkan Peninsula			Comprises the countries of south-eastern Europe (Albania, Bosnia-Hercegovina, Bulgaria, Croatia, Greece, Macedonia, Moldova, Romania, Slovenia and Yugoslavia.) In use since the early 19th century, the name means 'Mountains' from the Turkish *balkan* and was coined to describe those lands that had been under the direct control of the Ottoman Empire since the Treaty of Karlowitz (now Sremski Karlovci, Yugoslavia) (1699). Apart from the break-up of Yugoslavia (1991). today's national frontiers derive from the peace settlement after the First World War. Gave its name to 'balkanization', which is taken to mean the disintegration of an area into a number of smaller mutually hostile states.

Baltic States

Comprises Estonia, Latvia and Lithuania, an area which may have taken its name from the *Mare Balticum*, the Baltic Sea, so named by the Romans after an island (which never existed) which they called Baltica. From the middle of the 19th century the people generically became known as Balts, although the Estonians are not Balts and speak an entirely different language from the Finno-Ugric group. The name, however, may be derived from *bælt*, a Danish word used to describe some of the narrow passages, or belts, between the Danish islands. The Russian name for the area is Pribaltika, meaning 'adjacent to the Baltic Sea'.

Benelux

A collective name for Belgium, the Netherlands and Luxembourg. Also called the Low Countries because significant amounts of land are either just below or just above sea level. The Benelux customs union came into effect (1948) and this was followed by the Treaty of the Benelux Economic Union (signed 1958, operative 1960).

Gibraltar

Named after the Muslim commander Tariq ibn Ziyad who captured the 'Rock' (711). Corrupted from the Arabic *Jebel Tariq*, Mount Tariq. Spain captured it from the Moors (1462). Captured by British troops (1704) during the War of the Spanish Succession and formally ceded to Britain by the Treaty of Utrecht (1713), becoming a crown colony (1830).

Iberian Peninsula

Comprises Portugal and Spain, the name being derived from the Iberian tribe who took their name from the Iberus, the ancient name for the River Ebro.

Liechtenstein

Principality of Liechtenstein (Fürstentum Liechtenstein). Named after the family Liechtenstein, Light stone, who came from the castle of Liechtenstein near Vienna. Present boundaries drawn (1434), the territory belonging to the

Vaduz Count of Vaduz. The principality was created within the Holy Roman Empire (1719) from the two independent counties of Schellenberg and Vaduz. Part of the Rhine Confederation (1805–15) and the German Confederation (1815–66). Independent since 1866.

The name has evolved from Valdutsch: the Latin *vallis*, valley, and the Old German *Dutsch*, German. It is situated in the Rhine Valley. Capital.

Malta Melita Republic of Malta (Repubblika ta' Malta) since 1974. Consists of three inhabited and two uninhabited islands. Means 'refuge' in Phoenician. Occupied by Phoenicians, Carthaginians and Romans. Became part of the Eastern Roman Empire (395). Captured by the Arabs (870) and Normans (1091). Ceded (1530–1798) to the Knights of St John of Jerusalem (later known as the Knights of Malta). Invaded by Napoleon (1798) who ejected the Knights. Ceded to Britain at the Treaty of Paris (1814). Independent (1964).

Valletta Named after Jean Parisot de la Valette (1494–1568), Grand Master of the Knights of St John, who built the town after an Ottoman siege (1565). Capital (1570).

Monaco Principality of Monaco (Principauté de Monaco), a sovereign state within French territory. Possibly derived from a Phoenician word signifying a place to rest; or it may come from *monegu*, rock. Ruled by the Grimaldis since 1297. Annexed to France (1793–1814), protectorate of Sardinia (1815–61). Present borders from 1848 when Menton and Roquebrune were lost to France, thus reducing the principality's area to 1.9 sq km. A Franco-Monegasque treaty (1861) confirmed the loss but restored Monaco's independence. Became a constitutional monarchy (1911).

Monte Carlo

Means 'Mount Charles', so named (1866) by and after Prince Charles III (1818–1889).

San Marino

Republic of San Marino (Repubblica di San Marino); also known as the Most Serene Republic of San Marino. Named after a Christian stonemason and later saint, Marinus, who is alleged to have fled Dalmatia to escape persecution by the Emperor Diocletian (245–316) and who founded a hermitage on Monte Titano. Although surrounded by Italian territory, San Marino claims to be the world's oldest republic (301). However, the first evidence of independence dates only from 885 (the Montefeltro Decree) since when sovereignty has been maintained. Independence recognized by the Congress of Vienna (1815).

Scandinavia

Generally agreed to comprise Denmark, Finland, Iceland, Norway and Sweden, although the peninsula only includes Norway and Sweden. Derived from the Roman name Scandia.

Vatican City

Vatican City State (Stato della Città del Vaticano). Name derived from the Latin *mons vaticinia*, hill of prophecies. The smallest state in the world (0.4 sq km/0.2 sq miles). Within the city of Rome. The Italian government recognized Vatican City's independence and papal sovereignty in the Lateran Treaty (1929). This also delineated the territorial extent of the temporal power of the Holy See.

The Roman Empire 350

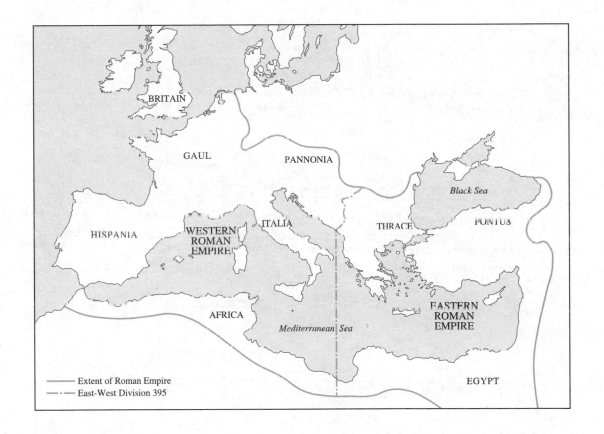

BRITAIN

GAUL

PANNONIA

Black Sea

HISPANIA

WESTERN
ROMAN
EMPIRE

ITALIA

THRACE

PONTUS

AFRICA

Mediterranean Sea

EASTERN
ROMAN
EMPIRE

EGYPT

—— Extent of Roman Empire
—·—·— East-West Division 395

The Frankish Empire 843

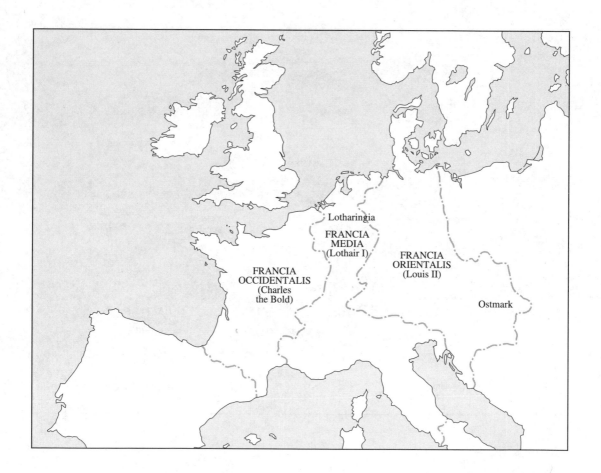

Lotharingia

FRANCIA
MEDIA
(Lothair I)

FRANCIA
ORIENTALIS
(Louis II)

FRANCIA
OCCIDENTALIS
(Charles
the Bold)

Ostmark

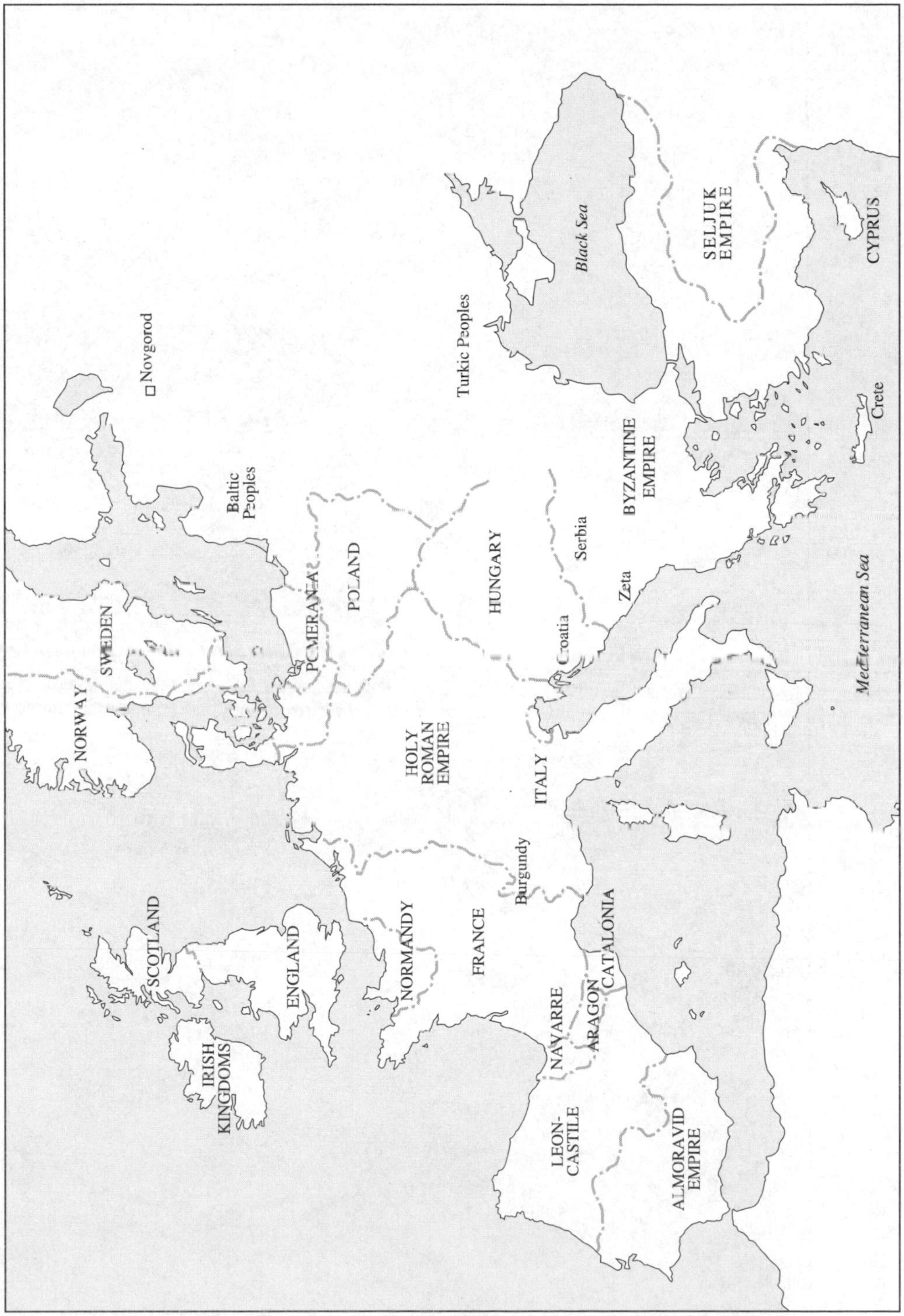

Europe 1000

Novgorod

Baltic Peoples

Turkic Peoples

Black Sea

SELJUK EMPIRE

CYPRUS

BYZANTINE EMPIRE

Crete

SWEDEN

NORWAY

POMERANIA

POLAND

HUNGARY

Serbia

Zeta

Croatia

HOLY ROMAN EMPIRE

ITALY

Mediterranean Sea

SCOTLAND

ENGLAND

NORMANDY

FRANCE

Burgundy

CATALONIA

ARAGON

NAVARRE

IRISH KINGDOMS

LEON-CASTILE

ALMORAVID EMPIRE

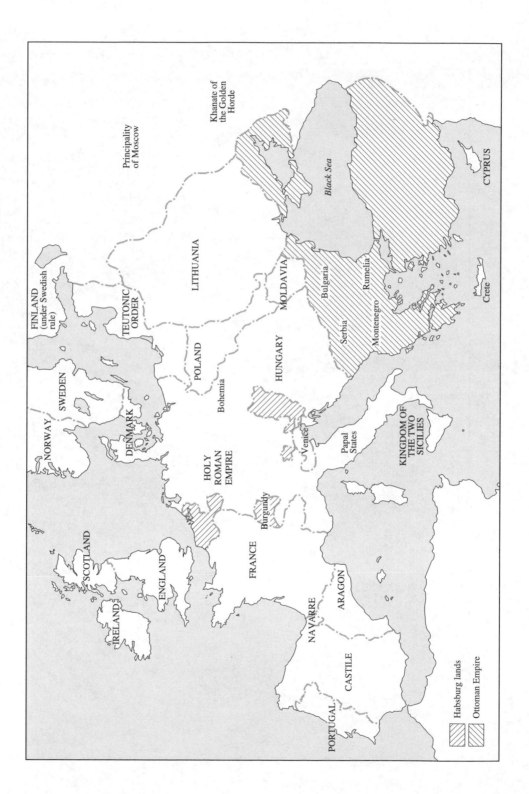

Europe 1500

Principality of Moscow

Khanate of the Golden Horde

Black Sea

CYPRUS

FINLAND (under Swedish rule)

TEUTONIC ORDER

LITHUANIA

MOLDAVIA

Bulgaria

Rumelia

Serbia

Montenegro

Crete

SWEDEN

POLAND

HUNGARY

NORWAY

DENMARK

Bohemia

Venice

Papal States

KINGDOM OF THE TWO SICILIES

HOLY ROMAN EMPIRE

Burgundy

SCOTLAND

ENGLAND

FRANCE

IRELAND

NAVARRE

ARAGON

PORTUGAL

CASTILE

Habsburg lands

Ottoman Empire

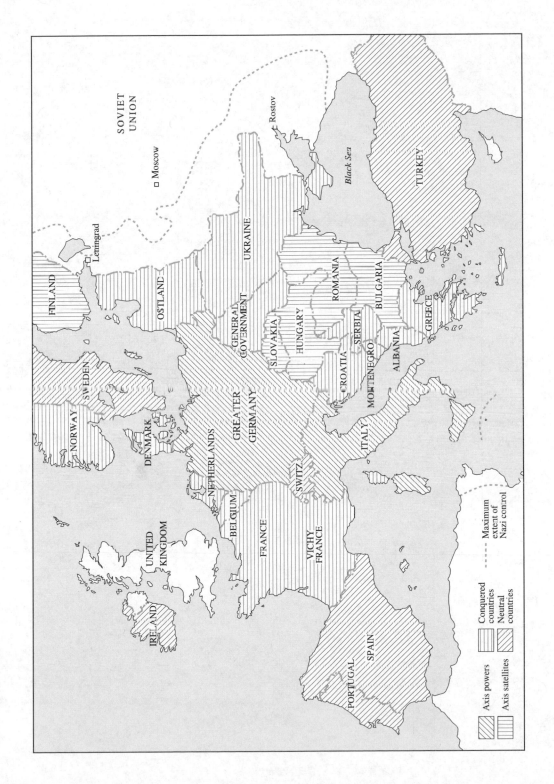

Hitler's Europe 1942

SOVIET UNION

□ Moscow

Rostov

Black Sea

TURKEY

Leningrad

FINLAND

OSTLAND

UKRAINE

ROMANIA

BULGARIA

GREECE

SWEDEN

GENERAL GOVERNMENT

SLOVAKIA

HUNGARY

SERBIA

CROATIA

ALBANIA

MONTENEGRO

NORWAY

DENMARK

NETHERLANDS

GREATER GERMANY

BELGIUM

FRANCE

SWITZ.

VICHY FRANCE

ITALY

UNITED KINGDOM

IRELAND

PORTUGAL

SPAIN

Maximum extent of Nazi control

Conquered countries

Neutral countries

Axis powers

Axis satellites

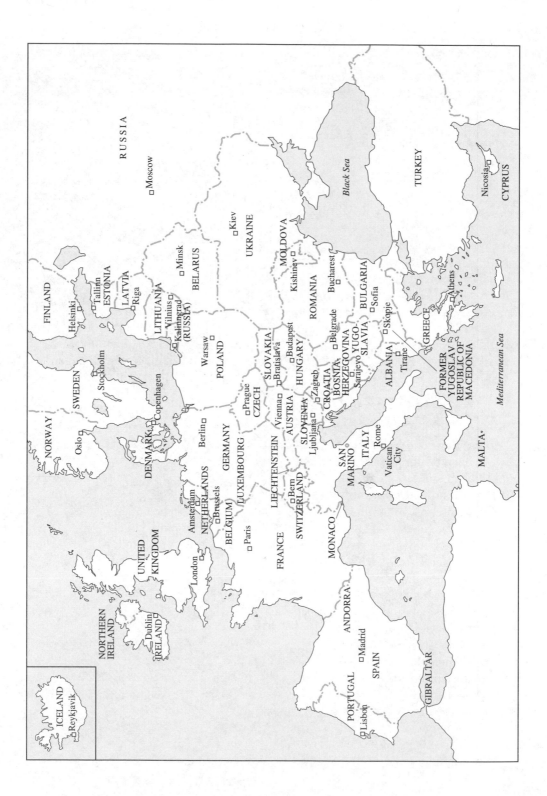

Europe 2000

ALBANIA

By 168 BC the Romans had subdued the southern Illyrians and Greeks living in modern-day Albania and it had become the Roman province of Illyricum. When the Empire was divided in 395 AD Illyria found itself a part of the East Roman (Byzantine) Empire. From then until the seventh century it was submerged under waves of barbarian invaders: Visigoths and Ostrogoths, Huns and Slavs. Beginning in the ninth century the country became a magnet for Normans, Angevins, Venetians and Serbs; in 1096 the First Crusade devastated the country.

A century later, with the death of the Byzantine Emperor Comnenus, the Albanians were able to establish the Principality of Arbania (or Arbanon), the first Albanian state with its capital at Kruja. It comprised nearly half of today's northern Albania, but it did not include the towns on the plain. In the 1340s the Serbs invaded and occupied the country. However, the death of the Serb Emperor, Stephen Dušan, in 1355 was the signal for revolt; within a few years the Serbs had been evicted.

The Ottoman invasion of the Balkans in the 1370s brought the Turks to the frontier of Albania in 1384. The decisive Battle of Savra the following year forced the Albanian lords to become vassals of the sultan. Byzantine rule was brought to an end. By 1430 the Turks were in control of most of the country; opposition, however, began to grow. In 1443 Gjergj Skanderbeg, lauded today as Albania's national hero, assumed leadership of the spreading resistance movement which lasted until his death in 1468. Only ten years later were the Turks able to impose control; and in the next century they began to implement a policy of Islamisation.

In the 19th century the Turks divided the country into seven *sandžaks* (military district) which formed the *vilayet* (province) of Rumelia. Despite various uprisings and their inability to subdue completely the mountain regions, they were able to maintain their rule until the end of 1912.

In an attempt to unite the four Albanian *vilayets* of Shkodër, Monastir (now in Macedonia), Kosovo (now Serbia) and Janina (now Ioánnina, Greece) in a single large province, the Albanian League of Prizren was founded in 1878 following the rebellions in Bosnia-Hercegovina in 1875 and in Bulgaria in 1876. A large Ottoman army advanced on, and occupied, Prizren in 1881, forcing the disbandment of the League. Nevertheless, it had succeeded in putting Albania on the international map and focusing Albanian aspirations. These alarmed its neighbours and when the First Balkan War (q.v.Macedonia) broke out in October 1912 they invaded Albania. Fearing that their lands would be incorporated into Greece, Montenegro and Serbia at the end of the war, the Albanians declared their independence a month later. Although this was recognized in principle, Albania was made a protectorate of the Great Powers; with Kosovo given to Serbia and Chamouria to Greece, about half the Albanian population was actually living outside the country.

During the First World War Albania was occupied by Austro-Hungarian, French, Greek, Italian, Montenegrin and Serb troops. Only when the Italians, the last to leave, departed in 1920 was Albania's independence confirmed internationally to the dismay of Serbia, which wanted a port on the Adriatic, and Greece, because of the Greek community in the south.

In 1924 Ahmet Beg Zogu, a warlord from the north, overthrew the government with Yugoslav help and assumed power. A year later Albania was declared a republic with Zogu as president; in 1928 he proclaimed himself King Zog I. During the 1930s Italian political and economic influence in Albania grew. However, in April 1939, Mussolini, increasingly irritated by Zog's independence, ordered an invasion. Zog was forced into exile and Albania and Italy were united; Kosovo and Chamouria were reintegrated. When Italy capitulated in September 1943 German forces occupied the country. Harried by the communist-dominated National Liberation Front (NLF), they withdrew in November 1944. The NLF took power the same month.

The People's Republic of Albania was proclaimed in January 1946 with Enver Hoxha, leader of the Albanian Communist Party since its foundation in 1941, as president. Hoxha's rule was characterized by military and economic alliances with Yugoslavia (1944–48), the USSR (1948–61), and China (1962–78). Thereafter, Albania withdrew into itself, making a virtue out of isolationism. Hoxha died in 1985 and gradually Albania emerged from its self-imposed purdah. Communist rule finally ended in April 1992.

As a result of the fighting in Kosovo (q.v.Yugoslavia) Albanian refugees flooded into Albania during 1998 and 1999. Strained relations with Yugoslavia, a lack of political stability and an ineffective administration mean that Albania will be dependent on the West for its security for some years to come.

Area:	28 748 sq km (11 100 sq miles)
Population:	3.49 million; Albanians, Greeks
Languages:	Albanian, Greek
Religions:	Muslim, Albanian Orthodox, Roman Catholic
Capital:	Tirana
Administrative Districts:	26 districts
Date of Independence:	28 November 1912
Neighbouring Countries:	Yugoslavia, Macedonia, Greece

English-speaking Name	Local Name	Former Names	Notes
ALBANIA	Shqipëria	Arbania/Arbanon Epirus	The Republic of Albania (Republika e Shqipërisë) since 1991. Previously the People's Socialist Republic of Albania (1976), the People's Republic of Albania (1946) and the Kingdom of Albania (1928). Shqipëria is generally taken to mean 'Land of the Eagles', a name gradually adopted (16th and 17th centuries) to replace Arbania/Albania which took its name from the Albanoi (the Byzantine Greek name; in Latin, Arbanenses) tribe which in turn took its name from the Indo-European word *alb*, mountain (cf. Alps). Epirus was an ancient country formed from modern southern Albania and north-western Greece. It existed as a Byzantine principality (1204–1337). Today it is an administrative region of Greece, much of northern Epirus having been united with Greece (1913).
Bajram Curri			Named after an Albanian nationalist who, when surrounded by enemy forces during the struggles of the 1920s, committed suicide in the mountains here.
Berat		Antipatria, Albanorum, Pulcheriopolis, Beligrad	The Byzantines named the city, *polis* in Greek, after Pulcheria, sister of the Eastern Roman Emperor Theodosius II (401–50, r.408–50). The present name is derived from Beligrad, 'White City', the name adopted by the Serbs after the city had been rebuilt (13th century).
Durrës		Epidamnus, Dyrrhachium, Durazzo	May mean 'Dangerous cliffs'. Capital (1013–20).
Elbasan		Skampis, Albanopolis	Albanopolis means the 'City of the Albanoi'. However, the Turks built a fortress here (1466) and the present name means 'The Fortress' from the Turkish *El Basan*.

Gjirokastër	Argyrókastron, Ergeri	Named after the Greek Princess Argyro who threw herself from the tower of the citadel (*kastron*), built by the Argyres (15th century), rather than surrender to a besieging Turkish army. It may, however, have been named simply after the Argyres tribe.
Korçë	Pelium, Koritsa	Derived from the Slav word *gorica*, small town. Occupied by the Greeks (1912), by the French (1916) and finally awarded to Albania (1920).
Krujë (Kruj)		Means 'Spring' (water). Skanderbeg's stronghold for 25 years. The siege of Krujë (1450) by the Ottoman Sultan Murad II (1404–51, r.1421–44, 46–51) was lifted by Skanderbeg (1405–68), which made him Albania's national hero and a hero throughout Europe.
Kuçovë	Qyteti-Stalin	Previously called 'City of Stalin' at a time after the Second World War (1939–45) when relations between the two countries were friendly.
Sarandë	Onchesmos	Derived from the Greek *saranta*, 40, from Agii Saranta, Forty Saints, the name of the Byzantine monastery which was dedicated to 40 saints.
Shkodër	Scutari	Also known as Shkodra or Skadar. Capital of the Illyrian King Gentius who surrendered to the Romans (168 BC). Capital (1913). The Italian Scutari comes from the Latin *scutarii*, shield-makers. Scutari was also one of the names for the present Üskûdar, a suburb of Istanbul (Turkey) on the Asian side.
Vlora (Vlorë)	Aulôn, Avlona, Valona	Situated in a bay, it means 'Hollow between the hills'. The National Convention, led by Ismail Qemal, proclaimed independence here (28 Nov 1912).

AUSTRIA

Invading the eastern Alps around 400 BC, the Celts founded the kingdom of Noricum. Some 200 years later the Romans arrived and by 15 BC they had occupied the area south of the Danube. They remained there until the fifth century when they were driven out by Germanic tribes, principally the Bavarians, who thereafter contested the area with the Slavs and their Avar masters. The Frankish Emperor Charlemagne became embroiled at the end of the eighth century when he defeated the Bavarians and the Avars and established a frontier district or *mark,* march, on this eastern border of his empire as a defence against potential invaders. The area became Christianized and ethnically German, but it fell under the control of Great Moravia (q.v.The Czech Republic) and then of the Magyars when they destroyed Great Moravia and defeated a Bavarian army near Pressburg (now Bratislava) in 907.

The distinct political entity that was to become Austria emerged in 976 when Leopold of Babenberg received the land as an imperial fief from the Holy Roman Emperor Otto II, whose father had loosened the Magyar grip. The eastern frontier was extended to Vienna. New marches were created in Styria and Carniola. Austria was made a duchy in 1156 and in 1192 the Babenbergs gained new lands with the inheritance of the Duchy of Styria. When the dynasty died out in 1246, the Austrian duchy was offered to a Bohemian prince, Přemysl Otakar II[1]. He proceeded to acquire other Alpine duchies to the south – Styria in 1260 and Carniola, Carinthia and Istria in 1269 – and set his eyes on the imperial crown. Instead, Count Rudolph IV of Habsburg[2] was chosen by the German electors in 1273 to be King Rudolph I of Germany and Holy Roman Emperor. He set about the overthrow of Otakar by claiming the duchies as imperial fiefs. Rudolph invaded Austria in 1276 and Otakar was killed in battle two years later on the Marchfeld. Rudolph now proceeded to take control of the former Babenberg lands for the Habsburgs, thus initiating the link between them and Austria.

Emperor Frederick III (1440–93) was the first Habsburg to use the motto *Austriae Est Imperare Urbi Universo* – Austria Shall Rule the World. When King Louis II was killed and the Hungarian army crushed by the Turks at the Battle of Mohács in 1526, the empty thrones of Bohemia and Hungary produced a dangerous vacuum in East Central Europe. The Habsburgs rushed to fill it, Ferdinand I completing the union of the Austrian lands with those of Bohemia (q.v.The Czech Republic) and Hungary when he became king of both. Thus was consolidated the Habsburg Empire, an empire that was to last until 1918. By marriage (Frederick III's grandson married the daughter of the Spanish monarchs Ferdinand and Isabella) more territory was acquired to the extent that the Habsburg lands became too large to govern and the dynasty split into the Austrian and Spanish branches in 1556. Ferdinand I retained the eastern lands, including Austria, and in 1558 received the imperial title.

Anticipating the extinction of the Spanish branch of the Habsburgs, Emperor Charles VI promulgated the Pragmatic Sanction in 1713, whereby he declared that the Habsburg lands were indivisible and that, should the male line die out, female members of the House should be eligible to succeed as empress. The Sanction was generally acknowledged throughout Europe and when Charles died in 1740 he was succeeded by his daughter Maria Theresa. Within two months she was challenged by Frederick II (later the Great) of Prussia who invaded Silesia and defeated the Habsburg army. This encouraged other countries to join him in the hope of weakening Austria. The War of the Austrian Succession lasted until 1748 and resulted in the loss of Silesia and the emergence of Prussia as the strongest rival to the Habsburgs in Germany. But Maria Theresa proved her mettle, demonstrating that she presided over a cohesive territorial unit. Nevertheless, an attempt to win back Silesia during the Seven Years' War (1756–63) failed. During the next three decades, however, Austria acquired great swathes of territory as a result of the three partitions of Poland in 1772, 1793 and 1795.

When Napoleon declared himself Emperor of France in 1804, Archduke Francis II of Austria (and King of Bohemia and Hungary) realized that he would lose his position of Holy Roman Emperor and so created the Austrian Empire with himself as Emperor Francis I. A year after Napoleon's defeat of a joint Habsburg, Prussian and Russian force at the Battle of Austerlitz in 1805 the Holy Roman Empire was dissolved. Francis now ruled an empire which consisted of Austria, Bohemia, Galicia, Hungary, Slavonia and Transylvania; the other states of the former empire were organized by Napoleon into the Confederation of the Rhine (q.v.Germany). Under the terms of the Treaty of Pressburg in 1805 and the Treaty of Schönbrunn in 1809, following its defeat at the Battle of Wagram and the French seizure of Vienna, Austria lost huge amounts of territory in Germany, Italy, Slovenia and Croatia, including access to the sea, and its acquisitions gained under the Third Partition of Poland.

After Napoleon's defeat in Russia in 1812 Austria watched cautiously from the sidelines until Metternich, the Austrian chancellor and foreign minister, felt bold enough to declare for the Fourth Coalition. The Austrian Army then participated in the 'Battle of the Nations' at Leipzig in 1813 and in the succeeding battles that led to the entry of the Allies into Paris in 1814 and Napoleon's exile.

The Congress of Vienna in 1814–15 restored to Austria much of the territory it had lost earlier and gave it new territory in Italy. No attempt was made to revive the Holy Roman Empire, but due to Metternich's diplomatic skills, the Habsburg Empire emerged stronger than before and with its prestige enhanced. It was he who established the Concert of Europe – the pledge of the victorious powers of Austria, Great Britain, Prussia and Russia to uphold the peace settlement.

However, in triumph lay the seeds of disaster 100 years later. Growing discontent with conservative rule and the rise of nationalism led to revolution, which in 1848 engulfed the Austrian Empire and much of Europe. Although the revolution achieved little in Austria itself, major upheavals took place elsewhere and it was to have major repercussions. The Habsburg monarchy only just survived.

Rivalry between Austria and Prussia for supremacy in Germany had been an important issue since 1815. In the view of Bismarck, the Prussian chancellor, Austria had become the main obstacle to Prussia's growth and he was keen to exclude its influence on Germany to allow Prussia to become the principal German state. He pursued a policy to isolate Austria from any potential allies. After the two powers became involved in war with Denmark over Schleswig-Holstein in 1864 Bismarck engineered the Seven Weeks' War against Austria in June 1866. The Prussians prevailed. Bismarck demanded no major territorial concessions apart from the surrender of Austria's last Italian possessions. Nevertheless, one of the consequences of the war was the 1867 'Compromise' (*Ausgleich*):

renewed Hungarian demands for self-rule led to the division of the Empire which came to be called the Dual Monarchy of Austria-Hungary. Both parts were constitutionally autonomous. Francis Joseph remained as emperor of the Austrian lands (Austria, Bohemia, Slovenia) and King of the separate Kingdom of Hungary. The two countries shared common foreign and defence policies.

Concerned that their invasion of Bosnia-Hercegovina in 1878 might draw in the Russians, the Habsburgs became intent on acquiring an ally. This suited Bismarck who was keen to ensure the new Germany's security. An alliance was concluded in 1879 and Bismarck established the Triple Alliance by including Italy in 1882. Germany was the dominant partner.

To demonstrate its authority in the Balkans, and to the dismay of Russia, Austria-Hungary annexed Bosnia-Hercegovina in 1908. Immediately Germany backed Austria-Hungary, indicating that it would not shy away from war if Russia were to challenge the annexation; Russia was too weak to do so. However, following the assassination of Archduke Francis Ferdinand, heir to the Austro-Hungarian throne, in Sarajevo in June 1914, the First World War was unleashed. At the end of it the Austro-Hungarian Empire was dissolved and Austria became an independent republic. South Tyrol was ceded to Italy and southern Styria to Yugoslavia (Slovenia).

Deflated from its grandeur as the nucleus of a huge empire to a small German-speaking rump, many Austrians now favoured union with Germany. This, however, was forbidden by the post-war peace treaties – in clear contravention of the principle of self-determination espoused in those treaties. Hitler, however, dreamed of war and his first target was the land of his birth, an easy picking. In March 1938 Nazi forces invaded the country and Austria was annexed (the *Anschluss*), becoming a province of the Greater German Reich.

Following Germany's surrender and the Soviet liberation of Vienna in 1945, the republic was restored. However, it, and Vienna, were divided into zones to be occupied by American, British, French and Soviet forces. Ten years later the four powers signed a treaty with Austria, recognizing its independence and pre-1938 borders. They agreed to withdraw their forces on the understanding that Austria would remain neutral and promise not to enter a union with Germany or restore the Habsburgs.

1. Přemysl meant the House of Přemysl indicating that Otakar belonged to the Přemyslid dynasty.

2. One of the greatest dynasties of Europe, it is also called the House of Austria. The name comes from a contraction of Habichtsburg, Hawk's Castle, the name of the ancestral residence in Aargau, Switzerland. The dynasty began with the Counts of Habsburg. It came to the fore when Count Rudolf IV was elected German king in 1273. In 1282 he bestowed Austria and Styria on his two sons. Habsburgs then ruled Austria until 1918 and also Bohemia, Germany, Hungary and Spain at different times.

Area:	83 853 sq km (32 376 sq miles)
Population:	8.29 million; Austrians, Yugoslavs, Turks
Language:	German
Religions:	Roman Catholic, Protestant, Muslim
Capital:	Vienna
Administrative Districts:	Nine states
Date of Independence:	15 May 1955
Neighbouring Countries:	Czech Republic, Slovakia, Hungary, Slovenia, Italy, Switzerland, Liechtenstein, Germany

English-speaking Name	Local Name	Former Names	Notes
AUSTRIA	Österreich	Noricum, Ostarichi	The Republic of Austria (Republik Österreich) since 1918. Its name is derived from the fact that it became a military district on the eastern border of Emperor Charlemagne's (742–814) Frankish kingdom. It was his 'Eastern Province', *Ostmark* or *Ostarichi* (996), which became Österreich, Eastern Empire.
Bregenz		Brigantium	Lying at the foot of the Pfänder Mountain, it is derived from the Celtic *briga*, height.
Burgenland			A state meaning the 'Land of castles'. Part of Hungary until ceded to Austria (1920).
Carinthia	Kärnten		A state named after a Celtic tribe living in Karantanija from *kar*, rock, a reference to the mountainous area in which they lived. The centre of the Celtic Kingdom of Noricum which became a Roman province (16 BC). Austrian crownland (1335).
Eisenstadt			Means 'Iron town'. Ceded from Hungary (1920).
Graz			Derived from the Slav *gradec*, small fort.
Hallein			Close to a saltworks, the first syllable *hail* means salt or salt works and is found in many names.
Innsbruck		Veldidena	Means 'Bridge over the River Inn'.
Klagenfurt		Virunum	According to legend, it is named after the *Klagefrau*, weeping woman, who supervised the *furt*, ford, over the River Glan.

Linz	Lentia	May be derived from the German *linde*, lime (tree).
Salzburg	Juvavum	Named after the nearby salt mines, *salz*, salt, and *burg*, *castle*. Salt was 'white gold' and people were paid in salt, thus the word salary.
Steyr		Named after the River Steyr, being located at the confluence of the Enns and Steyr rivers. Steyr means 'Stream'.
Styria	Steiermark	A state named after the town of Steyr and *marcha*, march or frontier district.
Tirol	Tirol	A state whose name was copied from the Castle of Tyrol which itself was named after a local family.
Vienna	Vindobona, Wenia, Wienis	Wien	The Roman name means 'Fort on the River Wien' although it may be derived from the Celtic *vondomina* or *vindo*, white, and *bona*, fort. Garrison for the 13th, 14th and then 10th Roman legions. Capital of the Holy Roman Empire (1558–1806), the Austrian Empire (1806–67), the Austro-Hungarian Empire (1867–1918) and the Republic of Austria. Occupied by the Hungarians (1485–90). Unsuccessfully besieged by the Ottoman Turks (1529, 1683). Occupied by the French (1805, 1809). Congress of Vienna (1815) which settled the fate of Europe after the cataclysm unleashed by Napoleon. Occupied by the Nazis (Mar 1938) and by the Russians (Apr 1945) and thereafter subject to quadripartite Allied administration (until 1955).

Vorarlberg

Raetia

A state meaning 'Before the Arlberg', a mountain pass linking the Danube and Rhine river systems. The people of Vorarlberg voted (1919) to join Switzerland, but the Peace Conference in Paris would not allow the union.

Wiener Neustadt

Means 'Viennese New Town' although it was founded (1194) as a frontier fortress against the Magyars. Allegedly paid for with the ransom money for the English King Richard I (1157–99).

BELARUS

Belarus was one of the three nations that evolved from Kievan Rus (q.v.Russia), but it is quite different from Russia and Ukraine in that it was never, and had never tried to be, independent until the 20th century.

After the Mongol Tatar hordes had taken Kiev in 1240 and sacked a number of Byelorussian towns the territory gradually passed to Lithuania during the 14th century. Nevertheless, a fair degree of autonomy was allowed and the population began to develop their own nationality and language. Following the personal union between Lithuania and Poland in 1386 as a result of the marriage of the Lithuanian Grand Duke Jogaila and Queen Jadwiga of Poland, the two countries were formally united in 1569 at the Union of Lublin.

Polish influence and culture began to take precedence, but during the 17th and 18th centuries Poland's strength dissipated and Byelorussia languished. The First Partition of Poland in 1772 resulted in Russia acquiring the eastern part of modern Belarus; the Second in 1793, the central region, including Minsk, the capital; and the Third in 1795 brought the rest of what became known as Byelorussia into the Russian Empire. Regarding Byelorussia as 'western Russia', a policy of Russification was pursued.

In 1839 Tsar Nicholas I abolished the Uniate Church (which acknowledged the supremacy of the pope), prohibited the use of the Byelorussian language and banned use of the name Byelorussia. Instead, the region was referred to as the Northwest Territory.

Byelorussia's first opportunity to become independent came in 1917 after the Bolshevik Revolution in Russia. Indeed, an independent Byelorussian Democratic Republic was proclaimed in December by the All-Byelorussian Congress, but within four days the Bolsheviks had disbanded it. Launching a new offensive, German troops occupied much of the country in early 1918. Under the terms of the short-lived Treaty of Brest-Litovsk in March of that year the Russians surrendered a large section of western Byelorussia to Germany. Three weeks later Byelorussia again declared its independence as the Byelorussian National Republic, but it remained subservient to German military control.

With Germany's defeat in November 1918 the Treaty of Brest-Litovsk was abrogated and as the Germans withdrew the vacuum was filled by Bolshevik troops. A Byelorussian Soviet Socialist Republic (BSSR) was proclaimed on 1 January 1919.

Three months later Polish troops invaded and declared Byelorussia to be part of Poland. From then until 1921 it was the scene of much fighting. The Treaty of Riga in March 1921 ended the Russo-Polish war and gave to Poland large areas of western Byelorussia which it was to keep until 1939. This had the effect of returning Poland's eastern frontier to its position prior to the First Partition. In this newly-won region efforts were made to 'Polonize' the Byelorussians. Central Byelorussia formed the BSSR while the eastern region became part of Russia.

When the Soviet Union was created on 30 December 1922, Byelorussia was one of the four founding members.

Byelorussia doubled in size during the 1920s when Russia gave up territory around Vitebsk, Mogilev and Gomel. Those areas of western Byelorussia lost to Poland at the Treaty of Riga were regained when Soviet troops invaded Poland following the

Nazi-Soviet Non-Aggression Pact in 1939. They were lost again when Germany attacked the Soviet Union in June 1941, but when the tide turned they once more became a part of the USSR in 1944. In 1945 a Treaty between Poland and the Soviet Union formally recognized Soviet possession and the Polish population was expelled. Like Ukraine, despite being a constituent republic of the USSR, Byelorussia was awarded its own seat in the United Nations.

The Byelorussians did not evince much enthusiasm for independence in August 1991 and there has been little change since. By the time the Soviet Union had been dissolved at the end of 1991 the new country had changed its name to the Republic of Belarus and joined the Commonwealth of Independent States[1]. Without the desire to build a separate national identity, Belarus has pursued a policy of close co-operation and eventual integration with Russia. On 8 December 1999 a Treaty of Union was signed. The two countries will form a confederation, governed by a Supreme State Council. They will retain their sovereignty, national identity (in the UN and other international institutions) and territorial integrity.

1. The Commonwealth of Independent States (CIS) was formed on 8 December 1991 to replace the Soviet Union. Initially it comprised the Slav troika of Belarus, Russia and Ukraine. It was joined later by all the other former Soviet republics except for the three Baltic States. In July 1992 its permanent headquarters was established in Minsk.

Area:	207 600 sq km (80 154 sq miles)
Population:	10.28 million; Belarusians, Russians, Poles, Ukrainians, Lithuanians, Jews
Languages:	Belarusian, Russian
Religion:	Eastern Orthodox
Capital:	Minsk
Administrative Districts:	6 provinces and the city of Minsk
Date of Independence:	25 August 1991
Neighbouring Countries:	Russia, Ukraine, Poland, Lithuania, Latvia

Eastern Europe 1948-91

English-speaking Name	Local Name	Former Names	Notes
BELARUS	Belarus	Belaya Rus, Byelorussia	The Republic of Belarus (Respublika Belarus) since Sep 1991. Previously the Byelorussian Soviet Socialist Republic within the USSR and earlier still known as Belaya Rus (White Ruthenia). Why 'white' is not known. It may be because one of the three major branches of the Slavs settled here and they were predominantly of fair complexion. Many Byelorussian towns were sacked by the Tatars so the theory that the 'white' means 'free', that is free from the Tatars, is dubious. However, 'free' may have the sense of freedom of spirit, the people always trying to defend their language and culture whatever the odds.
Bobruysk	Babruysk		Named after the small River Bobruyka, itself named after the Russian *bobr*, beaver.
Borisov	Barysaw		May indicate the site of an ancient battle, the Slavonic *bor* meaning battle or be named after a Polotsk prince, Boris Vselavich (12th century). Just north of the city a vicious battle was fought as Napoleon crossed the River Berezina (Nov 1812) during his retreat from Moscow. Renamed (1993).
Brest		Berestye, Brest-Litovsk	The name is derived from *berëst*, birch bark, of which there are many locally. Litovsk means 'Lithuanian', indicating that the city belonged to that country (14th century) before becoming Polish. Sacked by the Tatars (1241). Passed to Russia at the Third Partition of Poland (1795). The Treaties of Brest-Litovsk (1918), signed by the Central Powers with a nationalist Ukrainian delegation and Soviet Russia, concluded hostilities on the Eastern Front during the First World War; Russia lost Ukraine, Finland and its Polish and Baltic territories. Within Poland (1919–39), the 'Litovsk' being

Name	Variant	Meaning / History
Dzerzhinsk	Koydanovo / Dzyarzhynsk	removed (1921). Became a Soviet frontier town when Germany and the USSR divided Poland after the Nazi-Soviet non-aggression pact (1939). Originally named after Koydan, a Tatar leader, defeated here (1241). Renamed (1932) after Felix Dzerzhinsky (1877–1926), a fanatical Polish communist who founded (1917) the Soviet secret police, the Cheka (The All-Russian Extraordinary Commission for Combatting Counter-revolution and Sabotage). He was born on the Dzerzhinovo Estate (*derzhat*, to hold) from which he probably took his name.
Gomel	Homyel	Derived from the Slavonic *gom*, hill. Became Lithuanian (14th century), then Polish and Russian after the First Partition of Poland (1772).
Grodno	Hrodna	Means 'City' from *hrad/grad*. Sacked by the Tatars (1241).
Minsk	Mensk, Menesk	Means 'exchange', i.e. a market place, from *menyat*, to change. At first Russian, it passed to Lithuania (14th century) and then Poland before being regained by Russia at the Second Partition of Poland (1793). Sacked by the Tatars (1505), occupied by the French (1812), Germans (1918, 1941–44) and Poles (1919–20). Capital since 1919.
Mogilev	Mahilyou	Mogila means 'Grave' and Lev 'Lion'. According to legend, a young peasant was deeply in love with a beautiful girl, but the *pan*, local squire, refused permission for them to marry. The peasant died but his grave became known as the 'Tomb of the Lion'. Originated as a Lithuanian town, was passed to Poland and then Russia on the First Partition of Poland.
Polotsk		Probably named after the River Polota. The city lies at the confluence of this river and the Western Dvina. Polota may

Slavgorod	Slawharad	be taken from the Slavonic *pal*, marsh, or from the Polochane tribe.
		Means the 'Town of the Slavs'.
Vitebsk	Vitsyebsk	Takes its name from the River Vitba.

BELGIUM

Until 1579 the history of Belgium is inextricably linked with that of the Netherlands and Luxembourg (all together known as the Low Countries). Their common history is described in the section on the Netherlands.

In 1568 the Revolt of the Netherlands began in the ten provinces of the Southern Netherlands. Although it was successful in the seven provinces of the North, it failed in the South. In 1579 Artois and Hainaut formed the Union of Arras, committed to Catholicism and pledging allegiance to the Spanish king. Henceforth the name Spanish Netherlands was used to refer only to the southern provinces (later to become Belgium and Luxembourg). The seven Calvinist northern provinces signed the Union of Utrecht, which established a military league to resist the Spanish and from which was formed the United Provinces of the Netherlands. The Low Countries were no longer homogeneous.

Only in 1585, however, when Antwerp surrendered, could the Spanish claim to have subdued the South completely and gained its recognition of Philip II as king. The Thirty Years' War broke out in 1618 and the Spanish resumed their efforts to recover the North three years later. They were unavailing; the United Provinces had been lost for good and the Spanish were forced to recognize their independence. The Spanish lost more territory during the course of the war so that at the Peace of Westphalia in 1648 a new northern boundary for the Spanish Netherlands was drawn. The second half of the 17th century was characterized by regular French invasions to which the Spanish had no answer. Artois and parts of Flanders were lost. So the borders of modern Belgium do not coincide with the 1648 borders; the independent bishopric of Liège was acquired when it and Belgium were assigned to The Netherlands in 1815.

In 1700 the Spanish Habsburg dynasty ended with the death of the childless Charles II. He named Philip, Duke of Anjou, as his successor and as a result the Spanish Netherlands came under the control of Bourbon France for the next six years. However, during the War of the Spanish Succession, which broke out in 1701, the Duke of Marlborough and Prince Eugene of Savoy secured a series of victories over the French. After the Battle of Ramillies in 1706 the British and Dutch assumed joint rule of the Spanish Netherlands for the next seven years. At the Treaty of Utrecht in 1713 the French abandoned all claims to the Spanish Netherlands which passed to the Holy Roman Emperor Charles VI, head of the Austrian branch of the Habsburgs; they became the Austrian Netherlands. Nevertheless, Charles did not enjoy complete sovereignty because, under the terms of the 1715 Barrier Treaty, the Dutch were allowed to hold a line of military fortresses in the south for which the local inhabitants had to pay. Thus possession of this territory was considered something of an imposition. The Treaty of Utrecht also marked the southern border of modern Belgium.

The death of Charles VI and the accession of his daughter Maria Theresa in 1740 provoked the War of the Austrian Succession. The Barrier fortresses were easily overcome by the French who went on to occupy the Austrian Netherlands for four years. Austrian rule was only restored at the Treaty of Aix-la-Chapelle (now Aachen, Germany) which brought the war to an end in 1748.

An uprising against the Austrians followed the outbreak of the French Revolution in 1789. An independent republic, the United States of Belgium, was proclaimed, but it was short-lived, largely because the two principal leaders could not agree on the sovereignty of the provincial states. A year later the Austrians decided to re-impose imperial control and little resistance was encountered.

In 1792 France declared war on Austria and by the end of the year had occupied the whole of the Austrian Netherlands. Hailing the French as liberators in the hope that they would regain their independence, the Belgians quickly became disillusioned when they realized they had lost their autonomy and their traditional way of life. For 15 months between March 1793, when the French were beaten at the Battle of Neerwinden, and June 1794, when they defeated the Austrians at the Battle of Fleurus and drove them out for good, imperial power was restored in Belgium. After they returned, the French treated the country as conquered territory. In 1795, however, attitudes changed and in October Belgium and Liège were annexed and became a part of France. The Austrians finally acknowledged this at the Treaty of Campo Formio in October 1797.

Napoleon's defeat in Russia and at the Battle of Leipzig in 1813 heralded the end of French occupation. As conquered territory the Allies were not pre-disposed to consult the Belgians about their future. The British, in particular, were determined to remove Belgium from France and at the same time strengthen the Low Countries as a counter to any further French ideas of expansion. Thus, it was decided to re-unite the United Provinces and Belgium to form the Kingdom of the United Netherlands. Before the Congress of Vienna (Sep 1814 to Jun 1815) could confirm this arrangement, Napoleon escaped from Elba and began to receive a rapturous welcome wherever he went in France. Seizing the moment, Prince William of Orange proclaimed himself King of The Netherlands on 16 March 1815 as William I.

Perhaps it was inevitable that the new kingdom would split apart. Since 1579 when it had last been one country, the two halves of The Netherlands had followed different courses of development. The North was Dutch-speaking, Protestant and trade-oriented; the South was predominantly French-speaking[1], Roman Catholic, partly agrarian and partly industrialised. Under the new regime Dutch became the official language, Catholicism was downplayed in the educational system and both North and South had the same number of representatives in the States General, despite the fact that Belgium's population was double that of the North.

Conflicting interests and the feeling that they were not receiving equal treatment encouraged a protest movement among young catholic and liberal Belgians. By the end of 1828 a national opposition campaign had been launched. This came to a head in August 1830 at a time when William chose to mark the 15th anniversary of his accession. Opposition became rebellion in September and on 4 October a provisional government declared Belgium to be independent; a month later the Belgian National Congress announced that Belgium was to be a parliamentary monarchy.

William did not yield and prepared for war. The great powers of Europe met in London with the immediate task of preventing war. On 20 December the London Conference officially recognized the division of the kingdom. In January 1831 the Conference confirmed Belgium as an independent, neutral, kingdom. The search for a suitable king, however, proved more difficult than expected. Eventually Prince Leopold of Saxe-Coburg (a German principality), who had declined the Greek crown the previous year, agreed to his election as King of the Belgians. He was already the widower of Princess Charlotte (daughter of the future George IV of Great Britain); in 1832 he married the eldest daughter of the French king and Belgium's position in Europe, guaranteed by Britain and France, was assured.

Still William would not accept the situation. Within a month of ascending the throne in July 1831 Leopold had to contend with a Dutch invasion. With the help of French troops this was brought to a halt and the Dutch withdrew. However, the London Conference stepped in and awarded to The Netherlands the province of Limburg while Luxembourg was split. The French-speaking, western, part was to become a Belgian province. Since William refused to acknowledge these awards, they were not implemented immediately. Despite their gains in recent years, the Belgians were only too well aware that they had no real control over their foreign policy and were little more than pawns among the bigger players in Europe.

Suddenly, in March 1838, William agreed reluctantly to sign the peace treaty giving effect to the partition. It was ratified in April 1839 and the Limburg and Luxembourg settlements were put into effect. When Luxembourg became an independent state in 1867 its border with Belgium was finally fixed.

Dismissing Belgian neutrality, the Germans invaded in August 1914 and proceeded to occupy much of the country for the next four years. At the Treaty of Versailles in 1919 Belgium's official neutrality was abolished, but Eupen and Malmédy on the eastern border were regained after their loss in 1815. During the Second World War, from May 1940 to September 1944, Belgium was again occupied by the Germans.

Only in 1993 did Belgium become a truly federal state with the three regions of Brussels, Flanders and Wallonia.

1. Flanders (now the provinces of East and West Flanders, Antwerp, Limburg and Flemish Brabant) is the historic Dutch (Flemish) speaking part of Belgium. The five southern provinces of Wallonia are predominantly French-speaking. The language divide between the Romance and Germanic languages dates back to Roman times.

Area:	30 153 sq km (11 781 sq miles)
Population:	10.26 million; Belgians, Italians, Moroccans
Languages:	Dutch (Flemish), French, German, Italian
Religions:	Roman Catholic, Muslim, Protestant
Capital:	Brussels
Administrative Districts:	3 regions with 10 provinces
Date of Independence:	1830
Neighbouring Countries:	The Netherlands, Germany, Luxembourg, France

English-speaking Name	Local Name	Former Names	Notes
BELGIUM	België, La Belgique, Belgien	Spanish Netherlands, Austrian Netherlands	Kingdom of Belgium (Koninkrijk België (Dutch), Royaume de Belgique (French), Königreich Belgien (German)) since 1831. Named after the Belgae, inhabitants of northern Gaul in Roman times. The name Belgium came into use during the Thirty Years' War (1618–48).
Antwerp	Antwerpen, Anvers		Possibly derived from *aan het werp*, at the wharf, in reference to its location on the River Scheldt, or from the Germanic prefix *anda*, against, and some noun connected with *werpen*, to throw, to indicate some defensive fortification. According to legend, a giant, Druon Antigonus, exacted tribute from passing boatmen; he severed the hands of those who refused to pay and threw them into the river until Silvius Brabo, a Roman soldier, cut off one of the giant's own hands and threw it into the river.
Brabant			A province named after Silvius Brabo. The modern Brabant was formerly the southern part of the Duchy of Brabant which was split in two during the Eighty Years' War (1568–1648), the northern part going to the Dutch.
Bruges	Brugge	Municipium Brugense	Possibly named by the Vikings from *bruggia*, place of disembarkation; or from the Flemish *brug*, bridge, around which the city developed.
Brussels	Bruxelles, Brussel		Derived from *bruocsella* from *broec*, marsh and *sele*, village, meaning 'Settlement in the marshes'. Capital of the Spanish Netherlands (1531–1713) and Austrian Netherlands (1713–1795). Lost its status when part of France (1795–1815). Shared status as capital of the Kingdom of the Netherlands with The Hague (1815–30) until becoming

		capital of Belgium (1830). Gave its name to Brussels sprouts.
Charleroi	Charnoy	A village developed into a fortress (1666) by the Spaniards and named after Charles II (1661–1700), King of Spain (1665–1700).
Dinant		Probably derived from Diana, the Roman goddess and huntress.
Flanders, West and East	Flandre, Vlaanderen	Two provinces with a name probably meaning 'Lowland' or 'Land subject to flooding'.
Ghent	Gent, Gand	Probably derived from the Celtic *condate*, confluence, an acknowledgement of its position at the junction of the Rivers Lys and Scheldt. Site of the Pacification of Ghent (1576) which tried to unite the Low Countries against Spain.
Liège	Liège, Luik / Leodium	May be derived from the Legie brook, now an underground tributary of the River Meuse (Maas); or it may come from the word *leudi*, people. Annexed to France (1795); became part of the Kingdom of the Netherlands (1815) and Belgium (1830).
Limburg	Limbourg	A province, once part of the Duchy of Limburg and named after the House of Limburg. Split (1839) between Belgium and The Netherlands. Derived from the Germanic *lindo*, lime tree, and *burg*, castle.
Louvain	Leuven, Louvain	May be derived from the Low German *loo*, bushy hill, and *veen*, swamp.

Malmédy	Malmundarium	Means 'Cleansed from evil ones'. Awarded to Prussia after the Congress of Vienna (1815) and to Belgium after the Treaty of Versailles (1919).
Mons	Bergen, Mons	Means 'hill' from the Latin *mons*. The Flemish Bergen means the same.
Nieuwpoort	Nieuport	Founded as a 'new port' (12th century) for Ypres.
Ostend	Oostende	Originally Oostende-ter-Streepe, meaning the 'East end of the strip' – the seaside resort.
Oudenaarde	Audenarde	Means 'old earth' or 'old landing place'. Site of the Battle of Oudenaarde (Jul 1708), when the Duke of Marlborough and Prince Eugene of Savoy defeated the French during the War of the Spanish Succession (1701–14).
Pepinster		Named after the Pepin family, some of whom became the Mayors of the Palace in Austrasia (the eastern part of northern France, Belgium, Luxembourg and western Germany) during the Merovingian dynasty (476–750).
Tournai	Doornik, Tournai / Turnacum	Believed to be named after Turnus, a Roman leader.
Wallonia		A region named after the Walloons whose name is derived from the Germanic *walhon*, stranger.

BOSNIA-HERCEGOVINA

Having been a part of the Eastern Roman province of Illyria since 168 BC, Bosnia did not become a magnet for the Slavs until the sixth century AD. The following century two new Slav tribes arrived: the Croats and the Serbs. The Croats settled in the north, centre and the west, while the Serbs moved into the Drina River Valley in the east and further south into Hum (now Hercegovina). Between the tenth and 12th centuries Bosnia experienced Serb, Croat, Byzantine and Hungarian domination. In 1180, when the Byzantine emperor died, an independent Bosnian state was established for the first time by Kulin, who assumed the Hungarian title of *Ban* (Governor).

During the next 280 years the power and independence of his successors ebbed and flowed, very often according to pressure exerted by Hungary which maintained its claim to sovereignty. Two more powerful rulers emerged in the 14th century. Under *Ban* Stephen Kotromanić (1322–53) the neighbouring principality of Hum was incorporated into Bosnia. In 1353 he was succeeded by Stephen Tvrtko who expanded Bosnia still further southwards and acquired a part of the Dalmatian coast. For more than a decade Tvrtko was faced with a revolt by the Bosnian nobles, but in 1377 he felt able to proclaim himself 'King of the Serbs, and of Bosnia, and of the Coast'. Five years later he took control of most of the Dalmatian coast and some of the Adriatic islands. To his title he added 'King of Dalmatia and Croatia'.

Bosnia was at the height of its power when Tvrtko died in 1391. Thereafter, rivalry between noblemen, Hungarian interference and impending Turkish invasion all conspired to weaken Bosnian power. The Turks captured part of central Bosnia in 1448. In the same year the ruler of Hum, Stephen Vukčić, adopted the title of '*Herceg* (duke) of St Sava' to demonstrate his independence and Hum became Hercegovina. The fall of Constantinople in 1453 and the conquest of Serbia in 1459 left Bosnia open to a full Turkish onslaught. It came in 1463 and Bosnia was incorporated into the Ottoman Empire; Hercegovina followed in 1482.

The Turks regarded Bosnia as the assembly area for their campaigns against Hungary and Venice. In contrast to Serbia, where the Muslim Turks were seen as alien occupiers, some Bosnian landowners, keen to keep their land and privileges under Ottoman rule and to escape further persecution by the Catholic Croats and Orthodox Serbs, welcomed the Turks as liberators and voluntarily converted to Islam. It was a gradual process, taking over a hundred years for the majority of the population to become Muslim. The Ottoman aim was not to force the conversion of the people to Islam, but rather to keep the country under control and collect sufficient money and men to meet its commitments further afield. To this end, in 1580, it was organized in eight military-administrative districts (in Turkish, *sancàk*, banner; in Serbo-Croat, *sandžak*) and Bosnia, including some adjacent areas of Croatia and Serbia, was elevated to the status of a province within the Ottoman Empire.

The Turkish defeat at the gates of Vienna in 1683 brought an end to Turkish expansion and a reversal of roles: the defence of the lands they had conquered against the Christian advance. War between Austria and Turkey continued sporadically for the next 100 years. The Treaty of Passarowitz (now Požarevac, Serbia) in 1718 fixed part of the south-western border of Bosnia. The Treaty of Belgrade in 1739 restored to Turkey much of the territory, principally Hercegovina, gained by Austria in 1718. The Austrians also surrendered their claims to the lands south of the River Sava, thus fixing Bosnia's northern border.

A Christian peasant revolt in 1875 spread and Serbia and Montenegro declared war on the Ottoman Empire the following year; they were joined by Russia in 1877. The war ended in 1878 and at the Congress of Berlin, although still under Turkish suzerainty, Bosnia-Hercegovina was put under the control of Austria-Hungary and, in particular, the finance minister, Benjamin Kallay, to maintain order. His great aim was to develop a national Bosnian identity where the people would think of themselves as Muslim, Roman Catholic or Orthodox Bosnians and not as Croat or Serb Muslims. He failed and the results of his failure have been seen in the events of the 1990s.

On 5 October 1908 Bulgaria declared its independence from Turkey. The next day Austria-Hungry announced the annexation of Bosnia-Hercegovina, a move facilitated by the presence of Austrian troops. Besides causing alarm throughout Europe, this angered the Serbs, who formed secret societies and an organisation called *Mlada Bosna* (Young Bosna) to foment opposition to Habsburg rule and spread the concept of South Slav (Yugoslav) nationalism. To soften opposition the Austrians evacuated the *sandžak* of Novi Pazar, a small strip of territory which prevented Serbia and Montenegro having a common border.

Tension, nevertheless, grew and on a visit to Sarajevo on 28 June 1914 the Archduke Franz Ferdinand, heir to the Austrian and Hungarian crowns, was assassinated by a young Bosnian Serb revolutionary from *Mlada Bosna*. The Austro-Hungarian government held Serbia responsible and issued an ultimatum that no country could accept. Serbia agreed to all conditions except for one; this would have entailed the surrender of its sovereignty. Austria-Hungary immediately declared war. In less than two weeks the great European empires had been sucked into the dispute and the First World War had started.

When the Kingdom of the Serbs, Croats and Slovenes came into existence on 1 December 1918 it included Bosnia and Hercegovina. On King Alexander I's reorganization of the Kingdom into *banovine* (provinces) in 1929 (q.v. Yugoslavia), the major parts of Bosnia became the Vrbaska and Drinska *banovine* while Hercegovina was incorporated into the Primorska and Zetska *banovine*. Within the new Kingdom of Yugoslavia Bosnia-Hercegovina thus ceased to exist.

Shortly after Yugoslavia was invaded in April 1941 the territory of Bosnia-Hercegovina was incorporated into the Independent State of Croatia. This did not stop Tito and his Partisans, and indeed the Serbian Chetniks as well, making the new state the scene of much of the fighting against the invaders and each other. It was on Bosnian territory that the National Army of Liberation of Yugoslavia was founded.

The 1946 federal constitution of the new Yugoslavia recognized Bosnia-Hercegovina as one of its six republics. It was populated by three major ethnic minorities: Muslims (44 per cent), Serbs (33 per cent) and Croats (17 per cent). In 1971 the Bosnian Muslims (Bosniaks) were recognized as one of the six nations of Yugoslavia[1]. When Yugoslavia fractured in 1991 Bosnia-Hercegovina found itself in a difficult position: the Bosniaks and Croats wanted independence, the Serbs wanted to remain in Yugoslavia. The latter formed several 'Serb Autonomous Regions' in Bosnia-Hercegovina where they were in a majority. After a referendum boycotted by the Serbs, Bosnia-Hercegovina was declared independent of Yugoslavia on 3 March 1992 and this was internationally recognized on 7 April. The Serbs, keen to partition the republic along ethnic lines by joining Serb-occupied areas, protested violently and the bloodiest war to have taken place in Europe since the end of the Second World War began. The division of Bosnia-Hercegovina was at stake – and soon became fact with the proclamation of the Bosnian Serb Republic and the Croatian Community of Herceg-Bosna. Neither the Serbs nor the Croats conceded the fact that the Bosniaks were also entitled to a state of their own. The UN intervened and sent troops to Sarajevo in July.

By early 1993 the Bosnian Serbs held nearly 70 per cent of Bosnia-Hercegovina. After 18 months of bitter fighting in Central Bosnia Bosniak and Croat forces agreed to a ceasefire and this led to the creation of a Muslim-Croat Federation of Bosnia-Hercegovina on 1 March 1994. The aim was to terminate all fighting between Bosniak and Bosnian Croat forces and foster political, economic and military co-operation between them. In reality, it merely papered over the differences between the two sides.

The war between the Federation and the Bosnian Serbs, however, intensified in March 1995. The Serbs responded to Federation gains by indiscriminately shelling Bosnian towns. The war was effectively brought to an end by NATO air strikes against Bosnian Serb targets which began in May 1995. Muslim and Croat forces took advantage of the Serb discomfort to capture some 20 per cent of the territory held by their enemy. The Bosnian Serbs were brought to the negotiating table. The Peace Accords[2], initialled on 21 November 1995 in Dayton, Ohio, USA, followed. The warring parties agreed to recognize the legal status of Bosnia-Hercegovina as a single state with two distinct entities within it: the Bosnian Muslim-Croat Federation with 51 per cent of the territory and the Republic of Serbia (Republika Srpska) having 49 per cent. It was partition in all but name. The Muslims controlled two main areas in the north-west and the centre, the Croats land in the west adjacent to Croatia and the Serbs territory in the shape of a horseshoe: in the north along the Croatian border and to the east and south-east along the Yugoslav border (Serbia and Montenegro)[3].

With their own armies and police, and in the face of different ethnic loyalties and ambitions, the task of developing a common Bosnian identity, so desired by Kallay over a hundred years earlier, is daunting. The presence of NATO-led peacekeeping troops has prevented a return to violence, but were they to leave, the temptation to renew hostilities might prove difficult to resist. Meanwhile, Bosnia-Hercegovina is a *de facto* UN protectorate.

1. The term 'muslim' has ethnic and cultural connotations as well as religious ones. Over a century after the Turks left, the Bosniaks are the most secular of Islamic peoples and, contrary to Serb and Croat claims, there is little evidence to indicate the spread of Islamic fundamentalism.

2. Their full title is the General Framework Agreement for Peace in Bosnia and Hercegovina; they are usually known as the Dayton Accords. The final agreement was signed in Paris on 14 December 1995.

3. Bosnia-Hercegovina is not land-locked; it has a coastline of 12 miles but without a natural harbour.

Area:	51 129 sq km (19 741 sq miles)
Population:	4.34 million; Muslims, Serbs, Croats
Language:	Serbo-Croat
Religions:	Muslim (Sunni), Serbian Orthodox, Roman Catholic, Protestant
Capital:	Sarajevo; Banja Luka is the capital of the Serb Republic
Administrative Districts:	2 self-governing regions: the Croat-Muslim Federation, with 10 cantons, and the Serb Republic
Date of Independence:	7 April 1992
Neighbouring Countries:	Croatia, Yugoslavia

English-speaking Name	Local Name	Former Names	Notes
BOSNIA-HERCEGOVINA	Bosna i Hercegovina	Bosona, Hum, Zachumlje	Republic of Bosnia-Hercegovina (Republika Bosna i Hercegovine) since March 1992. A republic within Yugoslavia (1946). Bosnia takes its name from the River Bosna (or, in ancient times, Bosanius) while Hercegovina (which comprises 18 per cent of the country) means the 'Territory of a duke', or 'Duchy', from *herceg*, duke, *ov*, to make the genitive case of *herceg* to indicate possession, and *ina*.
Banja Luka			Means 'Baths of Luke' (St Luke). Capital (1583–1639). At the Battle of Banja Luka (1738) the Turks defeated the Austrians. The consequent Treaty of Belgrade (1739) delineated the present northern border of Bosnia. Became (Feb 1998) the 'capital' of Republika Srpska, the Serbian region of Bosnia-Hercegovina.
Bosanski Brod			Here *brod* means 'Crossing place' (over the River Sava) from Slavonia into Bosnia. See Slavonski Brod in Croatia.
Donji Vakuf			During the Ottoman occupation *vakuf* was a piece of land, the income from which was given to a religious charitable foundation. *Donji* means 'lower'. Fairly close by is Gornji, 'upper', Vakuf.
Drvar		Titov Drvar	The Bosnian town named after President Tito (1892–1980), Partisan leader during the Second World War, Prime Minister (1945) and President (1953) of Yugoslavia. Renamed (1992). In an attempt to destroy Tito's Supreme Headquarters here and capture or kill him, the Germans made a surprise, but unsuccessful, airborne assault (25 May 1944).

Name	Alternative names	Description
Jajce		Possibly named (14th century) by Hrvoja, King of Bosnia and Duke of Spalato, after the Neapolitan Castel dell' Ovo (Egg Castle) (in Serbo-Croat, *jaje*, egg) to indicate his friendship with King Louis of Naples. Last stronghold of the kingdom to fall to the Turks (1463). The Second Session of the Anti-Fascist Council of National Liberation of Yugoslavia (in Serbo-Croat, AVNOJ for short) was held here (Nov 1943) to set the parameters for the new socialist state.
Mostar	Narona, Andetrium	*Stari Most* means 'Old Bridge' but Mostar is more likely to be derived from *mostari*, 'Keepers of the Bridge'. The stone bridge itself was built (1566) by the Turks in honour of their Sultan, Suleiman the Magnificent (1495–1566, r.1520–66), to replace a wooden suspension bridge. It was destroyed by Croatian shelling (Nov 1993). Capital of Hercegovina.
Sarajevo	Vrhbosna, Bosna Saray	The Slavs built a castle on a hill (*vrh*, summit) to the east. This was captured by the Turks (1428) who built the present city (1462–89). Among the Turkish buildings was a *saray*, a palace. Sarajevo takes its name from a corruption of *saray-ovasi*, the fields around the palace. The site of the assassination (28 Jun 1914) of the Habsburg Archduke Franz Ferdinand (1863–1914), heir apparent to the Austrian throne, which led to the outbreak of the First World War. Besieged by the Bosnian Serbs (1992–95). Capital of Bosnia-Hercegovina (1850).
Srebrenica	Argentaria	The name is derived from the silver (in Serbo-Croat, *srebro*) mining in the area.
Travnik		Means 'Green city' on account of the lush vegetation. Capital of the Turkish governor (pasha) of Bosnia (1699–1850).

Tuzla   Soli, Salinas   Named after the local salt (in Turkish, *tuz*) mines.

Vlasenica   May be named after the Vlachs, a people who settled in northern Albania and southern Serbia and who today are to be found in Albania, Bulgaria, northern Greece, Macedonia and Serbia.

BULGARIA

By 46 BC the Kingdom of Macedonia had been eclipsed and the Romans had become dominant throughout the Balkan Peninsula. When their empire was divided in 395 AD, Moesia, the area between the Balkan Mountains and the River Danube, and Thrace, from the mountains to the Aegean Sea, became part of the Eastern (Byzantine) Empire. Between the fourth and sixth centuries this empire was ravaged by Huns and Visigoths. They were followed by Slavs. Turkic nomadic warrior horsemen, the Bulgars, arrived in the seventh century from regions north of the Black Sea. Defeating a Byzantine army, they founded a kingdom in 681 astride the Danube and stretching north-eastwards to the River Dnieper.

During the next 300 years, as the Bulgars were gradually absorbed by the more numerous Slavs and converted to Christianity, Bulgaria – the First Bulgarian Empire – spread northwards at the expense of the Byzantine Empire to incorporate modern Romania and Serbia, southwards to the Aegean and westwards to the Adriatic. During his reign (893–927) Tsar Simeon I the Great tried but failed to usurp the Byzantine throne, but he could claim that only the Byzantine Emperor was more powerful than he. Within 100 years of his death, however, his empire had collapsed, the Bulgarians becoming subject to Byzantium in 1018.

In 1185 a general revolt against Byzantium took place and was so successful that the Byzantine Emperor was forced to give the Bulgarians control of Moesia. The Bulgarian Kingdom – the Second Bulgarian Empire – was resurrected. Tsar Kaloyan and then his brother, Tsar Ivan II Asen, extended their control to Albania, Epirus, Macedonia and Thrace, and re-established Simeon's title of Tsar of all the Bulgars and Greeks. After Ivan II's death in 1241, however, Bulgarian power again declined. The Mongols attacked from the north and when the Serbs defeated the Bulgarians in 1330 at the Battle of Velbuzhd (now Kyustendil) their fall from grace was almost complete. Ten years later the Turks began to push north-westwards, capturing the Bulgarian capital of Tûrnovo in 1393. Three years later Bulgaria became a province of the Ottoman Empire. It was to remain so for over 500 years.

A revolutionary movement did not get started in Bulgaria until the 1860s. When the Ottoman Empire and Serbia went to war in 1876 the revolutionaries decided to take advantage of the situation. The uprising against the Turks was a disaster, resulting in considerable loss of life and failing utterly to dent Ottoman power. But it did raise Bulgarian national awareness and bring Bulgaria to the attention of the rest of Europe. A year later the Russians declared war on the Ottoman Empire, one aim of which was to win independence for the Bulgarians. The Turks were overcome.

At the Treaty of San Stefano in March 1878 Russia proposed a 'Great Bulgaria' stretching across the southern Balkans from the Black Sea to the Vardar Valley in Macedonia and from the Danube to the Aegean. This did not find any favour with the Great Powers, particularly Austria-Hungary and Britain, and, at the Congress of Berlin four months later, the proposal was watered down. Eastern Rumelia and Macedonia were to remain in Turkish hands. Bulgaria emerged as an autonomous principality between the Danube and the Balkan Mountains but still subject to the Sultan. It was just 37.5 per cent of the size of the San Stefano proposal. However, seven years later Eastern Rumelia joined Bulgaria as a result of a coup without provoking any Turkish reaction. It did, however, provoke the Greeks and Serbs into demanding territorial compensation for this Bulgarian expansion. The Serbs invaded in the north-west, but were annihilated at the Battle of Slivnitsa.

A principality needed a prince and in 1879 Alexander Battenberg, a nephew of Tsar Alexander II of Russia, was chosen; he had fought with the Russians in the Russo-Turkish war of 1877–78. Alexander was deposed in 1886 and the following year Prince Ferdinand of Saxe-Coburg-Gotha was elected Prince of Bulgaria.

In July 1908 the Young Turks seized power in Constantinople (now Istanbul) and declared that they intended to unite all Ottoman territories; these included Eastern Rumelia and Bosnia-Hercegovina. On 5 October Ferdinand declared Bulgaria to be an independent kingdom and assumed the title of Tsar; the next day Austria-Hungary annexed Bosnia-Hercegovina.

Bulgaria, Greece and Serbia were all keen to take advantage of the progressive decline in the Ottoman Empire in Europe to expand their territories, casting particularly covetous eyes on Macedonia. Anti-Ottoman treaties were made with both Serbia and Greece in 1912: apart from Macedonia, a distribution of Balkan territory under Turkish control was agreed in the first; no such agreement was made with Greece. In October the First Balkan War (q.v. Macedonia) broke out and ended in triumph for the Balkan armies. At the Treaty of London in May 1913 Turkey gave up all but a small slice of territory in Europe: Bulgaria gained a stretch of Aegean coastline and the city of Adrianople (now Edirne), second only in importance to Constantinople.

The following month the Bulgarians, who had sustained the heaviest losses in the war, and the Serbs began to argue about the distribution of territory given up by the Turks. The main area of dispute was Macedonia where the Bulgars were intent on not allowing the Serbs to have the upper hand. The Second Balkan War began on 30 June when Bulgarian troops suddenly attacked their former Greek and Serb allies in Macedonia. Bulgaria was overwhelmed by an alliance of Greeks, Serbs, Montenegrins, Romanians and Turks. At the Treaty of Bucharest in August Bulgaria was awarded just 10 per cent of Macedonia (Pirin Macedonia) and eastern Thrace which included a Greek minority; the loss of southern Dobrudzha awarded to Romania after the First Balkan War was confirmed. Turkey regained territory in Thrace, including Adrianople, lost to Bulgaria after that war, thereby preserving its foothold in Europe.

After some hesitation Bulgaria sided with the Central Powers during the First World War in the belief that they would win. The Bulgarian aim was to achieve the state proposed by the Treaty of San Stefano. Thus the focus was on the acquisition of the rest of Macedonia held by the Greeks and Serbs, and southern Dobrudzha. Both areas were occupied by the Bulgarian army. At the Treaty of Neuilly in 1919, Bulgaria, on the losing side, was forced to cede the Thracian territory and stretch of Aegean coast it had gained after the Balkan wars to Greece, four pockets of land in the west to the Serbs, and southern Dobrudzha back to Romania. About one million Bulgarians now found themselves living outside Bulgaria.

At the outbreak of the Second World War Bulgaria declared its neutrality. However, southern Dobrudzha was regained in 1940 as a result of German pressure on Romania following Russian demands. Anticipating that the war would be won by Nazi Germany, and keen to reacquire Macedonia, Bulgaria joined the Axis Powers in March 1941; a month later it joined in the invasion of Yugoslavia. Most of Macedonia and a slice of Serbia was the reward. Emotional ties of Slav solidarity and a sizeable communist party deterred Tsar Boris III, who had established a royal dictatorship in 1935, from joining the Germans in their attack on the Soviet Union in June 1941. Nevertheless, on 5 September 1944, the Soviet Union declared war on Bulgaria and three days later the Bulgarians changed sides and declared war on Germany. The unopposed entry of Soviet troops into the country the same day quickly led to an armistice. Indeed, Bulgarian troops joined Marshal Tolbukhin's Third Ukrainian Front and subsequently fought their way into Hungary and Austria.

In 1946 the monarchy was abolished and a republic declared. It was to be run in accordance with communist principles. Soviet troops left the country in 1947, the same year that Bulgaria's possession of southern Dobrudzha was confirmed. However, the Allies were unimpressed with Bulgaria's change of sides in 1944 and it was not allowed to keep any other territory occupied during the war.

Attempts to 'Bulgarize' the ethnic Turks in the 1980s failed, but they comprise only 10 per cent of the population; the existence of a Macedonian nationality is denied. Communist rule lasted until 1990 when the wave of reform that was sweeping over the communist regimes in East Europe engulfed Bulgaria.

Area:	110 912 sq km (42 823 sq miles)
Population:	8.31 million; Bulgarians, Turks, Gypsies, Macedonians
Languages:	Bulgarian, Turkish
Religions:	Bulgarian Orthodox, Muslim (Sunni), Protestant
Capital:	Sofia
Administrative Districts:	9 regions
Date of Independence:	5 October 1908
Neighbouring Countries:	Romania, Turkey, Greece, Macedonia, Yugoslavia

The Balkans 1878

English-speaking Name	Local Name	Former Name	Notes
BULGARIA	Bulgariya		The Republic of Bulgaria (Republika Bulgariya) since Nov 1990. Previously the People's Republic of Bulgaria (1947) and Kingdom of Bulgaria (1908–46). Named after the Bulgars, possibly from the Turkic *bulga*, mixed, meaning a mix of Turkic and Slav tribes.
Asenovgrad		Stanimaka	Original name means 'Defence of the mountain pass' but renamed after Tsar Ivan II Asen in honour of his victory over the Byzantines (1230) at Klokotnitsa.
Balchik		Krunoi, Dionysopolis	The original Greek name was taken from the local springs while the Roman name honoured Dionysus (also known as Bacchus), the god of wine. Within Romania (1913–40).
Batak		Desposhovo, Desposhovo Mahallesi	Means 'Marsh' while the Turkish name means 'Place of the Despot'.
Blagoevgrad		Skaptopara, Dzhumaja, Gorna Dzhumaja	Named (1950) after Dimitur Blagoev (1856–1924), a Macedonian Marxist who founded the Bulgarian Communist Party. The previous names meant 'Mountain market'.
Burgas		Pyrgos	Name derived from the Greek *pyrgos*, fortress.
Dimitrovgrad			Built (1947) and named after Georgi Dimitrov (1882–1949), communist prime minister, who was accused (1933) of organizing the Reichstag fire in Berlin; not guilty, the Nazi prosecution was unable to get a conviction.

Dobrich	Hadzhioglu, Bazardchik, Tolbukhin	Originally named after the Turkish merchant who built the first house. Bazardchik (until 1878), then Dobrich after a local 14th century boyar; Bazargic (1913–40) when part of Romania. Tolbukhin (1949–91) after the Soviet Marshal, whose 3rd Ukrainian Army liberated the region (1944).
Dolni Chiflik	Georgi Traikov	Means 'Lower (Private) Estate', reflecting its Turkish connection. Was named after Georgi Traikov (1898–1972), titular head of state from 1964.
Dryanovo		Means the 'Place of dogwood'.
Karlovo	Levskigrad	Briefly named after Vasil Levski (1837–73), a leading revolutionary, whose birthplace this was.
Khisarya	Augusta, Toplitsa, Diocletianopolis	First named after Augustus Caesar (63 BC–14 AD), the first Roman emperor (27 BC–14 AD). Called Diocletianopolis, City of Diocletian, by the Byzantines. Diocletian (245–316) was the Roman emperor (284–305). The present name is derived from the Turkish word for fortress.
Kyustendil	Pautalia, Velbuzhd, Konstandili	Renamed Velbuzhd (1018) after a local chief until the Battle of Velbuzhd (1330) when the Serbs defeated the Bulgarians and killed the tsar. When the Turks captured the town they called it Konstandili, Konstantin's Land, after Konstantin Dragaš, the Serb leader. This became corrupted to its present name.
Montana	Montanensia, Golyama Kutlovitsa Ferdinand, Mikhaylovgrad	Named (1891–1945) after Tsar Ferdinand (1861–1948, r.1908–1918), the first King of modern Bulgaria. Previously named in memory of Khristo Mikhaylov, the leader of a failed uprising (1923) against the right-wing regime.

Pernik	Dimitrovo	Named (1949–52) after Georgi Dimitrov.
Pleven	Storgosia, Kajluka	Originally Thracian, it was renamed by the Slavs. Given its present name from the River Pleva shortly after becoming Hungarian (1266).
Plovdiv	Pulpudeva, Pulcin, Philippopolis, Trimontium, Filibe	An ancient Thracian site, Puldin being the Bulgarian name, rebuilt by and named (341 BC) after Philip II of Macedon (382–336 BC) as 'Philip's city'. The Roman Trimontium (46–1364) meant 'Three hills'. Taken by the Turks and renamed Filibe (1364–1919). Present name from 1919.
Preslav	Veliki Preslav, Eski Stambul	*Veliki* means 'great' and the city was so-called because it became the capital of the First Bulgarian Empire (893) and because of its great size. The Turkish name meant 'Old capital', *eski* meaning 'old'.
Samokov		Means 'Ironworks', the town being known historically for its working of local deposits of ore.
Shumen	Chumla, Kolarovgrad	Named (1950–65) after Vasil Kolarov, a revolutionary and communist ideologist.
Simeonovgrad	Maritsa	Named after Simeon I the Great (864–927), Khan of Bulgaria (893–927) and Tsar of the First Bulgarian Empire (925–27).
Sliven	Enidzhe Kariesi	Derived from *slivane*, confluence, the town lying at the confluence of the Novoselska and Asenovska rivers. The previous Turkish name meant 'New town'.

Place name	Alternative names	Description
Smolyan	Ezerovo, Pashmakli	Located among lakes and forests, the Slav name means 'Town of lakes'. The Turkish name remained (until 1934) when the town was renamed after a Slav tribe, the Smoleni.
Sofia (Sofiya)	Serdica, Sredets, Triaditsa	Originally named after the Serdi. The Slavonic *Sredets*, centre, acknowledges that the city is roughly equi-distant from the Black Sea and the Adriatic and between Belgrade and Istanbul. Given the Greek name Triaditsa when Bulgaria was incorporated into the Byzantine Empire (1018). Occupied by the Turks (1382–1878) who named it Sofia after the Church of *Sveta Sofia*, Holy Wisdom. Capital since 1879.
Sopot	Vazovgrad	Briefly named after Ivan Vazov (1850–1921), poet and novelist whose work glorified Bulgarian history.
Sozopol	Apollonia, Susopolis	First Greek city-state. Named Apollonia because the town had a large statue of Apollo until it was stolen by the Romans (72 BC). The present name means 'Town of salvation'.
Stara Zagora	Beroea, Augusta Trajana, Vereia Irenopolis, Burue Eski Zagra	*Zad Gora* means 'Across the mountain'. *Stara* means 'old' although, having been sacked by the Turks (1877), the town is not old. Irenopolis, meaning 'Irene's town', is named after Irene Ducas (c.1066–1120), wife of the Byzantine Emperor Alexius I Comnenus. Its penultimate name meant 'Old fertile land' in Turkish.
Targoviste	Eski Dzhumaya	Its former Turkish name meant 'Old Friday' from *cuma*, Friday, to signify a Friday market. The town was famous for its cattle fair. The present name (from 1934) has much the same meaning from the Bulgarian *targovets*, merchant.

Varna	Odessos, Stalin	May be derived from the Slavic root word *vran*, the black one, from the fact that it is located on the Black Sea. A Greek city-state (585 BC). Sacked by the Avars (586). Named Varna (681). The Turkish victory at the Battle of Varna (1444) ended any European hopes of saving Constantinople (Istanbul) from the Turks. Named (1949–56) after Stalin (1879–1953), Soviet leader.
Veliko Turnovo	Turnovo	Veliko added (1965) to mean 'Great place of thorns'. Capital of the Second Bulgarian Empire (1185–1393). Sacked by the Turks (1393). Independent Kingdom of Bulgaria proclaimed here (1908).
Velingrad		Named after Vela Peeva, a Resistance hero who was betrayed and executed (1944). Previously Velingrad had consisted of three villages.
Vidin	Dunonia, Bononia	The original Celtic name gave way to the Roman Bononia. During the Second Bulgarian Empire it became known as Bdin.
Vratsa		A shortened version of *vratitsa*, small door, probably because the town is situated where the River Leva emerges from the Vratsata gorge. In Turkish times it was a meeting point for many trade routes.

CROATIA

Croatia has always been something of a frontierland: straddling the line between the Western Roman and Eastern Roman (Byzantine) Empires and the line between the Austro-Hungarian and Ottoman Empires; and, more recently, as part of Yugoslavia, lying as a buffer between NATO and the Warsaw Pact.

The Illyrians were the first known inhabitants of Dalmatia, some islands and a strip of land along the eastern shores of the Adriatic Sea. They were followed by the Greeks who began the colonization of the Adriatic coast in the sixth century BC; 300 years later the Romans arrived. The Roman province of Illyricum stretched along the coast from northern Albania to Istria and as far inland as the River Sava. In 395 AD the Roman Empire was divided and the land east of the River Drina became part of the Eastern Empire. Thus most of what was to become Croatia lay within the Western Empire.

The Slavs did not arrive in the Balkans until the sixth and seventh centuries. One of the Slav tribes, the Croats, settled between the Slovenes and the Serbs in what is roughly their present homeland. In 924 they established a kingdom centred on northern Dalmatia but with territory extending eastwards to include Slavonia. Susceptible to influences from the north and west rather than the east, they became Roman Catholic and adopted the Latin alphabet.

In 1091 Croatia and Slavonia were conquered by Hungary although allowed a degree of self-government. However, the *Pacta Conventa* of 1102, uniting the Kingdoms of Croatia and Hungary, was viewed differently by each side: the Hungarians believed it to sanction the direct incorporation of Croatia into Hungary, while the Croats saw it as a voluntary agreement whereby they could exercise autonomy in return for acknowledging the Hungarian king as ruler. Nevertheless, the link with Hungary was to last for over 800 years.

Between 1115 and 1420 Hungary and Venice vied for control of the Adriatic, and Dalmatian cities changed hands on a regular basis. Eventually, the greater part of Dalmatia (but not the city-republic of Dubrovnik) fell under Venetian control and remained in Venetian hands, and quite separate from Croatia, until 1797. It was then ceded to Austria and in 1805 to France, later becoming part of Napoleon's Illyrian Provinces until returning to Austria in 1815; it was to remain under Austrian control until 1918.

The advance of the Ottoman Turks and the defeat of the Serbian Army at Kosovo Polje in 1389 caused many Christian Slavs to move north into Croatia for security. The decisive defeat of the Hungarians by the Turks at the Battle of Mohács in 1526 effectively marked the end of the Hungarian monarchy. The following year the Habsburg Ferdinand of Austria was chosen as King of Croatia (and of Bohemia and Hungary), but it was not long before much of Croatia and practically all of Slavonia were also in Turkish hands. The line dividing the Islamic world and the Christian West now ran through Croatia.

The need to establish a barrier against the Turks led to the creation of the Military Frontier (in Serbo-Croat, *Vojna Krajina*) in 1578 under direct Austrian control. This further reduced the territory controlled by the Croatian *Ban* (Governor) and *Sabor*

(Assembly). To build up an effective defence large numbers of people had to be re-introduced into the area which had been de-populated by Turkish raids. Most of the newcomers were Serbs who had fled northwards ahead of the Turks. In return for military service they were given land. The possession of land in Croatia by the descendants of these Serbs was a factor that was to return time and again to bedevil relations between the Croats and Serbs, most recently in 1991 when war broke out between them.

During the 17th century the Turks began to lose ground. At the Treaty of Karlowitz (now Sremski Karlovci), Serbia, in 1699 they ceded practically all Croatia and much of Hungary to the Habsburg Emperor. They gave up more land at the Treaty of Passarowitz (Požarevac), Serbia in 1718. To the dismay of the Croats, much of this land was not returned to them but incorporated into the Military Frontier. They demanded that it should be brought under Croat administration. Not until 1873 was it finally abolished and the area absorbed into Croatia in 1881. By now the population of Croatia was nearly 25 per cent Serb.

As a result of the creation of the Austro-Hungarian Dual Monarchy in 1867, Croatia was placed under Hungarian jurisdiction. Although part of the Kingdom of Hungary, Croatia was granted considerable autonomy; Dalmatia, Istria and the Military Frontier continued under the direct control of Vienna. Autonomy was not enough for many, however, and Croat national feeling grew. However, during the First World War Croat troops fought as part of the Austro-Hungarian army. When the Habsburg Empire collapsed in 1918, Croatia declared the unification of Croatia, Dalmatia and Slavonia (Istria was seized by Italy in 1919), secession from Hungary and incorporation into the new South Slav Kingdom of the Serbs, Croats and Slovenes which was founded on 1 December 1918.

Overshadowed by the Serbs in the new kingdom, Croat disaffection grew until, in January 1929, King Alexander I, in trying to foster a sense of national identity, established a royal dictatorship and changed the name of the country to Yugoslavia; at the same time he abolished the regional names and reorganized the country into nine *banovine* (governorships), most of Croatia being divided into the Savska and Primorska *banovine*. This did not find favour with many Croats. A fascist terrorist movement, the *Ustaša* (Rebel), led by Ante Pavelić, was founded with Italian support and one of its members assassinated the king in 1934. By 1939 the Croats had managed to achieve a new and semi-independent status and agreement to a new *banovina* to be called Croatia.

Four days after the Axis Powers invaded Yugoslavia on 6 April 1941 the Independent State of Croatia was proclaimed. It included Slavonia, parts of Dalmatia not given to Italy, and Bosnia-Hercegovina. Though a puppet state underpinned by German and Italian arms, power was exercised by the *Ustaše*. Many Croats remained loyal to the Axis powers, massacring thousands of Serbs; anti-fascist Croats joined Tito's Partisans once his movement became established.

The 1946 federal constitution of the new communist Yugoslavia named Croatia[1], including Dalmatia, Slavonia and the previously Serbian part of Baranja, as one of the six republics. Its population still included a substantial minority of Serbs. As the concept of a unified Yugoslavia began to unravel after President Tito's death in 1980, Croat nationalism, regarded as tantamount to fascism by the communists, once again began to flower. Croatia's Serbs were dismayed. In October 1990 they created the Serb Autonomous Region (SAR) of Krajina. It was plain that the Serbs no longer wished to live under

Croatian rule. In a referendum in Croatia in May 1991 94 per cent of those who participated voted for an independent republic.

On 25 June 1991 Croatia declared its independence from Yugoslavia. Fighting broke out between Croats and Serbs in areas, particularly in Krajina and Slavonia, where Serbs were in a majority. They succeeded in driving the Croats out and occupying about a third of Croatian territory. In July the Croats (like the Slovenians) agreed to a three-month moratorium on the implementation of independence.

Two more SARs, Eastern and Western Slavonia, were formed and in December they were united in the so-called Republic of Serbian Krajina. An uneasy ceasefire was achieved in January 1992. The European Community recognized the independent Republic of Croatia on the 15th and the first UN troops arrived in February to keep the peace. Fear that the status quo might become permanent encouraged the Croats to re-structure and strengthen their Army. In May 1995 a successful attack was launched into Western Slavonia and in August most of the Serbian Krajina was recaptured. The Serbs, who had inhabited these regions since the 16th century, were expelled; only Eastern Slavonia with its large Serb majority remained under their control.

In August 1996 Croatia and Yugoslavia signed a 'normalization' agreement which included respect for each other's sovereignty within their international borders. Nevertheless, although the Dayton Accords of November 1995 confirmed Eastern Slavonia as an integral part of Croatia, it did not return to Croatian rule until January 1998. A century ago the Serbs constituted a quarter of Croatia's population; today the figure is no more than 3 per cent.

1. Along the Adriatic coast Croatia is split by 12 miles of Bosnian-Hercegovinian territory.

Area:	56 538 sq km (21 829 sq miles)
Population:	4.48 million; Croats, Serbs, Muslims
Languages:	Croat, Serbo-Croat
Religions:	Roman Catholic, Eastern Orthodox, Muslim
Capital:	Zagreb
Administrative Districts:	20 counties
Date of Independence:	8 October 1991
Neighbouring Countries:	Slovenia, Hungary, Yugoslavia, Bosnia-Hercegovina

English-speaking Name	Local Name	Former Names	Notes
CROATIA	Hrvatska		Republic of Croatia (Republika Hrvatska) since Jun 1991. Previously a republic within Yugoslavia (1946). Named after the Croats. The name may be derived from the Iranian *Choroatos*, nomads, from the Caucasus, or *hrbat*, mountain ridge, referring to the mountains along the Adriatic coast. Gave its name to the cravat, a form of scarf worn by Croat mercenaries (Old Slav, *khrovat*) in service with the French (17th century).
Bjelovar		Bjelovac, Wellewar	Means 'White fortress' from the Hungarian *vár*, castle or fortress.
Cavtat		Epidaurus	Captured by the Romans (228 BC), the name is derived from the Latin *Civitas Vetus*, indicating possession of the rights of a citizen of ancient Rome.
Crikvenica			Takes its name from *crkvica*, church, around which the first settlers made their home (14th century).
Dalmatia	Dalmacija		A region on the Adriatic coast including off-shore islands. Named after the Delmatae tribe, a name possibly taken from the Albanian *delmë*, sheep, to denote sheep breeders. During its recorded history under the rule of Illyrians, Greeks, Romans, Goths, Byzantines, Hungarians, Bosnians, Tatars, Croats, Serbs, Venetians, Sicilians, Normans, Turks, Austrians, Italians and Yugoslavs. Gave its name to the breed of dog, although the dogs were probably brought in to guard the borders and did not originate here.

Name	Other names	Description
Dubrovnik	Ragusium, Ragusa	Refugees fleeing from Cavtat (7th century) settled on the then island of Lava (which became Lausa, meaning 'rock' in Greek, then Rausa/Ragusium, Rhacusa, and Ragusa). A second settlement developed on the mainland which had a Slav name, Dubrava, meaning oak grove. The two settlements merged (12th century) when the channel separating them was filled in. It was a republic, the Republic of Ragusa, (1358–1808) steering a middle course between the Ottoman Empire and the Hungarian-Croatian kings until independence was ended by Napoleonic decree. By the Congress of Vienna (1815) ceded to Austria (1815–1918).
Hvar	Pharos, Lesina	An island. The Venetian name Lesina is derived from a Slavonic root word meaning woody.
Istria	Istra	A county named after the Illyrian tribe, the Histri. Seized (1919) by the Italians from Austria in an attempt to make it a permanent part of Italy, but surrendered after Italy's defeat in the Second World War. The Istrian peninsula extends into Slovenia.
Karlobag	Vegium, Bag	After Turkish destruction (1525), restored and named (1579) after Charles (Karl), the Habsburg Archduke of Styria and first commander of the Military Frontier against the Turks.
Karlovac	Karlstadt	Named after Archduke Charles of Styria and means 'Charles's town'. Became one of the three generalates of the Military Frontier.
Korčula	Korkyra Melaina, Corcyra Nigra, Curzola	An island. Means 'Black Korkyra', possibly because it was covered in dense pine woods.

Korenica	Titova Korenica	The Croatian town named (1948–92) after President Tito (1892–1980), Partisan leader, Prime Minister (1945), President (1953).
Kraljevica	Porto Re	The Austrians built a new 'royal' port here (18th century) from which the name is derived (in Serbo-Croat, *kraljevski*, royal).
Makarska	Muccurum, Macrum	Developed from the Roman village of Macrum which became Makar.
Opatija	Abbazia	Takes its name from the Benedictine Abbey of St James (in Serbo-Croat, *opatija*, abbey).
Pula	Polai, Pietas Julia, Polensium	According to legend, the name is derived from the Greek *polai*, the pursued, referring to Greek refugees from Colchis on the Black Sea. Renamed (c.40 BC) after Julius Caesar, its full name being Colonia Julia Pola Pollentia Herculanea. Became Austria's principal naval base (1848). Passed to Italy (1920). Administered by Anglo-American forces until passed to Croatia (1947).
Rijeka	Tarsatica, Trsat, Rika, Fiume	The Italian Fiume and Rijeka both mean 'River', the associated river being the Riječina. Seized (1919) by a small group of Italian irregulars led by the writer and adventurer Gabriele D'Annunzio (1863–1938) who remained in power for a year before being removed by the Italian government. Declared a free city in the Treaty of Rapallo (1920). At the Treaty of Rome (1924) the part west of the Riječina was assigned to Italy (1924–1945) and the part east to the Kingdom of the Serbs, Croats and Slovenes.
Šibenik	Castrum Sebenici, Sebenico	Named after a Slav tribe, the Sebenici.

Sisak	Segestica, Siscia	The Celtic settlement having come under Roman rule (2nd century BC), it was granted colonial status (1st century AD) as Colonia Flavia Siscia from which the present name is taken.
Slavonia		A region of eastern Croatia, named after its mainly Slav population.
Slavonski Brod	Marsonia	*Brod* means boat or, figuratively, crossing. Thus the name means a crossing place (over the River Sava) from Bosnia into Slavonia.
Split	Aspalathos, Spalatum, Spalato	Takes its name from the Latin *palatium*, palace. Emperor Diocletian (245–313, r.284–305) built (295–305) a huge palace in which to live after his retirement. Held by Venice (1420–1797) and Austria (1797–1918).
Trogir	Tragurion, Trau	May be derived from the Greek *tragos*, goat and *oros*, hill, meaning a settlement on Goats' Hill.
Varaždin	Garestine	Capital (1756–76).
Vinkovci	Colonia Aurelia Cibalae, Zenthelye	Previously called after the prophet Elijah, although Zenthelye translates as Saint Elijah.
Vis	Issa	An island. According to legend, named after Issa, a girl from Lesbos loved by Apollo.
Vukovar	Vukovo	Named after the River Vuka. Given the present Hungarian name (late 17th century). Destroyed by Serb and Yugoslav forces (Aug–Nov 1991).

Zadar	Idassa, Jadera Diadora, Zara	Sacked during the Fourth Crusade (1202). In Venetian hands (1409–1797) and then Austrian (1797–1920). By the Treaty of Rapallo (1920) given to Italy as an enclave on the Yugoslav mainland.
Zagreb	Andautonia, Agram	The first settlement was named after the Illyrian tribe, the Andautoni. May be derived from *zagrepsti*, to dig in, indicating a settlement surrounded by earthworks; or, possibly, from *za grebom*, behind the cliff, referring to the range of wooded hills to the north. Developed from two medieval and rival settlements: Gradec, meaning fortress, and Kaptol, bishopric. Gradec was accorded the rights and privileges of a royal and free city by Bela IV (1206–70), King of Hungary, after Kaptol had been sacked by the Tatars (1242). New building filled the narrow gap between them until they were united (1850). Agram (1526–1918). Capital of Croatia (possibly from 1557, except from 1756–76).

CYPRUS

An island with a dramatic history, Cyprus has been the target for a mix of civilisations that bequeathed their cultures and shaped its character. About 40 miles south of Turkey and nearly 500 from mainland Greece, it was first settled by Greeks in the second millennium BC. Its natural resources, particularly copper, attracted traders and settlers, but also adventurers and conquerors. After the Greeks came Phoenicians, Assyrians, Persians and Egyptians. By the middle of the fourth century BC the island was divided into ten city-kingdoms and, although the kings were subject to different conquerors, the links with Greece were maintained.

In 58 BC Cyprus combined with Cilicia in Turkey to become a Roman province. When the Roman proconsul, Sergius Paulus, was converted to Christianity in 45 AD Cyprus became the first province to have a Christian governor. The division of the Roman Empire in 395 made little difference to Cyprus which remained under Byzantine rule. A 250 year period of peace and stability ensued. However, Arab attacks began in 647 and control of the island was disputed between the Byzantines and Muslims for the next 300 years until the Byzantines finally gained control in 965.

One of the leaders of the Third Crusade was King Richard I the Lion-Heart of England. Heading towards the Holy Land in 1191, his fleet was scattered by storms off Cyprus. He came ashore, took control of the island and deposed the self-proclaimed Emperor of Cyprus, previously the Byzantine governor. After Richard had left the Cypriots rebelled and so the island was sold to the Knights Templar. They found the burden too great and asked Richard to buy it back; instead, in 1192, he sold it on to Guy de Lusignan, who had lost his Kingdom of Jerusalem in 1187. The Lusignans ruled until 1489, but not without difficulty in the face of increasing Genoese and Venetian control of the island's trade; in 1372 Famagusta was seized by the Genoese who held it until 1464. In 1426 the Egyptians conquered the island and reduced it to a vassal state of the Egyptian sultan.

The Venetians wanted Cyprus as a naval base and as a trading centre. When the Lusignan King James II married Catherine Cornaro, the daughter of a Venetian nobleman, in 1472, Cyprus became formally linked with Venice. James died in 1473 and his infant son in 1474, thus bringing the Lusignan line to an end. Cornaro ruled as queen, under tight Venetian control, until 1489 when she finally succumbed to pressure to cede Cyprus to Venice. It was during this century that the Greeks renewed their long association with the island as some sought to start a new life on an island of commercial and cultural wealth.

The 82 years of Venetian occupation were characterized by decay and neglect. The same period witnessed an expansion of the Ottoman Empire with Cyprus the target for many raids, rather than a full-blooded invasion. Finally, in 1571, the Turks attacked in strength and overcame the Venetians after a series of sieges and battles. More than 300 years of corrupt, oppressive and incompetent Ottoman rule followed. Ottoman neglect served only to unite the Greek ethnic majority and give mileage to *enosis,* the idea of union between Cyprus and Greece.

In 1878, as a result of the Congress of Berlin which amended the peace settlement arranged at the end of the Russo-Turkish war, responsibility for the administration of

Cyprus was given to the United Kingdom (UK) although it remained legally a part of the Ottoman Empire. The Cyprus Convention suited the Turks and the British. It gave the latter a base in the eastern Mediterranean from which to oppose any possible Russian expansion into what remained of the declining Ottoman Empire. Furthermore, with the opening of the Suez Canal in 1869, Cyprus had become strategically important. With control of Gibraltar, Malta and now Cyprus, the British dominated the Mediterranean. When Turkey sided with the Central Powers in 1914 the UK annulled the Convention and annexed the island. A year later Cyprus was offered to Greece provided it joined the Allies. The pro-German King Constantine declined the offer.

Under the terms of the Treaty of Lausanne in 1923, both Greece and Turkey acknowledged British sovereignty over Cyprus. It became a crown colony in 1925.

Well established during the 19th century, *enosis* now began to gather momentum. Violent pro-*enosis* riots took place in 1931. But the movement really began to advance after the Second World War. In 1955 Greek Cypriot extremists, led by Colonel (later General) Grivas, under the name EOKA (Ethniki Orgánosis Kyprion Agóniston – National Organization of Cypriot Fighters) began a campaign of terror to rid the island of the British; Turkish Cypriots and the Turkish government became alarmed. Sporadic fighting broke out between the two communities. Only in 1959 did the Greek and Turkish governments reach agreement which was then approved by the British government, and by the Greek and Turkish Cypriots. Cyprus was to become independent with a Greek Cypriot president (Archbishop Makarios) and a Turkish Cypriot vice-president in line with the population breakdown (80 per cent Greek, 20 per cent Turkish). Greece was allowed to station 950 troops and Turkey 650 troops on the island while the UK retained two sovereign base areas on the south coast. On 16 August 1960 Cyprus became a republic, its independence and territorial integrity guaranteed by the UK, Greece and Turkey. It was barred from any political or economic union with any other state and was not to be partitioned; thus Greece had sacrificed *enosis* and Turkey partition.

Very soon disagreements arose between the Greek and Turkish communities over a host of constitutional, political and economic issues. Agitation for *enosis* resumed and in December 1963 inter-communal violence broke out. The Turks were particularly vulnerable because they lived all over the island and were greatly outnumbered. In March 1964 UN peace-keeping troops (UNFICYP – UN Force in Cyprus) were deployed to the island; they are still there. Despite sporadic violence and terrorist activities by 'EOKA-B', relations between the two communities gradually improved. However, matters came to a head in July 1974 when the military junta, then in power in Athens, used the Cypriot National Guard to stage a coup to assassinate Makarios and achieve *enosis*. A former member of EOKA, Nikos Sampson, was proclaimed president.

The Turkish Cypriots called for British and Turkish military action to prevent *enosis*. The British refused to intervene and a Turkish invasion of northern Cyprus followed immediately. During the next three weeks the Turks seized all the territory north of what came to be known as the 'Attila Line', some 37 per cent of the island. In 1975 the Turkish Federated State of Cyprus was proclaimed, with the Turkish Cypriots stating that they did not want independence. In due course, however, an ethnic and political partition of the island took place. In 1983 a unilateral declaration of independence was made, the Turkish region being called the 'Turkish Republic of Northern Cyprus' (TRNC); its sovereignty was recognized only by Turkey.

Area:	9251 sq km (3572 sq miles)
Population:	793 000; Greeks, Turks
Languages:	Greek, Turkish, English
Religions:	Christian, Muslim
Capital:	Nicosia
Administrative Districts:	6 (2 in TRNC)
Date of Independence:	16 August 1960
TRNC:	
Area:	3355 sq km (1295 sq miles)
Population:	177 120 (1994)
Date of Independence:	15 November 1983

English-speaking Name	Local Name	Former Names	Notes
CYPRUS	Kípros (Gk), Kibris (Turk)	Makaria, Alashiya, Aeria, Iatanana	The Republic of Cyprus (Kypriaki Dimokratia (Greek)), Kibris Çumhuriyeti (Turkish)) since Aug 1960. The discovery of abundant deposits of copper gave the island one of its early names: Aeria probably comes from the Latin *aes*, copper. Rome's supply of copper came from Cyprus and was known as *aes Cyprium* which in time was shortened to *cuprum* to give the English word copper. Makaria means 'Blessed'. Cyprus also gave its name to the cypress tree.
Larnaca	Lárnax	Citium/Kition, Khittim, Salina	Name may be derived from the Greek *larnax*, sarcophagus, due to the large number unearthed in the area; Salina after the nearby salt lake.
Nicosia	Lefkosia (Gk), Lefkoza (Turk)	Ledra	Possibly named after Lefcon (280 BC), the elder son of Ptolemy I Soter, founder of the Ptolemaic dynasty, or after the tall cypress tree, *lefki*. Nicosia is a corruption of the Byzantine Lefkosia. Capital (since 10th century). Sacked by the Mamelukes (1426) and by the Turks (1570). Divided (1964) by the so-called Green Line to separate the Greek sector in the south from the Turkish sector.
Paphos	Páfos	Baffo, Basso	A combination of two cities, Nea (later Kato) Paphos and Ktima. Old Paphos (now called Kouklia) was located 16 km to the east. According to legend, named by the son of Paphos, the daughter of King Pygmalion, who fell in love with an ivory statue he had made of his ideal woman. The goddess Venus brought the statue to life and Pygmalion married her. Capital (3rd century BC). One of the city-kingdoms. Sacked by the Muslims (960).
Pólis		Arsinoe	Means 'City'.

Turkish Republic of Northern Cyprus	Kuzey Kıbrıs Türk Cumhuriyeti	A so-called republic, proclaimed (Nov 1983) and only recognized by Turkey.
Famagusta	Ammókhostos (Gk). Gazimağusa (Turk)	Arsinoe	The name is probably derived from *Fama Augusta*, possibly the 'Pride of Augustus' after the Roman Emperor (63 BC–14 AD). The Greek name is a corruption which means 'Buried in the sand', a reference to the silted mouth of the River Pedieos from the Greek *ammos*, sand, and *khostos*, built up. Seized by the Genoese (1372), the Venetians (1489) and after an eight-month siege (1570/71) devastated by the Turks.
Kyrenia	Girne (Turk)	Cerynia, Corineum	One of the city-kingdoms.

THE CZECH REPUBLIC

The Czech Republic, consisting of Bohemia and Moravia, lies at the heart of Europe. Its strategic position has ensured a turbulent history. From being a member of the Soviet-dominated Warsaw Pact, it joined the Pact's adversary, NATO, in 1999. It is the most westernized of all the Slav nations, and its capital, Prague, for a time capital also of the Holy Roman Empire, lies north and west of Vienna.

The Boii, a Celtic people, were the first known inhabitants of what is now the Czech Republic. They were followed by Germanic tribes. The first Slavs did not arrive until the sixth century and they disputed the territory with the Avars who had come from the Dnepr basin. By the close of the eighth century one of the Slav tribes, the Czechs, had become predominant in central Bohemia; another, the Moravians, had settled along the Morava River (part of which now forms the border between the Czech Republic and Slovakia) and come under the sway of Charlemagne, King of the Franks. Together, the Franks and Moravians expelled the Avars in 796.

Moravia was founded in the vacuum about 830 and by the end of the century Prince Svatopluk (870–894) had extended its borders to include much of modern Austria, western Slovakia, Bohemia, the southern part of modern Poland and the western part of modern Hungary; it came to be known as the Great Moravian Empire. However, it did not long outlive Svatopluk's death, succumbing to invasions by the barbarian Magyars; in about 907 the empire collapsed. The Christian Czechs to the west of the River Morava elected to ally themselves with the Frankish emperor, while the Slovaks to the east came under heathen Magyar subjection. For the next thousand years the Czechs and the Slovaks were to remain separated.

Under the Přemyslid dynasty (895–1306) Bohemia, protected by natural mountain frontiers, began to emerge as an influential political and cultural force in medieval Europe. Nevertheless, as the power of the new Saxon dynasty in Germany blossomed, Prince Boleslav I was forced to recognize Saxon suzerainty in 950. Five years later, led by Otto I the Great, the German king (and Holy Roman Emperor from 962), Czech, Bavarian and other troops annihilated the Magyars who, having been a threat for so long, never again attempted to invade Germany. Steady German immigration into Bohemia now began. In 1029 Moravia was incorporated into Bohemia, although it remained a separate province.

The expansionist ambitions of Prince Břetislav of Bohemia were anathema to the Holy Roman Emperor and in 1041 he forced Břetislav to swear allegiance to him. Bohemia-Moravia became an imperial fief (which it was to remain until the fall of the empire in 1806), although it was still considered to be the inherited property of the House of Přemysl. Although Břetislav's second son, Vratislav II, received the title of King of Bohemia in 1085, it was only in 1212 that the emperor finally confirmed the status of Bohemia as a kingdom. Vratislav encouraged German colonization as a counterweight to the increasing influence of the Czech nobility.

Various conquests brought in Silesia and Lusatia to the north and, under King Otakar II Přemysl, land stretching as far as the Adriatic Sea. By the time of Otakar's death in 1278, however, most of these territories had been lost and the Czech Lands of Bohemia and Moravia were in decline. In 1306 the assassination of the new, teenage, king brought the Přemyslid dynasty to an end. After four years of dispute the Czech nobility offered the throne to Duke John of Luxembourg, son of Emperor Henry VII.

His son, Charles I (whose mother was Czech), came to the throne in 1346 and nine years later was crowned Holy Roman Emperor (as Charles IV). He transferred the imperial administration to Prague and in doing so introduced a golden age for Bohemia. His Golden Bull of 1356 asserted that the Holy Roman Empire was a league of sovereign nations and that Bohemia was pre-eminent among the secular electors of the empire. Silesia and Brandenburg were brought into the kingdom.

Charles died in 1378, the same year that the schism in the Roman Catholic Church took place. Beginning to preach against the abuses of the Church in 1402, Jan Hus became the leader of a reform movement; the cause was later taken up by Luther with the eventual consequence that a large part of Europe became Protestant. After Hus was burnt at the stake as a heretic in 1415, the Hussite Wars broke out (1419–1437). The reform movement now became a more widespread protest movement directed against the Catholic Church and German ascendancy in Bohemia. Despite forcing the Germans to resettle in the border regions, the Czechs still remained inferior to them, the peasants being even worse off than before. A further consequence was to isolate Bohemia from the empire and, simultaneously, deprive it of an effective centre; weakened, it was to become prey to the Habsburgs.

To counter the growing Turkish threat, the Czech king, George of Poděbrady, suggested a royal union of Poland, Hungary and the Czech kingdom and recommended a Polish prince, Vladislas of the Jagiełłonian dynasty, as his successor. In 1471 George died and Vladislas was duly elected as King of Bohemia; George's rival, Matthias I Corvinus (Matyas Hunyadi), King of Hungary, retained Moravia, Silesia and Lusatia which he had captured a few years earlier. When Matthias died in 1490 Vladislas was elected to replace him as Ulászló II and Bohemia was reunited with its sister provinces.

After the Hungarian defeat by the Turks at the Battle of Mohács in 1526 during which the Bohemian/Hungarian king, Louis II, was killed, the Czech throne was claimed by the Austrian Habsburg, Archduke Ferdinand I. This claim was based on a treaty between his grandfather, the Holy Roman Emperor, and Vladislas and the fact that he was married to Vladislas' daughter, Anne of Hungary, the brother of Louis II. Only after Ferdinand had agreed to certain concessions did the Czech nobles elect him as king – but with no hereditary rights. Although Ferdinand became emperor in 1558, he valued Bohemia above all his other possessions.

The Hussite revolt against Ferdinand II's attempt to impose Catholicism throughout his realm in 1618 precipitated the Thirty Years' War. Two years later the Czech Protestants tried to depose Ferdinand, but were decisively defeated by Catholic forces at the Battle of the White Mountain. Although no country loses its independence as the consequence of a single battle, this defeat sounded the death knell of the Bohemian kingdom's independence. During the next ten years it was converted into a Habsburg province. The Habsburgs imposed a new constitution whereby they became a hereditary dynasty and Protestantism was outlawed. The Peace of Westphalia in 1648 confirmed the inclusion of the Czech crown lands (Bohemia, Moravia, Silesia; Lusatia had been ceded to Saxony in 1635) into the Habsburg Empire. A new wave of Germanization swamped Czech nationalism and reduced the Czech language to little more than a peasant dialect. The Czech nation all but evaporated.

A cultural revival began after the Napoleonic Wars, but an uprising against the Habsburgs in 1848 failed. When the Habsburg emperor was forced to establish the Dual Monarchy of Austria-Hungary in 1867 Bohemia became a province of Austria; Moravia was administered as a separate Austrian crown land. Henceforth, until the outbreak of the First World War, the political energies of the Czechs were directed at challenging German domination of Bohemia and Moravia. Relations with their Habsburg rulers became increasingly strained.

When Austria declared war on Serbia in July 1914 and subsequently on France and Russia, the Czechs found themselves in the tragic position of having to fight against two Slav states and for a cause which was anathema to them. Furthermore, it was obvious to them that were Austria-Hungary, in alliance with Germany, to win the war then their dreams of independence would vanish and their political position deteriorate. Thus they had no desire for victory. While many troops deserted, Thomas Masaryk, a politician who had been advocating union with the Slovaks, and Edward Beneš went abroad in 1915 to publicize the political aspirations of the Czechs and Slovaks.

The pressure for union between the Czechs and Slovaks grew, first to be an independent state within the Habsburg empire, and then, as that empire began to crumble, as a state totally separate from it. Fresh impetus was gained in January 1918 when Woodrow Wilson, the American president, proclaimed his Fourteen Points, following his country's entry into the war on the Entente's side. Wilson spoke of the need to grant autonomy to the minorities of Austria-Hungary – the 'right of self-determination for all nations'. On 30 May 1918 representatives of Czech and Slovak exiles in Pittsburgh, USA, issued a manifesto on the creation of an independent Czech-Slovak state. On 28 October 1918, as the war was drawing to a close, the Czech National Committee in Prague seized power and proclaimed the creation of the independent Republic of Czechoslovakia, uniting Bohemia, Moravia, 'Austrian' Silesia and Slovakia; on 14 November Masaryk was elected president. On 28 June 1919 the Treaty of Versailles granted international recognition to the new Czechoslovakia and confirmed its borders.

Yet this new state was artificial: a state of two equal halves that were not equal. The Czech half was industrialized and politically mature; the Slovak half, which had suffered considerable neglect under Magyarization, was agricultural and backward. These differences, to which was added cultural distinctions, soon gave rise to Slovak nationalist sentiments. Inevitably, demographic problems also confronted the Czech government. The population included three million Germans, one million Hungarians and half a million Poles out of a total of 15 million; none of these minorities welcomed the new state with much enthusiasm. The emergence of Nazism in Germany in the 1930s encouraged the German-speaking population of the border regions of Bohemia and Moravia, the Sudetenland, to agitate for autonomy.

As Germany rearmed the Sudeten Germans took heart. Their claims intensified; Hitler supported them; the Czech government made concessions; secession was demanded even though the Sudetenland had never been a part of Germany. In September 1938 the British, French, German and Italian governments, without consulting the Czechs, signed the Munich Settlement whereby Prague had to cede to Germany all areas in Bohemia and Moravia which had a German population exceeding 50 per cent, an area that exceeded a third of Czechoslovakia. In return, Germany promised that no further territorial claims would be made. The take-over was quick and peaceful, but it had the effect of making the rest of Czechoslovakia completely indefensible. Thus the balance of power in Central Europe was upset in favour of self-determination as espoused by Woodrow Wilson. Six months later, in March 1939, Nazi troops invaded the Czech Lands and established a German Protectorate of Bohemia and Moravia.

Czech resistance was largely passive, most of the population submitting to their oppressors. As the Red Army began Czechoslovakia's liberation from the east in 1944, the American Army drove in from the west, but no further than Pilsen. Only in May 1945 did the Czechs rise against the Nazis when Soviet troops were just a few days away from Prague. Soviet armies occupied most of the country until the end of 1945 while three million Germans were brutally deported. It was only in 1997 that the ill-feeling generated by Nazi excesses and Czech ethnic cleansing was laid to rest with the formal signing of a joint declaration cementing the postwar reconciliation between Germany and the Czech Republic.

At the end of the war Soviet territorial gains included a part of Slovakia, Subcarpathian Ruthenia. The consequent common border with Czechoslovakia paid dividends when the Central Group of Soviet Forces was established in the country in 1968. In February 1948 the Communist Party seized power in a coup d'état and in May Czechoslovakia was proclaimed a 'people's democracy'. Stalinization followed. In January 1949 both Bohemia and Moravia ceased to exist as political entities. A brief period of political tolerance, known as the 'Prague Spring' began in January 1968 when Alexander Dubček became First Secretary of the Communist Party. The Soviet Union and its Warsaw Pact allies felt threatened by Dubček's experimental 'socialism with a human face'. In line with the Brezhnev Doctrine[1], to stifle the reforms, they invaded Czechoslovakia in August 1968 and Dubček's government was removed. Separate Czech and Slovak republics within a Czechoslovak federation were established. Non-Soviet Warsaw Pact troops were quickly withdrawn, but Soviet troops remained for the next 23 years.

The policies of *perestroika* (re-structuring) and *glasnost* (openness) promoted by the Soviet leader, Mikhail Gorbachev, following his rise to power in 1985, encouraged the Czechoslovaks to protest against the policies of their government. Protests turned into mass demonstrations and stronger demands for reform. What came to be known as the 'Velvet Revolution' began with a huge demonstration in Prague on 17 November 1989. It was quickly followed by nation-wide demonstrations calling for an end to the Communist Party's monopoly of power and the introduction of parliamentary democracy. By the end of the year communist rule was over and Václav Havel had been elected as president.

Any hopes that the old Czech-Slovak antagonism would now wither were soon dashed. Cultural differences and mounting nationalism, together with political and economic problems, drove a wedge between the Czechs and the Slovaks. In the elections of 1992 parties favouring separation were strongly supported; it was generally agreed that this was preferable to the inevitable compromises that would otherwise be necessary. On 1 January 1993 the 'Velvet Divorce' took place and, only 74 years after its creation, Czechoslovakia ceased to exist.

---

1. So-called in the West after Leonid Brezhnev, the Soviet leader who devised it, this doctrine permitted the Soviet Union to intervene when "the essential common interests of other socialist countries are threatened by one of their number."

Area:	78 864 sq km (30 450 sq miles)
Population:	10.19 million; Czechs, Slovaks
Languages:	Czech, Slovak
Religions:	Roman Catholic, Protestant
Capital:	Prague
Administrative Districts:	7 regions and Prague
Date of Independence:	1 January 1993
Neighbouring Countries:	Poland, Slovakia, Austria, Germany

The Disintegration of the Austro-Hungarian Empire 1920

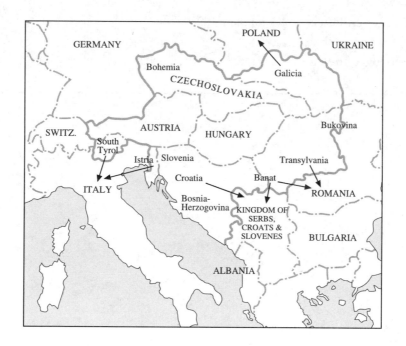

English-speaking Name	Local Name	Former Names	Notes
THE CZECH REPUBLIC	Česká Republika		The Czech Republic since 1 Jan 1993 when it separated from Slovakia. Part of the Czech and Slovak Federative Republic (Apr 1990), of the Czechoslovak Socialist Republic (Jul 1960), of the People's Republic of Czechoslovakia (Jun 1948) and of the Republic of Czechoslovakia (1918). Named after the Czechs, *češi*, a Slav tribe which came from the East and which may possibly have taken its name from *četa*, company or group of warriors or *Čech*, a legendary Slav chieftain who led his people to the region.
Bohemia	Čechy	Boiohemum	Named after the Celtic Boii tribe and means 'Home of the Czechs'. Encompasses five regions of the Republic. Became a kingdom (11th century–1620) within the Holy Roman Empire and thereafter a province in the Habsburg Empire. Part of Czechoslovakia (1918–39 and 1945–92). Čechy is taken from the chieftain Čech. Bohemian has come to describe a socially unconventional person – a 'gypsy' – since gypsies were believed at one time to come from Bohemia.
Brno		Brünn	Probably derived from the Celtic *brynn*, hill town or, less likely, from the Old Czech *brn*, clay; it lies at the confluence of the Svratka and Svitava rivers. Capital of Moravia. It gave its name to the famous Bren gun, developed in Brno and made in Enfield, England, hence the name Bren, and used during and after the Second World War.
Bystrice			Means 'Swift stream'.
České Budějovice		Budweis	Gave its name to Budvar or Budweiser beer.

Františkovy Lázně	Egerbrunnen, Franzenbad	Means 'Franz's springs'. Named after Francis (1768–1835, r. 1804–35) I of Austria and II of the Holy Roman Empire (r. 1792–1806).
Havlíčkuv Brod	Deutsch Brod	Renamed (1945) after the Czech poet and political journalist Karel Havlíček Borovský (1821–56). *Brod* means 'ford' (on the River Sazava) and is retained in memory of the German miners who first settled here.
Hradec Králové	Königgrätz	Means 'Castle of the Queen'. Associated with Queen Elizabeth of Poland, wife of Wenceslas (Václav) II (1271–1305), King of Bohemia (1278–1305) and King of Poland (1300–05). It became the seat of the widowed queens of Bohemia (14th century). Site of the Battle of Königgrätz, also known as Sadowa, (Jul 1866) at which the Prussians crushed the Austrians during the Seven Weeks' War.
Jablonec nad Nisou	Gablonz an der Neisse	According to legend, named after the only apple tree (in Czech, *jablon*) left standing after the area had been laid waste by the Catholic Lusatians during the Hussite Wars (1417–39). Nad Nisou means 'On the (River) Neisse'.
Jáchymov	Joachimsthal	The German name means 'Joachim's valley'. Founded (1516), it is probably named after Joachim I Nestor (1484–1535), Elector of Brandenburg. Silver from the local mines led to a silver coin, the Joachimsthaler, being minted (1519). The word dollar is derived from *thaler*.
Karlovy Vary	Karlsbad	Means 'Charles' Spa' (in Czech, *Var*, boiling). According to tradition, Holy Roman Emperor Charles IV (1316–78, r.1355–78) of Luxembourg (also King of Germany, 1347–78), discovered the spring when a stag he was chasing leapt into it. His personal physician declared the spring to have healing properties and so Charles built a hunting lodge nearby.

Moravia	Morava	North and South Moravia are two regions which take their name from the River Morava which itself comes from the German *mark*, march. Centre of the medieval Kingdom of Great Moravia before being incorporated (1029) into the Kingdom of Bohemia. Became an Austrian crown land (1848) and a province of Czechoslovakia (1918). Annexed by Germany (1939–45).
Olomouc	Mons Julii, Olmütz	Derived from Old Czech *holy mauc*, bare rock. Developed from either a Roman or Slav fort. At the Peace of Olomouc (1478) Moravia was ceded to Hungary. Capital of Moravia (1187–1648).
Pilsen	Plzeň	Name derived from the Old Czech *plz*, damp, a reference to the fact that it was founded at the confluence of four rivers. Famous for Pilsner beer.
Pisek		Means 'sand' from the gold-producing sand of the River Otava.
Prague	Praha	According to legend, some time in the seventh or eighth century, Prince Krok led his people south to a rocky hill that is now known as Vyšehrad, 'High castle'. His youngest daughter, Princess Libuše, who saw the glory of Prague in a vision, decreed that they should build a city at a place in the forest where they would find an old man constructing the threshold (in Czech, *prah*) of his house. He was discovered on Hradčany Hill. Later Libuše married a humble ploughman (in Czech, *přemysl*) and together they founded Prague and established the Přemyslid dynasty. The Czech Protestants were defeated at the Battle of Bílá Hora (White Mountain) (1620), thus permitting Ferdinand II (1578–1637, r. 1617–19, 1620–27, King of Bohemia, to regain his kingdom having been deposed. The Treaty of Prague

Slavkov	(Aug 1866) ended the Seven Weeks' War. Capital of the First Republic (1918). Hitler entered in triumph on the heels of his troops (15 Mar 1939). Liberated by the Soviet Army (9 May 1945). Invaded by Warsaw Pact armies (Aug 1968) to put an end to the 'Prague Spring'. A student demonstration (Nov 1989), savagely broken up by the police, led to the 'Velvet Revolution' and the downfall of the government.
Austerlitz	Nearby is the site of the Battle of Austerlitz (Dec 1805), also known as the Battle of the Three Emperors (Napoleon, Francis I of Austria and Alexander I of Russia), where the French overwhelmed the Austrians and Russians. Austerlitz is remembered as a railway station in Paris.
Tábor	Means 'Fortified camp'. Named after Mount Tabor, Israel, traditionally believed to be the place of Christ's Transfiguration (St Matthew, Chapter 17 although Mt Tabor is not specifically mentioned). Was a fortified base for the Táborites, a radical wing of the Protestant Hussites during the Hussite Wars. Was the first European commune and the inhabitants lived under military discipline.
Theresienstadt	Founded (1780) by Emperor Joseph II (1741–90, r.1780–90) as one of three towns fortified against the Prussians. Named in honour of his mother, Empress Maria Theresa (1717–1780, r.1740–80) as 'Theresa's town'. Became a Jewish ghetto (1941) and transit concentration camp for Jews bound for extermination camps, usually Auschwitz.
Terezin	
Gottwaldov	Previously named (1948–90) after Klement Gottwald (1896–1953), Prime Minister of Czechoslovakia (1946); engineered the Communist coup d'état (Feb 1948); President (Jun 1948).
Zlin	

DENMARK

About 500 AD a Swedish tribe calling themselves Danes moved south and established themselves in what later came to be called Denmark. Part of their possessions, however, was Scania (Skåne) in modern Sweden, which allowed them to control the Sound (between Helsingør and Helsingborg), the exit from the Baltic Sea into the North Sea and beyond. This contributed to their influence and power in the region for the next 11 centuries. The northward drive of Charlemagne's Franks in about 800 forced the Danes to defend themselves, but also inspired them to look beyond their own narrow concerns. The Vikings, pagan Scandinavian warrior/traders (from *vikingr,* pirate), began to dominate the Baltic and in 865 a strong force invaded England, the precursor to the establishment of a relatively short-lived Anglo-Danish empire which included southern Sweden.

During the 11th and 12th centuries Denmark was plagued by internal strife and civil war, but when Valdemar I the Great assumed the throne in 1157 a new era dawned. As Denmark gained in strength a policy of expansion was pursued: Norway was subjugated, Pomerania and Mecklenburg on the north German coast were annexed, Germany's largest island, Rügen, was seized and in 1219 Estonia was conquered. Denmark's southern boundary was redrawn further south at the River Eider, encompassing Schleswig, which became a dukedom and a fief associated with Denmark. Its population was largely German-speaking. When Valdemar II was captured in 1223 he was forced to surrender much of the territory that had been captured – but not Rügen or Estonia which was sold in 1346.

In 1375 Valdemar IV died and was succeeded by his five-year-old grandson, Olav. His father, King Haakon VI of Norway (and an earlier joint King of Sweden), died five years later and Olav succeeded to the throne, thus uniting the two kingdoms. This union between Denmark and Norway was to last until 1814. His mother, Margaret, daughter of Valdemar IV, ruled in his name. When Olav died in 1387, Margaret was elected as regent and given the right to nominate the next king. With no direct heir, she chose her nephew, Erik of Pomerania. In 1388 the Swedish nobles proclaimed her the 'rightful ruler' of Sweden and as a result the three Scandinavian countries, including that part of Finland which belonged to Sweden, were united under a single head – in Denmark Margaret reigned by inheritance, in Norway by marriage and in Sweden by election.

A formal union, the Kalmar Union, was proclaimed when Erik was crowned King of Denmark, Norway and Sweden in 1397. Erik favoured Denmark and Danes were appointed to important positions in Norway and Sweden without any reciprocity. The seeds of discontent were sown and Erik was deposed in 1438 and in both other countries by 1442. His nephew, Christopher of Bavaria, was elected to succeed him, but he died in 1448 without an heir. Count Christian of Oldenburg was then selected and he became King of Norway as well in 1449.

When the last male heir to the County of Holstein died Christian I concluded an agreement with the family in 1460 which allowed him to become Count of Holstein. Furthermore, it was agreed that Schleswig and Holstein were never to be divided, although Holstein remained a fief of the Holy Roman Empire. The seeds of the conflict 400 years later were sown.

The next 60 years were marked by sporadic conflict between Denmark and Sweden which came to a head in 1520 when the Danes attacked and defeated the Swedish army. Christian II was crowned hereditary King of Sweden and the Stockholm Bloodbath took place (q.v.Sweden). Three years later the Swedes rebelled and elected their own king. Jutland, Lübeck and Holstein-Gottorp joined the rebellion, all renouncing their allegiance to Christian who fled the country. The Union was dissolved. Frederick of Holstein-Gottorp, Christian's uncle, was crowned King of Denmark and Norway that same year.

The expansionist policies of both Denmark-Norway and Sweden brought them into conflict, first during the Nordic Seven Years' War (1563–70) and then during the Thirty Years' War (1618–48) when Sweden acquired the eastern Baltic and lands on Denmark's northern and southern borders. Christian IV reacted by contesting Swedish possession of the north German coast. This intervention between 1625–29 proved disastrous, Jutland being temporarily occupied by an imperial army under Wallenstein (commander of the Holy Roman Emperor's forces during the Thirty Years' War).

When Sweden became involved in Poland, Denmark took the opportunity to declare war in 1657. The Swedes immediately occupied Jutland and marched across the frozen sea to Funen, making for Copenhagen. The Treaty of Roskilde in 1658 brought hostilities to an end and the loss of Denmark's territory in Sweden. The Swedes, however, thought that the Danes might still meddle in the continuing war in north Germany. They broke the terms of the treaty and invaded Zealand with a view to taking over the entire country. The Dutch, frightened that this heralded Swedish control of the Sound, sent a fleet to relieve besieged Copenhagen. Combined with Polish attacks on Sweden and uprisings in Denmark, this had the effect of stopping the Swedes in their tracks and brought about the Treaty of Copenhagen in 1660. Although defeated, the Swedes were allowed to keep the Sound provinces taken at the Treaty of Roskilde (Scania, Blekinge and Halland), but the Sound itself would no longer be under the control of a single country. The treaty also largely established the present frontiers of Denmark, Norway and Sweden.

Denmark-Norway's participation in the Great Northern War (q.v.Sweden) ended in 1720. Denmark settled for a large sum of money instead of the return of its provinces in Sweden lost in 1658. When Denmark signed an Armed Neutrality Treaty with Sweden in 1794, later joined by Russia and Prussia, Britain was deeply angered. After Denmark broke a later agreement with the British, this anger took a practical form in 1801 when the Royal Navy was sent to Copenhagen to destroy the Danish fleet. The Danes withdrew from the Neutrality Treaty.

At the mercy of the British fleet, the Danes found themselves in an awkward position when Napoleon announced the imposition of the Continental Blockade in 1806. They played for time but the British, concerned that Denmark might join the continental powers and use its remaining ships to invade England, bombarded Copenhagen in 1807 and captured the Danish fleet. A few days later Denmark joined Napoleon. After the French defeat at the Battle of Leipzig in 1813, and on the losing side, Denmark was forced to cede Norway (but not Iceland, the Faeroes or Greenland) to Sweden at the Treaty of Kiel in 1814.

Intermittently, during the next 50 years, the Schleswig-Holstein question was to complicate Danish-German relations. When the Holy Roman Empire expired in 1806, Denmark brought Holstein (raised to a duchy in 1474) under the Danish crown, a reasonable move since the Danish king was also the Duke of Holstein. However, in 1815, German-speaking Holstein joined the newly-formed German Confederation, though the king remained its duke; the German-speaking element in southern Schleswig believed that Schleswig, a Danish fief, should join Holstein to form a single country within the Confederation. Danish nationalists, however, wanted to separate the two and annex Schleswig to Denmark. When the German-speaking population of Schleswig demanded independence from Denmark and affiliation to the German Confederation in 1848 it was resolutely opposed. However, even the Danish-speaking element was keen to preserve the historical unity between the two duchies and was not averse to joining the Confederation. A rebellion in support of independence broke out and a three year war between Denmark and Prussia ensued. Prussian troops occupied Schleswig-Holstein and Jutland, but were forced to withdraw under British and Russian pressure. Agreements in 1851 and 1852 stipulated that Denmark consisted of three parts: the kingdom, Schleswig and Holstein; that Holstein could continue membership of the German Confederation; and that no measures should be taken to bind Schleswig any closer to Denmark than to Holstein.

Trouble flared again in 1863 when a joint constitution was conferred on Denmark and Schleswig. This meant that the latter was annexed to Denmark in contravention of the earlier agreements. This was unacceptable to Prussia and Austria and the following year Denmark found itself at war. The Danes were defeated and Jutland was again occupied. At the Treaty of Vienna in 1864 Denmark was forced to relinquish both duchies to Prussia and Austria. As a consequence of the Prussian victory over the Austrians in the Seven Weeks' War in 1866 the duchies became part of Prussia. The Danish speakers of Northern Schleswig were dismayed.

Denmark remained neutral during the First World War. At the Treaty of Versailles it was decided that separate plebiscites should be held in the northern and southern parts of North Schleswig to decide to which country they should belong. In 1920 the northern part (north of Flensburg) voted for inclusion in Denmark, the southern half to remain in Germany. The Danish-German border drawn then has remained to this day.

Despite declaring its neutrality, Denmark was occupied by Nazi Germany during the Second World War. In 1944 Iceland declared its independence of Denmark and in 1948 self-government was granted to the Faeroe Islands. In 1953 Greenland was incorporated formally into the Kingdom of Denmark and in 1979 was granted home rule.

Area – Mainland:	43 069 sq km (16 629 sq miles)
Greenland:	2 175 600 sq km (840 000 sq miles)
Population:	5.32 million; Danes, Turks, Faeroese
Greenland:	56 000; Inuit (Eskimo)
Languages:	Danish, Inuit, Faeroese, German
Religions:	Evangelical Lutheran, other Christian, Muslim
Capital:	Copenhagen
Administrative Districts:	14 counties and 2 municipalities
Neighbouring Countries:	Germany

DENMARK

English-speaking Name	Local Name	Former Names	Notes
DENMARK	Danmark		The Kingdom of Denmark (Kongeriget Danmark). Named after the Danes, a tribe from Sweden. The provinces of Skåne and Halland (now in Sweden) formed the Danes' border territory, march or *mark*, in Danish.
Åbenrå		Obenroe, Apenrade	Derived from 'Åen ved Opnør', stream by Opnør, Opnør being an ancient village since disappeared. Prussian (1864–1920), Apenrade being the German name. Å is sometimes written as Aa.
Ålborg		Alabu, Alaburg, Aleburgh	Means 'Town on the river' from the Jutlandish Ål, channel, and *borg*, town. Ålborg lies on the south side of the Limfjord.
Århus		Arosel, Arusensis, Arus	Means ' At the river's mouth' from old Danish *å*, river, and *os*, mouth. The river, or rather stream, is no longer visible.
Bornholm		Burghaendeholm	An island and county with a name meaning 'Island, *holm*, of the Burgundians'. They emigrated to Poland (1st century AD) and later to central France.
Copenhagen	København	Hafn, Mercatorum Portus, Kopmannaehafn	Means 'Merchants' port' from *køber*, merchant, and *havn*, port. The Latin name was adopted (c.1200) before giving way to the Danish (1253). Capital since 1443. Treaty of Copenhagen (1660) largely established the present frontiers of Denmark, Norway and Sweden.
Elsinore	Helsingør		Possibly taken from the Norwegian tribal name Halsen (Hælsing) with *ør*, spit of land, added.

Esbjerg		Until the harbour was built (1868) the area was known as the 'Bank', *bjerg*, where the bait is put on the hook'. It is also possible that the name comes from Eskebjerg, ash tree bank.
Falster		An island whose name is derived from the old Danish *fiala*, hiding place.
Fredericia		Founded (1650) and named after Frederick III (1609–70), King of Denmark-Norway (1648–70).
Frederikshavn		Originally called the 'The beach, *strand*, in the parish of Flade'. Renamed (1818) after Frederick IV (1671–1730), King of Denmark-Norway (1699–1730).
Jutland	Jylland	North, South and West Jutland are counties named after the Jutes, a Germanic people who left Jutland (5th century) to invade southern England. The Battle of Jutland (1916) was the only major battle between the British and German fleets during the First World War. The result was indecisive, both fleets claiming victory.
	Gotland	
Langeland		An island whose name is derived from *lang*, long, and *land*, country.
Nyborg	Nyburg(h)	Named after the new, *ny*, castle, *borg*, built (1170) to protect the Store Strait.
Odense	Othenesuuigensem, Odansue, Othense, Othænsheret	Means the 'Shrine, or sanctuary, of Odin', the Norse god of war. Odin is also known as Woden to give Woden's day, Wednesday.
Ringkøbing	Rennumkøpingh	A county, and small town, deriving its name from the parish of Rindum and *købing*, market.

Roskilde	Roschald, Roschilde	Derived from Hroarskilde, Hroar, or Roir, being the name of its legendary founder, and *kilde*, sacred spring, thus 'Roir's spring'. Capital from c.1020–1443. The Treaty of Roskilde (1658) brought hostilities between Denmark and Sweden to an end and the loss of Denmark's Swedish territory.
Sønderborg	Synderburg, Sundherburg	Means 'Southern castle' after a royal castle probably founded by King Valdemar I the Great (1131–82, r.1157–82). Destroyed and annexed by the Prussians (1864) until returned to Denmark as a result of a plebiscite (1920).
Svendborg	Swineburgh	Name derived from *svin*, wild boar, and *borg*.
Thisted	Tystath, Thistedtt	Derived from the Germanic peoples' god of war and justice, Tyr, and *sted*, place. Tuesday is derived from the Anglo-Saxon version, Tiu.
Vejle	Wæthlæ	Name derived from *vadested*, wading place or ford.
Viborg	Wibergum, Vvibiaergh	Means 'Sacred hill', from *wi* and *borg*, a reference to the fact that it was once a centre of pagan worship and later of Christianity.
Faeroe Islands	Foroyar	Faereyiar	Means 'Sheep islands'. Colonized by the Vikings (c.800). Became a Norwegian province (1035) and, with Norway, passed to Denmark (1380). Under British protection (1940–45). Parliament declared independence from Denmark (1946), but the Danish king dissolved the parliament. Granted home rule (1948). Now seeking to sever all or most links with the mainland; independence is likely to be granted if the majority of the Faeroese desire it. The Faeroes may become a sovereign nation under the Danish monarchy.

Name	Former name	Meaning
Tórshavn		Means 'Thor's harbour', Thor being the Germanic word for thunder. Thor was a warrior god. Thor's day became Thursday.
Greenland	Kalaallit Nunaat	Named (982) by a Norwegian, Erik the Red, to deceive and thus encourage settlers from Iceland. Annexed by Norway (1261). After the original Norse settlements had gradually ceased to exist, colonized by the Danes (1721). Became an integral part of the Kingdom of Denmark (1953) and granted home rule (1979).
Aasiaat	Egedesminde	Originally named (1759) by Niels Egede after his father Hans, a missionary and trader, who founded the first trading company (1721). The present name means 'The spiders'.
Ilulissat	Jakobshavn	Means 'Ice mountains'. Originally named after Jakob Severin who overcame a small Dutch fleet in Disko Bay (1739).
Ittoqqortoormiit	Scoresbysund	Originally named after William Scoresby, a Scottish sea captain and whaler. The present name means 'People with peat'.
Qaqortoq	Julianehåb	Founded (1755) and named after Queen Juliana Maria of Denmark, originally of Brunswick-Wolfenbüttel, the second wife of King Frederick V (1723–66).
Nanortalik		Means 'Bear country'.
Nuuk	Godthåb	An Inuit (Eskimo) word meaning 'Summit or promontory'. Its original name (1721–1979) means 'Good hope'. Capital.
Paamiut	Frederikshåb	Originally named after Frederick V (1723–66), King of Denmark and Norway (1746–66), meaning 'Frederick's hope'. Now means 'Those at the mouth of the Kvanefjord.

Qasigiannguit	Christianshåb	The previous name meant 'Christian's hope' after Christian IV (1577–1648), King of Denmark-Norway (1588–1648). The present name means 'Small spotted seals'.
Qeqertarsuaq	Godhavn	Means 'Big island' with the earlier Danish name meaning 'Good harbour'.
Sisimiut	Holsteinborg	Originally named (1756) after the Danish Count Johan Ludwig von Holstein. The new name means the 'Place where foxes have earths'.

ESTONIA

Unlike the Latvians and Lithuanians who are both Indo-Europeans, the Estonians, like the Finns and Hungarians, are of Finno-Ugric stock. They originally came from the lands between the Ural Mountains and the River Volga, but little is known of the country before the first Viking invasions of the ninth century. Traders rather than settlers, they were squeezed between Danes and Swedes from the west and Russians from the east. Nevertheless, they were still able to maintain a certain territorial integrity. Attempts to Christianize them failed. In 1193 Pope Celestine III proclaimed that crusades against heathens should be pursued, not only in the Holy Land, but also in the north.

In 1202 a German military order of knights, known as the Sword Brothers, was formed to undertake the task. Resistance to them was fierce. Eventually, in 1217, the knights prevailed, although northern Estonia did not succumb. The Danish King Valdemar II was invited to help and he invaded Estonia from the north in 1219; the Estonians were finally defeated in 1227. Thereafter, the Sword Brothers went on to complete the conquest of Livonia (now Latvia and southern Estonia) by 1230. Defeated by the pagan Lithuanians in 1236, the Sword Brothers were absorbed into the Order of the Teutonic Knights as a subsidiary branch, the Livonian Order, the following year. Its aim was to bring the area between the River Vistula in Poland and the Gulf of Finland under German rule. For seven centuries the Estonian people, the majority peasants, were subject to a ruling class of Germans.

In the north, in 1343, the Estonians rose in revolt against their Danish masters. The knights rushed to quell it and took military control. This turn of events left the Danish king no choice but to sell his share of Estonia (modern northern Estonia), including the islands, to the Livonian Order in 1346. Estonia was now legally a part of the Holy Roman Empire and of the Confederation of Livonia; Germanization gathered pace.

The defeat of the Teutonic Knights at the Battle of Grünwald (Tannenberg) in 1410 by Polish-Lithuanian forces seriously weakened the Order and began its gradual collapse. Sweden seized northern Estonia in 1558. The same year, hoping to prevent domination of the area by Poland-Lithuania and gain access to the Baltic Sea at Narva, Tsar Ivan IV the Terrible invaded Estonia and Livonia. This signalled the start of the Livonian War (1558–83). The Livonian knights were too weak to offer much resistance and in 1561 disbanded their Order. Estonia was allocated to Sweden. Ivan now became hard-pressed to retain his gains in the face of opposition from the Swedes, Poles and Danes. In 1581 the Swedes recaptured Narva and drove the Russians from the Gulf of Finland. Finally, in 1582/3, Russia made peace, renouncing its claims to Livonia and Estonia.

The Swedes were not satisfied. After fighting the Poles for the Baltic seaboard, they took control of all Livonia by the Treaty of Altmark in 1629. Finally, by driving the Danes from the islands in 1645, Sweden became the undisputed ruler of modern Estonia, restraining the power of the German landowners and beginning what became known as the "good old Swedish times". Nine years later both the Poles and the Russians returned to the fray, but were beaten off after six years of fighting.

In 1700 the Russians, Poles and Danes combined to launch a military campaign against Sweden which was to become known as the Great Northern War (1700–1721). The Russian Tsar, Peter I the Great, keen to have an outlet to the Baltic Sea which would provide direct access to the West, considered Karelia and Ingria (q.v.Finland) to be ancient Russian lands. Furthermore, with the ports of Riga, Reval (now Tallinn), and Narva in Swedish hands, Russia was suffering severe economic loss. After defeating the Russians at Narva in 1700, Charles XII of Sweden invaded Russia, but disaster followed at Poltava (Ukraine) in 1709. The Swedish will collapsed and the next year Peter drove them out of Estonia and Livonia. The final result of the war was the complete destruction of Swedish power in the Baltic. In 1721, at the Treaty of Nystad, Sweden ceded Estonia and most of Livonia to Russia.

For the next 197 years Estonia remained a province of Russia although the influence of the German nobility remained strong. The unification of Germany in 1871 encouraged Russification for fear that the peasants might become more Germanized which, in turn, might encourage German expansion into the Baltic provinces.

In March 1917 revolution in Russia brought about the abdication of Tsar Nicholas II and Estonian demands for autonomy increased. In April the new Russian Provisional Government agreed to amalgamate Estonia and the Estonian-speaking northern part of Livonia and give it autonomy. Eight months later Lenin seized power in Russia and an Estonian communist puppet government was appointed. German forces moved in to eject the Bolsheviks and by late February 1918 had occupied the country. That same month the Estonian Diet proclaimed independence, a declaration the Germans ignored. At the Treaty of Brest-Litovsk the following month Soviet Russia gave up its Baltic territories to Germany. German occupation lasted until the armistice in November 1918.

The Bolshevik government in Russia declared the Treaty of Brest-Litovsk null and void and quickly filled the vacuum with its own troops. The Estonians resisted and, after 14 months of conflict, were able to free the country and advance into Soviet Russia. The Treaty of Tartu ended hostilities in February 1920. The Soviet government recognized the independence of Estonia, renounced 'for all time' all rights of sovereignty, and surrendered a little territory beyond the River Narva in the north-east and in Pskov province inhabited by Estonians.

The non-aggression pact signed by Germany and the Soviet Union in August 1939 included a secret protocol providing for the annexation by the USSR of Estonia (and Latvia). The following month the Soviet Union imposed a Treaty of Mutual Assistance on Estonia whereby 25 000 Soviet troops as well as air and naval units would be stationed in the country. In June 1940 Estonia was occupied by Soviet troops in violation of the pact. The government resigned and was replaced by Soviet appointees who promptly applied for Estonia to join the USSR. In August it was admitted into the Union as the Estonian Soviet Socialist Republic.

Soviet rule did not last long. In July 1941 German troops invaded and remained in control until ejected by Soviet troops in September 1944. Some 2 200 square kilometres of Estonian territory south of Lake Pskov were ceded to the Soviet Union. Sovietization began in earnest, spurred on perhaps by the population's truculence and sporadic attacks by an anti-Soviet guerrilla movement. It was only in the years following Mikhail Gorbachev's assumption of the post of General Secretary of the Soviet Communist Party in 1985 and the introduction of his policy of *glasnost* (openness) that the Soviet grip loosened in any meaningful way.

In November 1988 the Estonian Supreme Soviet declared the sovereignty of the Republic, but it was not until March 1990 that a transitional period towards the re-establishment of independence was announced. On 19 August 1991 an attempt was made in Moscow to overthrow Gorbachev. The next day the Estonian Supreme Council proclaimed full and immediate independence. This was recognized by the Soviet Union two weeks after the coup had collapsed. Thousands of Russian troops were based in Estonia but, by the end of August 1994, they had all been withdrawn.

The Russian interest in Estonia remains: it is considered to be in Russia's natural sphere of influence. Moreover, over a third of the population (of which the Russians contribute 30 per cent) are East Slavs, largely due to the policy of settling Russians in the non-Russian republics of the Soviet Union. Many Soviet military personnel who served in Estonia chose to stay there after retirement.

Area:	45 227 sq km (17 462 sq miles)
Population:	1.42 million; Estonians, Russians, Ukrainians, Belarusians, Finns
Languages:	Estonian, Russian
Religion:	Estonian Orthodox, Evangelical Lutheran
Capital:	Tallinn
Administrative Districts:	15 counties
Dates of Independence:	24 February 1918/20 August 1991
Neighbouring Countries:	Russia, Latvia

English-speaking Name	Local Name	Former Names	Notes
ESTONIA	Eesti	Estland	The Republic of Estonia (Eesti Vabariik) from 1918–1940 and since Aug 1991. The Estonian Soviet Socialist Republic (1940–41, 1944–91). Named after the Eestii, people living on the eastern shores of the Baltic Sea. 'Old Estonia' joined with the northern part of Livonia to form modern Estonia (1920).
Hiiumaa		Dagö, Khiuma	Means 'Giant's land' because, according to legend, a giant lived on the island. The previous Swedish name of Dagö means 'Day island' because it took a whole day to reach it from the mainland.
Kuressaare		Arensburg, Kuressaare, Kingisepp	Named (1952–91) after Viktor Kingisepp (1888–1922), an Estonian Bolshevik leader.
Paide		Paelinn, Weissenstein	Means 'Town of limestone'.
Saaremaa		Ösel	The largest island which, after occupation by the Danes (1559), Swedes (1645) and Russians (1721), only became Estonian in 1918.
Tallinn		Kolyvan, Reval	Originally meant 'Strong' or 'Brave' from the old Estonian *kaleva*. Then renamed Tanin Lidna, 'Danes' Fort' (in Estonian, *Taani Linn*), having been founded (1219) by the Danish King Valdemar II (1170–1241, r.1202–41). Sold to the Teutonic Knights (1346), passed to Sweden (1561), captured by the Russians and held by them (1710–1917). Shortened to Tallinn (17th century). Changed to Reval by the Germans after the ancient district of Rävala and back to Tallinn

(1917). Reval may come from the Danish word *refvall* alluding to the rock circle which rises sharply from sea level or from *rev*, sandbank. Capital since 1918.

Tartu   Tarbatu, Yuryev, Derpt, Dorpat

Originally a settlement to provide protection from invaders, it may be named after the ancient Estonian god Tar. Then named (1030) after its founder, the Grand Prince Yaroslav I the Wise of Kiev (1019–54) whose baptismal name was Yuriy, who built a fortress on the site. Then Derpt/Dorpat, whose meaning is unknown, (from 13th century until 1893) and Yuryev again (1893–1918). The Treaty of Tartu (Feb 1920) ended hostilities between Estonia and Soviet Russia which then recognized Estonian independence.

FINLAND

Following in the footsteps of their primitive ancestors and coming from the eastern shores of the Baltic Sea to the south, the first Finns, probably small groups of hunters and trappers, began to settle in south-west Finland in about 400. When, in the eighth century, Swedish Vikings also began to establish themselves on the south-west coast and open up trade routes to the east, the Finns moved inland and eastwards to Lake Ladoga (now Russia). From then until the 12th century Finnish history is obscure.

During the 12th century religious rivalry broke out: the west of Finland gravitated towards Swedish Catholicism while the east favoured the Russian Orthodox Church. In 1157 King Erik of Sweden launched an expedition, partly a crusade and partly an attempt to strengthen trade routes, into Finland. Swedish influence grew in the west, but further attempts to conquer Karelia in the east at the beginning of the 14th century were only partially successful. At the Treaty of Pähkinäsaari (now Petrokrepost, Russia) in 1323 a boundary was drawn, somewhat ill-defined through unpopulated wilderness and thus open to later dispute, from the eastern end of the Gulf of Finland north-west to the Gulf of Bothnia. Finland and the western part of Karelia were confirmed as part of the Swedish kingdom while the eastern part fell to the Russian principality of Novgorod.

From 1323[1] to 1809 Finland remained subject to Sweden and from then until 1917 to Russia. Only in 1917 did Finland gain its independence.

In 1362 Haakon, a joint King of Sweden (and Haakon VI of Norway), gave Finland a status equal to the other provinces in the Swedish kingdom and the right to vote in royal elections. The Finns were fortunate: they had not been subjugated and they had not lost their lands or their freedom. Thus there was no antagonism between the two peoples and the result was that there was no conflict between them during the five centuries that they were together.

In 1364 Haakon was overthrown by rebellious Swedish nobles and the crown was given to Albert of Mecklenburg. His unpopularity encouraged a Swedish nobleman, Bo Jonsson Grip, to undertake a campaign to win Finland for himself. By 1374 he had succeeded. Before he died in 1386 he had ensured that Finland's interests should be guarded by the Swedish nobility rather than the king. When Margaret, regent in Denmark and Norway, agreed to accept the rule of Sweden in 1388, Finland was included in the 1397 Union of Kalmar (q.v.Denmark).

Part of the Swedish realm, Finland played its part in the many wars that Sweden fought during the next three centuries. In 1570 hostilities broke out with Russia and they were not brought to a conclusion until 1595 at the Treaty of Täyssinä when Russia formally recognized a Russo-Finnish border, a little further to the east and up to the Arctic coast. In the meantime, in 1581, the Swedish king had made Finland a grand duchy and in 1634 it became an integral part of Sweden. The Peace of Stolbova in 1617 between Russia and Sweden gave to Finland Ingria (also known as Ingermanland, at the eastern end of the Gulf of Finland) and more of Karelia; this had the effect of excluding Russia from the Baltic. The population of these areas were of the Greek Orthodox faith which accounts for the presence of adherents in Finland today.

The Great Northern War, in which Swedish power in the Baltic was challenged by a coalition of Russia, Denmark-Norway and Poland, began in 1700. After capturing Viipuri (now Vyborg, Russia) in 1711, the Russians occupied Finland from 1713–1721. Under the Treaty of Nystad (now Uusikaupunki) in 1721 Sweden had to cede the south-eastern part of Finland – southern Karelia, including Viipuri and Kexholm, – Ingria and the Baltic provinces. With these acquisitions Russia regained a Baltic coastline and a protective zone round its new capital of St Petersburg. The Russians and Swedes went to war again in 1741. Two years later it ended with the Treaty of Turku which gave a further slice of south-eastern Finland to Russia.

Following the Peace of Tilsit in 1807 France and Russia became allies. Diplomatic efforts to get Sweden to join the alliance against England failed and in February 1808 the Russians invaded Finland. In accordance with their military strategy, Swedish forces withdrew into Sweden. The Finns, weary of tight Swedish control, were prepared to accept Russian assurances of autonomy. By the Treaty of Fredrikshamn (now Hamina) in 1809 Sweden ceded Finland, including the Åland Islands, to Russia. Finland became a grand duchy under Russian protection – Tsar Alexander I called himself Emperor and Autocrat of all the Russias and Grand Duke of Finland – and a state in its own right for the first time in its history. The Finnish territory ceded by Sweden in 1721 and 1743 was returned in 1812 and most of the border drawn then continues to the present day; Helsinki became the capital in place of Turku.

Gradually Finnish nationalism began to develop. Four-fifths of the population spoke Finnish while some 15 per cent spoke Swedish, the language of the administration, higher education and polite society. The movement to get Finnish to be the language of all Finns got off to a slow start, but in 1863 Tsar Alexander II decreed that Finnish was to be given equal status to Swedish in the administration in 20 years time. After Tsar Nicholas II declared in 1899 that he had the power to enforce laws in Finland without the Finnish Diet's approval, a policy of Russification began. Russian was to become the third official language and the Finnish Army was disbanded. Resistance, combined with a decline in Russian fortunes with defeat by Japan in 1905, led to growing internal dissent. When the Bolsheviks seized power in November 1917, the Finnish parliament decided to sever ties with Russia and independence was declared on 6 December 1917.

In January 1918, led by a revolutionary Marxist element, the Social Democratic Party attempted to bring down the government. They seized Helsinki and other towns in southern Finland. The Finnish Civil War continued until May by which time the Finnish communists had been trounced. Although not technically at war with the new Soviet regime, the hostility between the two countries could only be resolved by treaty. The Treaty of Tartu (q.v.Estonia) in February 1920 fixed Finland's eastern border: in the south it remained unchanged, but in the north the port of Petsamo (now Pechenga, Russia) was acquired.

After German and Soviet forces had dismembered Poland in 1939, the Soviet Union made various demands on neutral Finland, largely to "increase the security of Leningrad". The most important were to lease the Hangö peninsula and the surrounding area for 30 years, to cede various islands in the Gulf of Finland, and to redraw the border on the Karelian Isthmus in favour of the Soviet Union. The Finns refused, but did not believe that the Soviet Union would resort to war. They were sadly disabused of this notion, however, on 30 November when Soviet troops invaded and the ill-judged Winter War began. Without foreign assistance and despite heroic resistance, the Finns were forced to sue for peace and this was granted in March 1940 by the Treaty of Moscow. Finland had to cede territory in the south-east, not only those areas demanded the previous year, but up to the line agreed in 1721, and lease Hangö as a Soviet base for 30 years. The Treaty also included a clause whereby both states agreed not to undertake any aggression against the other or join any alliance directed against the other. By fighting so well, however, the Finns were able to prevent total Soviet occupation and the need to accept a conqueror's terms.

Although claiming not to be an ally of Germany but rather pitched into the war by unprovoked Soviet aggression, the Finns took the opportunity, after the German attack on the Soviet Union in June 1941, to regain the territory lost the year before. Their attempts to convince the world that their 'Continuation War' was a 'private' war with the Soviet Union – merely to win back lost territory – and that they were neutral as far as the World War was concerned were not wholly successful.

As the German onslaught on the Soviet Union was brought to a halt and then reversed, the Finns lost enthusiasm for the war and in February 1943 decided to pull out of it. Without guarantees of freedom and independence this proved impossible. In April 1944 Soviet terms were rejected on the grounds that they were so stringent that the Finns would be unable to fulfill them and would therefore lay themselves open to Soviet occupation. In June Soviet forces attacked in the Karelian Isthmus and, although the Finns were able to stabilize their lines in July, they began negotiations for a ceasefire. In September 1944 a peace accord was signed with the Soviet Union (confirmed by Treaty in Paris in February 1947) whereby Finland consented to the conditions of the 1940 Treaty of Moscow, agreed to expel all German forces from the country, cede Petsamo on the Barents Sea and lease the Porkkala peninsula (instead of Hangö) for 50 years; in the event it was returned in 1955. Thus some 46 600 square kilometres (18 000 square miles) – larger than Denmark – became Soviet and Finland lost its direct access to the ice-free Arctic Ocean.

1. Some argue that Finland became a part of the Swedish realm as a result of Erik's invasion in 1157.

Area:	338 145 sq km (130 558 sq miles)
Population:	5.18 million: Finns, Swedes, Lapps (Sami)
Languages:	Finnish, Swedish, Lapp, Russian
Religions:	Evangelical Lutheran, Greek Orthodox
Capital:	Helsinki
Administrative Districts:	11 provinces and one autonomous province
Date of Independence:	6 December 1917
Neighbouring Countries:	Norway, Russia, Sweden

English-speaking Name	Local Name	Former Names	Notes
FINLAND	Suomi		Republic of Finland (Suomen Tasavalta) since 1917. The local name may come from the Finnish *suo*, marsh, and *maa*, land, in reference to the country's many lakes. The English Finland, the land of the Finns, may be from the Germanic *finna*, fish-scale, which is *suomu* in Finnish. This could be a reference to the type of clothing worn by the primitive Finnish tribes. Another theory is that *suomi* comes from the old Swedish *somi*, mass, meaning that the Swedes at one time considered the Finns to be inferior, 'the masses'. The *Fenni* was the name given to the Lapps, or Sami, by Tacitus in his *Germania* (98).
Åland Islands			Means 'River land'. Though belonging to Finland with the Finnish name Ahvenanmaa, these islands were long inhabited by Swedes and Swedish language and culture are still predominant. Seized by Russia (1714) and included when the Grand Duchy of Finland was ceded to Russia (1809). The Ålanders sought reunion with Sweden (1919–21) but the Islands were allocated to Finland by the League of Nations (1923).
Hamina		Frederikshamn	Originally named (1753) after Frederick I (1676–1751), King of Sweden (1720–51). The Swedish *hamn* means harbour. Under the Treaty of Frederikshamn (1809) Sweden ceded Finland to Russia.
Helsinki		Helsingfors	The Swedish *fors* means waterfall, a reference to the fall at the original site of the city at the mouth of the River Vantaa, some three miles north of its present location which has better access to the sea. Helsing comes from Helsingi, a tribal name. Helsinki is the Finnish version of the Swedish name. Attacked by Russia (1808) and incorporated into the

		Grand Duchy of Finland (1809). Capital since 1812. To reduce tension between the Soviet Union and the West (by accepting the status quo in Europe – the boundaries agreed at Yalta), the Helsinki Final Act (or Helsinki Accords) was signed (Aug 1975) at the end of the first Conference on Security and Co-operation in Europe.
Kokkola	Karleby	Almost entirely Swedish-speaking, it was known by its Swedish name until the 20th century.
Lappeenranta	Villmanstrand	Founded (1649) by Queen Christina of Sweden (1626–89; queen 1644–54). On the coat of arms that she received was a somewhat primitive-looking man so the town was called Villmanstrand, 'Wild Man's Shore'. The Russians defeated the Swedish and Finnish armies here (1741) and the town passed to Russia under the Treaty of Åbo (now Turku) (1743).
Mikkeli	St Michel	Named after St Michael.
Oulu	Uleåborg	Means 'Water' with *borg*, fort, added for the Swedish name. It lies on the River Oulujoki.
Porvoo	Borgå	The Finnish and Swedish names both come from the River Porvoonjoki/Borgåå on which it lies. Finland was granted autonomy as a grand duchy after the Finnish Diet had sworn allegiance to Tsar Alexander I here (1809).
Tampere	Tammerfors	The second half of the name in both cases means 'rapids' (the Tammer rapids) which were close to where the town was founded.

Turku	Åbo	Means 'Market place'. At the mouth of the River Aura, the Swedish Åbo means 'Settlement by the river' from *å*, river, and *bo*, settlement. The Treaty of Åbo (1743) ended a two-year war between Sweden and Russia which gave Russia a slice of southern Finland. The capital before Helsinki but moved (1812) because it was considered too close to Sweden.
Uusikaupunki	Nystad	Both names mean 'New town' although Nystad was founded in 1617 by Gustav II Adolf (1594–1632) King of Sweden 1611–32). At the Treaty of Nystad (1721), which concluded the Great Northern War (1700–21), Sweden ceded Estonia, Ingria and Livonia 'in perpetuity' to Russia as well as a strip of Karelia, including Vyborg.
Vaasa	Korsholm, Nikolainkaupunki, Vasa	Re-named (1606) after the Swedish royal house (1523–1818). Destroyed by fire (1852), rebuilt and given the name Nikolainkaupunki, meaning the 'Town of Nicholas' after the Russian Tsar Nicholas I (1796–1855, r.1825–55).

FRANCE

The Gauls, a Celtic people, spread west and south from the Rhine valley into what is now France and Italy from about 1200 BC. In the second century BC the Romans began their conquest of Gaul which Julius Caesar completed between 58 and 52 BC. His successor, Augustus, then divided Gaul into four provinces: Aquitania, Belgica, Lugdunensis and Narbonensis (previously Provincia, The Province). Five centuries of Roman rule brought the Gauls Christianity, the Latin language, a legal system and a road network.

Frankish raids into Gaul from east of the Rhine frontier of the Roman Empire began in about 250 and thereafter Roman power began to decline. These onslaughts became more intense during the fifth century and encouraged the Franks to exploit their success and settle north of the River Loire; the Visigoths penetrated the south-west (Aquitaine and Provence), and the Burgundians moved into the south-east. As the Franks converted to Catholicism and mixed with and married the Gallo-Romans, they came to identify themselves with the local culture and language, although continuing to call themselves Franks. After the fall of the Roman Empire in 476, Clovis, a Frankish chieftain and grandson of Merovech after whom the Merovingian dynasty is named, united all the Frankish tribes under his authority, subdued Burgundy, annexed Aquitaine and pushed the Visigoths beyond the Pyrenees.

After his death in 511 Clovis's kingdom was divided between his four sons. During the next 20 years they succeeded in bringing practically the whole of Gaul under their control; and in 558 the kingdom was reunited under Chlotar I when the third of his brothers died. Three years later Chlotar died and his kingdom was again divided between his four sons. Rivalry and conflict ensued. Over the next 200 years Frankish hegemony was challenged. Weak and lazy kings were unable or unwilling to respond and in due course began to give up their power to a high official known as the Mayor of the Palace. At the Battle of Poitiers (sometimes Tours) in 732 Charles Martel, one of the Mayors, defeated the Moors and stopped further Muslim penetration into France. Over the next nine years he made himself master of most of the kingdom. Martel died in 741 and his two sons shared his inheritance. A new king came to the throne in 743, but one of his sons, Pepin III the Short, deposed him and in 751 usurped the throne and founded the Carolingian dynasty, the most famous member being one of his sons, Charlemagne, Charles the Great.

Charlemagne set out to extend his demesne. He conquered all of Gaul but was unable to subdue the Bretons of Brittany. He pushed eastwards to the River Elbe and into Spain, making the Frankish Empire the greatest power in Western Europe. The unpopular Pope Leo III sought his support and on Christmas Day 800 crowned Charlemagne Holy Roman Emperor. By this act Leo also hoped to demonstrate the supremacy of the Papacy. While Charlemagne was pleased to be recognized as an emperor, he felt that he had achieved the imperial title through his own qualities rather than coronation by the Pope. Charlemagne died in 814 and his empire did not long survive him. In 806 he had ordained that his empire should be divided between his three sons. Two of them, however, died before he did. At the Treaty of Verdun in 843 his grandsons divided the empire into three kingdoms: Louis II the German received a Teutonic nation in the east, called Francia Orientalis, later to be called Germany; Charles II the Bald was given a Romanic state called Francia Occidentalis, later to become France, in the west; and a Middle Kingdom, Francia Media, the northern part of which was named Lotharingia after Lothair I. When Lothair II died in 869 Lotharingia was divided between his uncles, Louis and Charles. This corridor of land was to be disputed intermittently (as Lorraine) between France and Germany until the end of the Second World War.

The Carolingian Empire continued to disintegrate in the years after 843. Largely independent states that owed only nominal allegiance to the king began to develop. The arrival of Vikings in the Loire region led to Charles III the Simple creating the Duchy of Normandy in 911. In addition, Anjou, Aquitaine, Brittany, Burgundy and Gascony became ever more powerful. When Louis V died in 987, Hugh Capet was elected king and the Carolingian dynasty ended. Capet's demesne centred on Paris, Orléans and Blois, later known as the Île-de-France.

In 1066 Duke William of Normandy invaded England. At the Battle of Hastings the English were defeated and the king killed. William proclaimed himself king. As Duke of Normandy he remained a vassal of the King of France. But as King of England he constituted a serious threat. In 1137 Eleanor inherited the Duchy of Aquitaine, which now also included Gascony, and in 1152 she married the grandson of Henry I of England, Henry Plantagenet. Already Count of Anjou and Duke of Normandy, he was more powerful than the King of France. In 1154 Henry became King of England as Henry II, thus uniting that country, Normandy and the west of France under his rule.

When Philip II Augustus came to the French throne in 1180 he was determined to expand the Capetian demesne and drive the Plantagenets from France. He was fortunate in that John, youngest son of Henry II and Eleanor, became King of England in 1199. By 1214 John had lost most of his continental possessions to Philip. By this time too, the French king's control had spread well beyond the Île-de-France. Expansion continued and by the time the Capetian dynasty came to an end in 1328, only Aquitaine, Brittany, Burgundy, and Flanders remained outside the French kingdom which had become the most powerful in Europe.

In 1328 the crown passed to Philip VI of Valois, a cousin of the last Capetian king. However, Edward III of England, whose mother, Isabella, was the daughter of Philip IV of France (who had died in 1314), also had a legitimate claim to the French throne. Initially Edward accepted Philip's accession, but when Philip moved to seize Guyenne (part of Aquitaine) in 1337, Edward, as Duke of Guyenne, renewed his claim and took an army to Flanders. Thus started the Hundred Years' War (1337–1453). After early English advances the pendulum swung towards the French when the 17 year-old Joan of Arc raised the siege of Orléans in 1429. By 1453 Normandy and Aquitaine had been conquered and English territory was confined to Calais (which was given up in 1558).

By the time Louis XI died in 1483 Anjou, Brittany, Burgundy, Picardy and Provence had been absorbed. The 17th century, when France was the dominant power in Europe under Louis XIV, and the first two-thirds of the 18th saw further acquisitions in the south-west (Armagnac and Navarre), in the north-east (Artois) and the east (Alsace, Lorraine and Franche-Comté); French borders became much as they are today. Flanders was an area of special interest, not only to France but also to England and Spain which controlled the Spanish Netherlands. No country wanted to see it occupied by the others.

In due course, Flanders was divided between France, Belgium and The Netherlands.

During the 18th century France was intent also on containing England and expanding its colonial empire. In 1789, however, revolution swept the monarchy away, a republic (the First Republic) being declared in 1792 and Louis XVI being executed the following year. The Reign of Terror followed during which thousands went to the guillotine. A new regime, the Directory, ruled from 1795 to 1799 when General Napoleon Bonaparte seized power. He became First Consul and, in 1804, Emperor (of the First Empire). A supreme military leader, Napoleon then proceeded to overrun most of Europe before his ill-advised adventure into Russia ended in disaster in 1812 and subsequent defeat at the Battle of Waterloo in 1815.

The monarchy was restored. But by the Treaty of Paris France lost several border districts, including Savoy and the Saar basin that had been annexed in 1789–92, and French borders returned to their pre-1789 positions. In 1848 the monarchy was overthrown again to be replaced by the Second Republic with Louis Napoleon, nephew of Napoleon Bonaparte, as president. Four years later, to popular acclaim, he became Emperor Napoleon III.

This Second Empire, however, was to be undone by the Franco-Prussian War, engineered by the German chancellor, Bismarck, in 1870. It proved to be a military disaster for the French. The Germans reached Paris and Alsace-Lorraine was lost; Napoleon was taken prisoner. He was deposed and the Third Republic was declared.

In 1904 the Entente Cordiale was signed with Britain and the French turned their attention to their colonial empire (renamed the French Union from 1944). The First World War was fought mainly on French soil, but after the Allied victory France regained Alsace-Lorraine.

The Third Republic survived until 1940. After surrendering to invading German forces in May, northern France came under German occupation while the French collaborationist government, headed by Marshal Pétain and called the French State, moved to Vichy; at the same time General de Gaulle formed a Free French Government-in-Exile in London. France was liberated by Allied Forces in 1944 and France's pre-war borders were restored; additionally, small areas of Italy were ceded to France. The Fourth Republic was instituted in 1946 and the Fifth in 1958.

Area:	547 026 sq km (211 207 sq miles)
Population:	59.06 million; French, Germans, North Africans, Indo-Chinese, Basques
Language:	French
Religions:	Roman Catholic, Muslim
Capital:	Paris
Administrative Districts:	22 departments
Neighbouring Countries:	Belgium, Luxembourg, Germany, Switzerland, Italy, Monaco, Spain, Andorra

The Plantagenet Lands 1154-1300

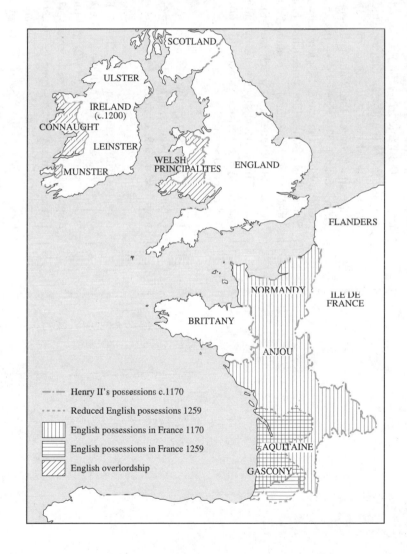

Legend:
- — · — Henry II's possessions c.1170
- ‑ ‑ ‑ ‑ Reduced English possessions 1259
- English possessions in France 1170
- English possessions in France 1259
- English overlordship

Map labels: SCOTLAND, ULSTER, IRELAND (c.1200), CONNAUGHT, LEINSTER, MUNSTER, WELSH PRINCIPALITES, ENGLAND, FLANDERS, NORMANDY, ILE DE FRANCE, BRITTANY, ANJOU, AQUITAINE, GASCONY

English-speaking Name	Local Name	Former Names	Notes
FRANCE	La France	Gaul, Francia	French Republic (République Française) since 1792. However, the monarchy was not abolished until Emperor Napoleon III (1808–73) was deposed (1870). Name derived from the Germanic tribe, the Franks, who conquered Gaul after the fall of the Roman Empire. Under the first Frankish dynasty, the Merovingians (476–750), Francia was confined to the area between the Rhine and Seine rivers (see Île-de-France below). The Franks gave their name to the French monetary unit, the franc. Belgium, Luxembourg and Switzerland also use the same name (since 1 Jan 1999 a sub-division of the euro). The name Frank probably meant 'fierce' and in time 'free' which gave the English word 'frank'. The wooden shoes worn by the Gauls were called *gallicae* by the Romans, hence the word galoshes (now a waterproof overshoe).
Aix-en-Provence		Aquae Sextiae	Founded (123 BC) by the Roman proconsul Sextius Calvinus, the waters, *aquae*, were named after him. The town became the capital of Provence (see below).
Alsace		Elsass	A region with a name meaning 'Those living on the outside', that is, west of the River Rhine. A Frankish duchy (496), subsumed into Lotharingia (870) and thereafter a much disputed territory. Ceded to Germany (1871) after the Franco-German war; returned to France (1919), occupied by Germany (1940) and once more restored to France (1945). Gave its name to the Alsation breed of dog, also known as the German shepherd.
Amiens		Samarobrivia, Ambianum	Named after the Ambiani. Samarobrivia means 'Bridge over the River Somme'.

Angers	Juliomagus	Named after the Gallic tribe, the Andecavi. Under the Romans it was called 'Julius (Caesar)'s market'. Capital of the former Duchy of Anjou. Geoffrey IV Plantagenet, Count of Anjou, married (1128) Matilda, daughter of Henry I of England. Their son became Henry II of England (1154) thus founding the English House of Plantagenet which provided 14 kings (1154–1485).
Antibes	Antipolis	Means the 'City on the opposite side of the bay', i.e. from Nice.
Aquitaine	Aquitania	A region with a name associated with *aqua*, water; it lies on the Bay of Biscay. When Eleanor (c.1122–1204), daughter of the Duke of Aquitaine, married (1137) Louis VII (1120–80, r.1137–80), Aquitaine and France were united. After their divorce Eleanor married the Count of Anjou (1152) who two years later became Henry II, King of England (1133–89, r.1154–89). Aquitaine remained an English possession (until 1453).
Arles	Arelate	Located where the Rhône River delta begins, the city's name is probably derived from the Indo-European root word *ar*, water or river.
Arras	Nemetacum/Nemetocenna	Named after the Atrebates, a tribe in north-east Gaul. The Peace of Arras (1482) fixed the northern borders of modern France.
Auvergne		A region named after the Celtic Arverni tribe.
Auxerre	Autessiodurum	Probably taken from a personal name, meaning 'Fort of Autessio'.

Place	Roman/Old name	Description
Avignon	Avennio	The name is based on the Indo-European root word *ab*, water. The present name is derived from the Roman name. Papal seat of the Catholic Church (1309–1377). Factionalism in Rome led the French king, Philip IV the Fair (1268–1314, r.1285–1314), to invite the pope to set up his seat in Avignon. Bought (1348) by Pope Clement VI (1291–1352) and remained papal property until it was annexed by France (1791).
Bayeux	Baiocasses, Augustodorum, Civitas Baiocassium	Derived from the name of the Celtic Baiocasses tribe. The famous Bayeux Tapestry, depicting the Norman conquest of England (1066), was probably woven here (c.1092).
Bayonne	Lapurdum	The name is taken from the Low Latin *baia*, bay, and the Basque *on*, good. First produced (17th century), the bayonet took its name from the city.
Beauvais	Caesaromagus, Civitas de Bellovacis	Named after the Bellovaci tribe whose capital it was; after Caesar captured it (52 BC) it became 'Caesar's market' and then 'City of the Bellovaci'.
Besançon	Vesontio	Derived from its original name which is based on the Indo-European root word *ves*, mountain. The town was developed at the foot of a high rock on which the Romans built a fort.
Bordeaux	Burdigala	May be taken from the name of the local Celtic tribe, the Bituriges Vivisci.
Boulogne, Boulogne-Sur-Mer	Gesoriacum, Bononia	Name derived from the Gaulish *bona*, fort.
Bourges	Avaricum	Names after the Bituriges whose capital this was.

Brittany	Bretagne	Armorica was the Romanized version of the Celtic word for seaside. Named after the Britons who fled to this area of France (5th century) to escape the invading Angles, Saxons and Jutes. An independent duchy, it did not become part of France (until 1532). Now a region.
Burgundy	Bourgogne	A region named after the Burgundians, a Scandinavian tribe who migrated from the southern shores of the Baltic Sea (1st century). During its history parts of Burgundy have been a kingdom, a duchy and a county. Upper Burgundy was known as Jurane Burgundy and Lower Burgundy as the Kingdom of Provence. When these two were united they became known as the Kingdom of Arles (from 13th century). Cisjurane Burgundy referred to the County of Burgundy (see Franche-Comté below) while the Duchy was that part of the realm west of the River Saône; it was annexed by France (1477). The colour burgundy is probably derived from the colour of the famous red Burgundy wine produced in the region.
Caen	Catumagos	Means 'Battlefield' from *catu*, battle, and *mago*, field. Caen was critical to the German defence after the Allied invasion of Normandy (Jun 1944); consequently it suffered considerable damage.
Calais	Calesium	Named after the Gaulish Caleti tribe. Withstood an English siege for nearly a year (1346). Was the last piece of English territory to fall to France (1558).
Cannes	Canois	Means 'Cane harbour', a reference to the reeds growing in the surrounding area.
Centre		A region, whose name describes its position in the approximate centre of France.

Champagne-Ardenne			A region. Derived from the Latin *campania*, plain. Known world-wide for its wine, champagne.
Chartres	Autricum		Named after the Celtic tribe, the Carnutes. Sacked by the Normans (858).
Cherbourg	Coriallum, Carusburc		Name may be derived from the Germanic *hari*, army, and *burg*, fort.
Clermont-Ferrand	Augustonemetum		Originally named after the first Roman emperor, Augustus (63 BC-14 AD). Created by the union (1731) of Clermont and Montferrand. Clermont means 'clear mountain', i.e. clearly visible, and Montferrand 'Ferrand's mountain'.
Dieppe			Probably derived from the Saxon word *deop*, deep, a reference to the depth of the mouth of the River Arques here.
Dijon	Dibio, Castrum Divionense		Derived from the Latin *divus*, divine or godlike, from an unknown Divius. Famous for its mustard.
Dunkirk	Dunkerk	Dunkerque	Name derived from the Flemish 'Church of the Dunes'. Site of the evacuation of some 350,000 British, Belgian and French troops to Britain (May-Jun 1940).
Franche-Comté	Cisjurane-Burgundy		A region meaning 'Free county' and was the name given to the County of Burgundy (1137) after the count had refused to pay allegiance to the German king. After a victorious struggle he became known as the Free Count.
Fréjus	Forum Julii		Founded by, and named after, Julius Caesar as a naval base and trading post.

Gascony	Novempopulana, Vasconia	An historical duchy named after the Basques who overran it (from 561). Belonged to the English (1154–1453). The Spanish for Basque is *Vasco*; the Basques were at one time called the Vascones.
Grenoble	Gratianopolis, Grelibre	Evolved from the original name meaning the 'City of Gratian' after the Roman Emperor Flavius Gratianus (359–83, r.375–83). Grelibre was used for a short time during the period of the French Revolution because *noble* was considered to show a connection with the nobility.
Île-de-France	Francia	A region with Paris as its capital. The area round Paris was known as Francia. *Île*, island, signifies an area between rivers. The Capetian dynasty (987–1328) began to spread territorially from the centre of this area into what eventually became France.
Languedoc-Roussillon		A region whose name is derived from *langue d'oc*, the traditional language of southern France where *oc* means 'yes'.
Le Havre	Le Havre-de-Grace	Means the 'Haven of Grace' to refer to the harbour built (1517) from a fishing village by King Francis I (1494–1547, r.1515–47).
Le Mans		Named after a Gallic tribe, the Cenomani. World-famous for its 24-hour motor race held in June each year. The car that covers the greatest distance in 24 hours is the winner.
Lille	Lisle	Possibly corrupted from *L'Île*, 'The Island', which it was often called since it was originally founded in marshland. Gave its name to lisle, a fine and smooth cotton thread invented here.

Limoges	Lemovices, Augustoritum	Named after the Lemovices, a Gallic tribe.
Limousin		A region named after the Lemovices. The people used to wear distinctive hoods. Early motor cars resembled these hoods and thus became known as limosines.
Lorraine	Lotharingia	A region whose name has evolved from Lotharingia, the Kingdom of Lothair II (835–69, r.855–69). Ownership repeatedly disputed between France and Germany. To France (1766), passed to Germany (1871) after which Alsace-Lorraine together became known as Reichsland; returned to France (1919), captured by Germany (1940) and restored to France (1945).
Lyon	Lugdunum	Means 'Lug's fort'. Lug was one of the main Gallic gods. Capital of Gaul. Sacked by the Romans (197).
Mâcon	Matisco	Derived from the Roman name which may be associated with the root word *mat*, mountain.
Marseilles	Massilia, Villa-sans-Nom	Marseille For its support of the monarchy the city temporarily lost its name (1793) and was called Ville-sans-Nom (Town without a Name). The French national anthem, La Marseillaise, was named after the city (1795). Composed (Apr 1792) in Strasbourg as the Battle Song for the Army of the Rhine, it was quickly adopted by volunteer militia units from Marseille as they marched north to Paris (1792). As they entered the capital singing the song, it was immediately dubbed La Marseillaise.
Mulhouse		Takes its name from the German *mühle*, mill, and *haus*, house, indicating a settlement by a mill. Joined the French

Nancy	Nantiacum	Republic (1798), ceded to Germany (1871), and returned to France (1918). Means the 'Place, *acum*, of Nantio', an obscure individual.
Nantes	Condevincum, Civitas Namnetum	Named after the Namnetes. The Edict of Nantes (Apr 1598) ended the Huguenot Wars (1562–98), giving the Huguenots religious and political freedom, but denied them permission to try to convert Catholics.
Narbonne	Narbo	Named after the first Roman colony in Gaul, Narbo Martius, founded (118 BC).
Nevers	Noviodunum Aeduorum, Nevirnum	The name is linked to the River Nièvre, having been contracted from the Roman names.
Nice	Nike, Nikaia	Probably means 'Victory' from the Greek *nikē*, after the Phoenicians had defeated a neighbouring tribe.
Nîmes	Nemausas	Named after the 'spirit' of a local spring or fountain. Gave its name to denim, a hard-wearing cotton twill fabric, *de Nîmes*, of Nîmes.
Normandy	Normandie	Now divided into two regions, Basse-Normandie and Haute-Normandie, it is named after its inhabitants, the Normans, themselves originally Norsemen or 'North men'. Previously (5th century) part of the Frankish Kingdom of Neustria. The site of the 'D-Day' landings, the Allied invasion of the Continent (6 June 1944).
Orange	Arausio	Takes its name from Arausio, a Gaulish god. Developed into an independent county (11th century), became a principality

Orléans	Genabum, Aurelianum	Named after Aurelian (c.215–75, r.270–75), Roman Emperor. Genabum comes from a combination of words, *gen*, bend (in a river) and *apa*, water. Last French stronghold during the Hundred Years' War (1337–1453) and the site of Joan of Arc's victory over the English (1429). The Massacre of St Bartholomew's Day (1572) during which some 1000 Protestants were killed took place in the city. Occupied by the Prussians (1870).
Paris	Lutetia	Sometimes known as Lutetia Parisorium. The Celtic *lut* means 'marsh'. This was the chief town of the Parisii and thus the full title means 'Marsh town of the Parisii'. Capital since 987. The storming of the Bastille (1789) triggered the French Revolution. Surrendered to the Prussians (1871) and occupied by the Germans (1940–44). Many treaties: 1763 – ended the Seven Years' War; 1783 – ended the American War of Independence; 1814/15 – ended the Napoleonic Wars; 1856 – ended the Crimean War; 1898 – ended the Spanish-American War; 1919/20 (sites around Paris) – ended the First World War; 1947 – confirmed the 1944 Russo-Finnish armistice; 1973 – engendered a ceasefire in the Vietnam War and the withdrawal of all US forces although fighting soon resumed. The Founding Act on Mutual Relations, Co-operation and Security (May 1997) provided for a new security partnership between NATO and Russia.
Périgueux	Vesuna	Named after a Gaulish tribe, the Pterocorii. The Romans named it after a local spring.

and passed to the Dutch House of Nassau (1544), thus creating the Dutch House of Orange. At the Treaty of Utrecht (1713) ceded to France.

Picardy	Picardie		A region whose name is derived from *pique*, pike, a weapon which was used to great effect by the men of this area.
Poitiers		Limonum	Named after the Pictones or Pictavi, a Gallic tribe. The Battle of Poitiers (1356) resulted in a decisive victory for the English over the French. (See Tours for another Battle of Poitiers (732)).
Poitou-Charentes			A region named after the Pictones.
Provence		Gallia Transalpina, Provincia	Means 'Province' because it was the first Roman province to be established beyond the Alps – Gallia Transalpina. Now part of the region of Provence-Alpes-Côtes d'Azur.
Quimper	Kemper		Name derived from the Breton *kemper*, confluence, since it lies on the confluence of the Rivers Odet and Steir.
Rennes			Named after the Redones, a Gallic tribe whose capital this was.
Rheims	Reims	Durocortorum	Named after a Gallic tribe, the Remi. Site of the German surrender to the Allies (7 May 1945).
Rouen		Rotomagus	The original name may have meant the 'Market of Roto', a local individual. In English hands (1066–1204 and 1419–49) during which time Joan of Arc was burnt at the stake (1431).
St Lô		Briovera	Renamed after St Lô, bishop of Coutances (6th century).
St Malo			Named after the Welsh missionary monk Maclovius (d. c.640) who fled to Brittany (6th century).

Place	Roman/earlier name	Notes
St Tropez		According to legend, the separated head and body of the martyred Torpes, a Christian centurion, were put in a boat and cast adrift. St Tropez lies at the point where the boat made its landfall.
Soissons	Noviodunum, Augusta Suessonium	Named after the Suessiones, a Gaulish tribe who made this their capital (3rd century).
Strasbourg	Argentoratum, Strateburgum	Means 'Fortress on the street' from the Old German *straza*, street, and *burg*, fort. Occupied by the Germans (1871–1918, 1940–44).
Toulon	Port de la Montagne	Name may be derived from the Celtic *tol*, hill. Given its former name during the French Revolution. Both names refer to high ground, in this case the nearby Faron Mountains.
Toulouse	Tolosa	As with Toulon, may come from *tol*, a reference to the nearby Pyrenees.
Tours	Caesarodunum, Civitas Turonum	Named after the Turons and, for a time, 'Caesar's Fort'. The Battle of Tours also known as the (First) Battle of Poitiers (732) was fought close by and resulted in victory for Charles Martel, the Frankish leader, over the Moors.
Troyes	Augustabona, Civitas Tricassium	Named after a Gallic tribe, the Tricasses. Gave its name to the system of troy weights: the grain, the pennyweight (24 grains), the ounce (20 pennyweights) and the pound (12 ounces); the common avoirdupois pound has 16 ounces.
Valenciennes		May be named after Emperor Valentinian I (321–75, r. 364–75); or it may be derived from *val des cygnes*, valley of

the swans. Ceded to France at the Treaty of Nijmegen (1678).

Vichy	Vicus Calidus	Derived from its Latin name meaning 'Warm settlement'. Capital of Vichy France (July 1940-Sep 1944) which was formally known as *État Français*, French State, France having been divided into two parts (Jun 1940): one occupied by the Germans and the other, the French State, in the south-east, under nominal French sovereignty. However, its leaders, Marshal Philippe Pétain (1856–1951) and Pierre Laval (1883–1945), collaborated with Nazi Germany. The inhabitants of the town are known as *Vichyssois* while the supporters of the regime were known as *Vichystes*. Bottled Vichy water is known internationally as is Crême Vichyssoise, a soup created by the French chef of a New York hotel (1910).
Corsica	Corse	Kyrnos	An island and region of France. Became a province of France, having been sold by the Republic of Genoa (1768).
Ajaccio		Ajax	Name may be derived from the Low Latin *adjacium*, resting place, where shepherds stopped with their flocks. Capital (1811).
Bastia		Marina di Cardo	Took its name from *bastiglia*, the fortress which was built by the Genoese to guard the town. Capital (1380–1811).
Bonifacio		Giola	Named after the Genoese Count Bonifacio who built a castle here (828).

GERMANY

As Roman legions penetrated into modern Germany they were met by Germanic tribes from the north and from Scandinavia. In 9 AD, under Hermann (Latin: Arminius), they defeated three Roman legions, thus preventing Roman domination east of the River Rhine. Incursions by nomadic Huns from the east encouraged some of the German tribes to migrate. Between the fourth and sixth centuries the Franks, one of the largest of the tribes, moved west to France, the Anglo-Saxons north-west to Britain, the Lombards south to Italy and the Visigoths south-west to Spain. The tribes from Friesland, Saxony, Lorraine, Thuringia, Franconia, Swabia and Bavaria that remained in central Europe had little in common politically and no cohesion.

In 774 the Frankish king, Charlemagne, conquered the Lombards and in 800 established the Carolingian empire when he was crowned as emperor of all the Romans (Holy Roman Emperor) by the pope. However, 29 years after his death in 814, his empire was divided between three of his grandsons at the Partition of Verdun and the seven tribes found themselves grouped together under a Frankish king, Louis (Ludwig) the German. This East Frankish kingdom was later to evolve into Germany and Austria. Louis gained further territory in the west when the Middle Kingdom, Lotharingia (q.v.France), was divided in 870 at the death of his nephew.

The seven tribal regions remained largely independent duchies. When King Louis the Child died without an heir in 911, the Duke of Franconia became the first truly 'German' king – chosen by the Germans themselves and not imposed upon them. When he died seven years later the crown passed to the Duke of Saxony and a century of Saxon rule began. The kingdom was expanded to the east and south, and the rest of Lotharingia, to which the imperial title was linked, was acquired. Otto I the Great became king in 936 and King of Italy in 951. Victory over the Magyars in 955 did much to establish the south-eastern border. The march of Austria was established as a buffer state between the Bavarians and the Magyars who were never able again to threaten the Upper Danube or Rhine valleys.

Having subdued defiant vassals and defeated the Magyars, Otto founded the First German Reich. To consolidate his authority he was crowned Holy Roman Emperor in 962. However, he failed to establish the principle of hereditary succession, thus leaving the monarchy open to selection by the German princes. While the Salian (1024–1125) and Hohenstaufen (1138–1254) emperors managed to maintain the state of Germany, the territorial princes and other nobles were able to preserve their autonomy. It was they and the Teutonic Knights[1] who were largely responsible for extending the boundaries of the kingdom eastwards as far as the River Oder and establishing a springboard for further advances. By 1400 Brandenburg, Pomerania, Silesia, Lusatia, Prussia, Austria and Styria had all been settled by Germans.

The Hohenstaufen emperors turned their attention southwards to Italy. By marriage the Kingdoms of Italy and Sicily were acquired. However, their attempts to impose direct imperial rule over these regions met with opposition from the papacy and led to a struggle which weakened the emperors' authority; in consequence, the power of the princes waxed. Following the death of the last Hohenstaufen, the old German state fell apart and numerous petty statelets came into existence. Yet more were created when they were divided as a result of inheritance, grant or conquest. Only in 1273 was an obscure count from Swabia, Rudolph of Habsburg, elected king. Within ten years the acquisition of Austria and Styria had elevated the Habsburgs to a prominent position among the German princes.

To bring stability to the German monarchy and regulate the election of the Holy Roman Emperor a constitutional document known as the Golden Bull was promulgated in 1356. Only seven prince-electors were to form the electoral body: the King of Bohemia, the Prince-Archbishops of Cologne, Mainz and Trier, the Count Palatine of the Rhine, the Margrave of Brandenburg and the Duke of Saxony. The Bull also confirmed the territorial sovereignty and powers of the electors. These powers, however, began to decline due to the practice of dividing territory between surviving sons rather than passing it on the eldest.

Sigismund (of the House of Luxembourg), Holy Roman Emperor and King of Germany, Bohemia and Hungary, was less interested in Germany than he was in his other possessions. Before he died without a male heir in 1437 he named his son-in-law, Albert of Habsburg, Duke of Austria, as his successor. In 1438 Albert was chosen as king, beginning, with one exception, a period of unbroken Habsburg domination until the dissolution of the empire (and the First Reich) in 1806. In due course the empire was to be referred to as the Holy Roman Empire of the German Nation, an acknowledgement that its main component was the German lands and that the emperor ruled other lands outside the confines of the empire.

In 1517 a little-known friar and theology professor, Martin Luther, wrote his Ninety-five Theses against indulgences and unleashed the Reformation. His criticism of the papacy and its teachings led to his eventual excommunication, but found enthusiastic support in parts of Germany. Social and religious ills, and a lack of political unity and vision, rightly or wrongly, were put at the door of an unreformed and intransigent papacy. Many succumbed to the appeal of Lutheranism. By the mid-1520s some German states and cities had cut their ties with Rome and become Lutheran.

Religious upheavals followed, but it was not until 1546 that Emperor Charles V, a Roman Catholic, had time to turn his attention to his Protestant territories which included Brandenburg, the Palatinate, Saxony and a number of bishoprics. By this time Lutheranism had become more than just a religious movement; the Protestant princes had developed a political agenda and military strength. The war that Charles started against the princes was militarily inconclusive. It ended with the Peace of Augsburg in 1555. Lutheranism was legalized and rules formulated whereby the ruler of each territory or city could choose between Roman Catholicism or Lutheranism; everyone subject to that ruler then had to follow the religion he had chosen. The Protestant Reformation further disunited Germany.

Germany was the main battlefield for the Thirty Years' War (1618–48). A series of savage wars, it was principally a contest between the Habsburg emperor, supported by the Catholic states and princes, and a group of Protestant principalities and towns with help from Sweden and the United Netherlands. France, although Catholic, also intervened on the side of the Protestants to ensure that the House of Habsburg did not become too powerful. The German countryside was devastated and the population decimated. The Peace of Westphalia in 1648 set the seal on the reformed Europe. It confirmed the Peace of Augsburg and therefore earlier Protestant successes, and preserved almost complete territorial sovereignty for the German princes, although they still owed nominal allegiance to the emperor. Such sovereignty was enjoyed by some 300 principalities, 50 imperial cities and almost 2,000 imperial counts and knights, whose territory was sometimes measured in acres. As a result of the Peace, Sweden received Western Pomerania (a region, now in Germany, along the Baltic coast) and Bremen, while France assumed sovereignty over Alsace; Brandenburg, a small state in northern Germany, got Eastern Pomerania (now in Poland).

Within 100 years of the Peace of Westphalia, Austria, the most powerful German kingdom, had become a major state in Europe. It was followed closely by Prussia which had been inherited by the Hohenzollern Elector of Brandenburg in 1619 although remaining under Polish suzerainty. By 1648 Brandenburg was a conglomeration of scattered territories which, in the view of Frederick William the Great Elector, could only be held together by a powerful standing army. In 1657 events conspired to allow him to become the sovereign Duke of Prussia. In 1701 his son, who became Frederick I, was given the title of King of Prussia by the emperor in Vienna and the name Prussia was soon used to describe Brandenburg-Prussia. Frederick's son, Frederick William I, proceeded to almost triple the size of his army, making it the fourth largest in Europe.

It was the Great Elector's great-grandson, Frederick II the Great, who wanted to put this army to use. In 1740, on becoming Prussian king, he began the War of the Austrian Succession (1740–48), the catalyst for which was the death of the Habsburg emperor Charles VI without a male heir. Frederick offered to accept the Pragmatic Sanction (q.v.Hungary) if Austria were to cede Silesia, a province within the Kingdom of Bohemia, to him. The Austrians refused, Frederick attacked and conquered Silesia.

The Seven Years' War (1756–63) involved the great powers of Europe fighting not only in Europe but also in America and India. It broke out when Austria seemed ready to try to win back Silesia. In the German theatre Frederick attacked first, invading Saxony. Although coming close to defeat later, the Treaty of Hubertusburg in 1763 confirmed Frederick's hold on Silesia but forced him to give up Saxony. By the first partition of Poland in 1772, however, he gained West Prussia (except for the port of Danzig (now Gdansk) which only came with the second partition in 1793) which enabled him to link East Prussia with his lands to the west. More territory was gained at the third partition in 1795. Prussia had become the leading German state and a major force in Europe.

Before 1806 Prussia played little part in the Napoleonic wars and Napoleon was able to manipulate the smaller German states in pursuit of his territorial aims. In 1805, crushed by successive Napoleonic victories over Russian and Austrian armies, Emperor Francis II was forced to sign the Peace of Pressburg (now Bratislava). The following year 16 middle German states formed the Confederation of the Rhine under Napoleon's protection. When they seceded from the 1000 year-old Holy Roman Empire Francis abdicated and the Empire was dissolved. Napoleon now proceeded to tighten his grip on Germany and forced 36 of the 38 states (excluding Austria and Prussia) to join the Confederation. He founded the Kingdom of Westphalia, the nucleus of which was Prussia's western provinces, which he gave to his brother Jérôme, and extended his realm as far east as Lübeck.

Napoleon's mastery of Germany was consolidated with the rout of Prussian armies and the occupation of Berlin in October 1806. At the Treaty of Tilsit (now Sovetsk) in

July 1807 Prussia lost almost half its territory. So dominant was France that Napoleon could not be challenged until his army was humiliated in Russia in 1812. The Battle of Leipzig in 1813 resulted in further defeat for the French, withdrawal from Germany east of the Rhine and the collapse of the Confederation of the Rhine. Having been lost to the French in 1793, the Rhineland was regained in 1814.

The Congress of Vienna from September 1814 to June 1815 did not result in German unification, but nor did it attempt to turn the clock back and resurrect the Holy Roman Empire or the conglomeration of statelets in Germany. It created instead a new German Confederation comprising 39 sovereign states. These included the great powers of Austria and Prussia, minor kingdoms such as Bavaria, Saxony and Württemberg, duchies such as Baden and Hesse-Darmstadt, small principalities, and free cities such as Hamburg and Lübeck. The King of Hanover, the fourth largest German state after Austria, Prussia and Bavaria, was also the King of Great Britain. Neither Austria, Prussia nor the German kings wanted German unification for fear of losing their status. Austria, particularly, was determined to maintain its historic position as first among equals.

With the creation of a Customs Union, *Zollverein,* in 1834 the commercial, if not political, unification of many of the Confederation's states was achieved. Austria and Hanover were the most notable absentees. By 1867 all the states except for Austria, Bremen and Hamburg had joined.

In 1862 Otto von Bismarck became the Prussian prime minister. He believed unification – with Prussia dominant – to be both desirable and inevitable. His first opportunity to pursue this aim came the next year when the Danish government, in contravention of earlier agreements, moved to annex Schleswig, the northern half of which was populated by Danes, the southern by Germans (q.v.Denmark). Schleswig and Holstein were both linked to the Danish crown, but Holstein was populated by Germans and was a member of the German Confederation. In 1864 Bismarck seized the opportunity to detach both duchies from Denmark and invited the Austrians to join in their occupation.

Austria and Prussia, however, had different aspirations for the duchies. But more important to Bismarck was the future of Germany from which he was keen to eject Austria; if this proved impossible, then he might have settled, at least temporarily, for a Hohenzollern sphere of influence in the north and a Habsburg one in the south. Francis Joseph was not amenable and relations deteriorated. Intent on challenging Austria even at the risk of war, Bismarck offered Austrian Venetia to Italy on the proviso that the Italians would go to war with Austria if Prussia did so. Further diplomatic manoeuvring provoked the Austrians into declaring war and the Seven Weeks' War broke out in June 1866. The Prussians were victorious, but magnanimous in victory. Nevertheless, the Habsburg emperor had to agree to the loss of Venetia, the dissolution of the Confederation and the establishment of a North German Confederation (which excluded Austria) under Prussian leadership. This incorporated 21 German states, including Schleswig-Holstein, though none from Austrian territory. The Kingdom of Prussia stretched in the east from what is now Kaliningrad (Russia) to west of the River Rhine.

Prussia's growing power in Germany was seen as a threat to French security, while France had come to be regarded as Germany's traditional enemy. A successful war could meet their different needs. For Bismarck these included completing the unification of Germany and annexing Alsace-Lorraine which had been traditionally German territory until given to France at the end of the Thirty Years' War in 1648. In 1870 Bismarck was able, by means of a leak to the press, to provoke the French into declaring war. Within ten months the Germans had occupied Paris, won the war and signed the Treaty of Frankfurt. Alsace-Lorraine was ceded to Germany, a loss which caused huge resentment in France. More importantly, the southern states (principally the Kingdoms of Saxony, Bavaria and Württemberg, and the Grand Duchies of Baden and Hesse) and Mecklenburg joined the Confederation to form a united German nation – the Second Reich – under the Prussian Hohenzollerns in January 1871. The Prussian king, Wilhelm I, was proclaimed German emperor (*kaiser*) and crowned in Versailles; Bismarck became the first chancellor of the German Empire.

Having defeated Denmark, Austria and France in the space of seven years and thus generated a sense of insecurity throughout Europe, Bismarck now became preoccupied with the possibility of encirclement. He thus negotiated an alliance in 1879 with the Habsburg Austro-Hungarian Empire and in 1882 with Italy to form the Triple Alliance. Thereafter, he was concerned solely with peace in Europe. However, when he retired from the scene in 1890 a new assertiveness, combined with a massive programme to expand the German Navy, encouraged an opposing alliance: by 1907 the Triple Entente of Britain, France and Russia was ranged against Germany. Italy's reliability had proved ephemeral and Germany's only ally now was Austria-Hungary. When the Archduke Francis Ferdinand was murdered in Sarajevo in 1914, the Austro-Hungarians delivered a sharp ultimatum to the Serbs. This was strongly supported by the Germans who believed that if Russia was to support its Slav compatriots and war was to ensue, then it was better sooner rather than later. The Austro-Hungarian Empire was in danger of falling apart and if it did so then Germany would be completely isolated.

The loss of the First World War brought an end to the Second Reich and the abdication of the kaiser, ushered in the Weimar Republic and forced Germany to accede to the punitive 1919 Treaty of Versailles. The terms were harsh, if not totally unexpected. Alsace-Lorraine had to be returned to France, the Saarland was taken over by the League of Nations (until 1935 when the inhabitants voted overwhelmingly to return to Germany), and the Danes of northern Schleswig, allowed the choice, elected to become part of Denmark. The Rhineland was demilitarized, German forces being allowed no closer than 50km east of the river. In the east, West Prussia was ceded to Poland and Danzig was declared a Free City under the League of Nations. The 'Polish Corridor' was thus created, giving Poland access to the Baltic Sea and severing East Prussia from the rest of Germany. Also gained by Poland were substantial parts of the province of Posen (Poznan) and Upper Silesia. Thus, millions of Germans came under Polish, and indeed Czech, rule. This was directly contrary to Woodrow Wilson's concept of self-determination[2] and infuriated the Germans. Finally, Germany lost its overseas colonies.

In 1933 Adolf Hitler became chancellor. The Weimar Republic gave way to the Third Reich and a national socialist (Nazi) totalitarian state. Determined to end the indignities of the Versailles Treaty and gain space (*Lebensraum)* for the Germans in the east, Hitler was not averse to achieving his aims by means of war. One of the shortcomings of the Versailles Treaty had been the lack of any international mechanism to verify German disarmament. Rearmament began secretly in the 1920s and, after he had come to power, Hitler accelerated it. By 1936 Germany was strong enough to move troops into the Rhineland in violation of the Treaty. Neither Britain nor France, nor even the League of Nations, offered any resistance. That same year Hitler forged the Berlin-Rome axis with Mussolini. In March 1938 Germany peacefully annexed Austria; in October Britain and France approved the annexation of the largely German Sudetenland (q.v.The Czech Republic) in return for a promise of no more German expansion. Six months later Hitler seized the rest of Czechoslovakia. To ensure that Germany would not have to fight a war on two fronts as it had in the First World War, Germany signed a non-aggression pact with the Soviet Union in August 1939.

The Second World War began on 1 September 1939 when German forces invaded Poland. After three years of success in both the west and the east the tide turned and by May 1945 Germany had been overrun and the Third Reich destroyed. As a result of various Allied conferences during the war, the German territories east of the Oder and Neisse rivers, together with Danzig, became Polish; and the northern part of East Prussia was transferred to the USSR. Prussia ceased to exist. The Allies then divided what remained of the country and Berlin, separately, into four zones of occupation. In 1949 the era of formal German unity came to an end after only 78 years. The American, British and French zones were consolidated into the Federal Republic of Germany, usually known as West Germany; the Soviet zone, the eastern third of Germany, became the German Democratic Republic, otherwise known as East Germany. The Soviet sector of Berlin, or East Berlin, became its capital; West Berlin became an isolated self-governing exclave of West Germany within East Germany. In 1954 East Germany was given nominal sovereignty. The following year West Germany became a sovereign state and joined NATO, while East Germany became a founding member of the Warsaw Pact.

Allocated to France at the end of the war and virtually autonomous since 1947, the Saarland rejoined West Germany in 1957, becoming a state.

In 1961 the physical division of the two countries was emphasized by the erection of the Berlin Wall which separated the western and eastern sectors of the city. This closed the gap in the fortified Inner German Border which ran from the Baltic south to the Czech border. Mutual hostility between the two Germanies ensured uneasy stability and the situation remained locked until the end of 1989 when the communist regimes in Eastern Europe began to fall. On 9 November the Wall was opened and on 3 October 1990 West and East Germany reunited to become one country within the European Community (now European Union). Having been Europe's principal battlefield for centuries, a new era has dawned in which Germany can no longer threaten the peace of Europe by being either too weak or too strong.

1. The Teutonic Knights had their origins in a German field hospital at the siege of Acre in 1189–90 during the Third Crusade, their Order being named the Hospital of St Mary of the Teutons in Jerusalem. In 1198 it was militarized, the knights having two functions: fighting and caring for the sick. In the 13th century they transferred their principal activities to eastern Europe, but their headquarters remained in Palestine until it fell to Islam in 1291.

2. On 8 January 1918 the American President, Woodrow Wilson, put forward to a joint session of Congress his ideas for the post-war settlement. Of the Fourteen Points, eight he considered to be obligatory and six to be desirable. He made specific proposals concerning the territory of Austro-Hungary, Belgium, France, Germany, Italy, Montenegro, Poland, Romania, Serbia, and Turkey.

Area:	357 041 sq km (137 854 sq miles)
Population:	82.69 million; Germans, Turks, Yugoslavs
Language:	German
Religions:	Lutheran, Roman Catholic, Muslim
Capital:	Berlin
Administrative Districts:	16 states
Date of Independence:	1949 (unification: 1990)
Neighbouring Countries:	Denmark, Poland, Czech Republic, Austria, Switzerland, France, Luxembourg, Belgium, Netherlands

The German Confederation 1815

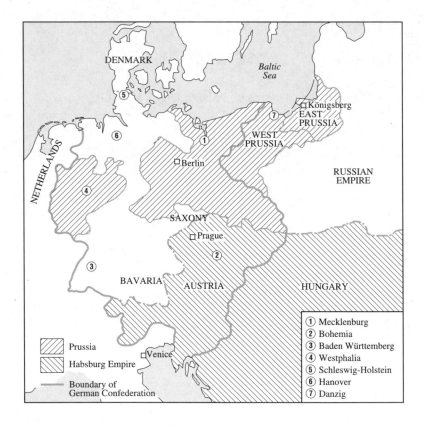

English-speaking Name	Local Name	Former Names	Notes
GERMANY	Deutschland		The Federal Republic of Germany (Bundesrepublik Deutschland) since May 1949. The German Democratic Republic inaugurated in the Soviet zone of the former Germany (Oct 1949). The two republics united as the Federal Republic of Germany (Oct 1990). Deutschland means the 'Land of the Dutch'. *Lingua theodisca*, which meant the 'popular tongue of the German tribes', in due course became *Deutsch* which indicated people who spoke in the vernacular rather than using the more academic Latin of the church. The origin of the name Germany is not clear, although the Romans called the country Germania. Germany is unlike most countries insofar as its name is not universal: the French call it *Allemand* and the Spanish *Alemania* after the Alemanni tribe, the Italians *Germania* but their word for German is *tedesco*, the Slavs use variations based on the root word *nem*, mute, to mean 'not able to speak our language' (in Russian, *Nemetchina*, in Serbo-Croat, *Nemačka*) and the Finns call it *Saksa*, the land of the Saxons.
Aachen		Aquis Granum, Aix-la-Chapelle	The springs of Grannus were named after the Celtic god of healing and later used as baths by the Romans. The Treaty of Aix-la-Chapelle ended the War of the Austrian Succession (1748) and the Congress of Aix-la-Chapelle (1818) aimed to direct European affairs after the Napoleonic Wars. The 'la-Chapelle' of the French name refers to the fact that Charlemagne (742–814), King of the Franks (768–814) and Holy Roman Emperor (800–14), is buried here and is used to differentiate it from other towns beginning with 'Aix'. The name Aachen is derived from the Old German *aha*, water, a reference to the springs here. Annexed by France (1801–15).

Augsburg	Augusta Vindelicorum		Named by the Roman generals Drusus and Tiberius after their stepfather as 'The fortress (*burg*) of Augustus in the land of the Vindelicians', a Celtic tribe the Romans had subjugated. The Peace of Augsburg (1555) recognized only two religious denominations, Roman Catholicism and Lutheranism; the ruler of each territory was to choose one which would then be binding for his subjects.
Baden-Baden	Aquae Aureliae		Means 'Baths' and to distinguish it from any other baths in Germany it has added a second Baden to show that it is in the state of Baden-Württemberg.
Bavaria		Bayern	A state, *Land*, named after a Germanic tribe, the Baiovarii, who settled in the region (c.500).
Berlin			Situated among lakes and rivers, the name may be derived from the Slav *brl*, marsh or swampland or even sluice gate. Capital of the Duchy of Brandenburg (1486) which became the Kingdom of Prussia (1701), then Germany (1871–1945, since 1990). The German parliament held its first plenary session in the refurbished Reichstag (19 Apr 1999) having moved from Bonn and this marked the dawn of the so-called 'Berlin Republic'. From 1945 until the reunification of Germany (1990), divided into American, British, French and Soviet sectors. East Berlin became the capital of the German Democratic Republic (1949) and West Berlin, an enclave within that Republic, a state of the Federal Republic of Germany. A fortified wall separating the American, British and French sectors from the Soviet sector was erected (1961). The collapse of the communist regime (1989) was accompanied by the dismantling of the wall. The Treaty of San Stefano (Mar 1878), which caused an international crisis at the end of the Russo-Turkish war, was overturned at the Congress of Berlin (Jun/Jul 1878): Russian gains were reduced.

Place	Older names	Description
Bonn	Castra Bonnensia, Bonnburg	Derived from the Gaulish *bona*, fortress or city. Garrison of the Roman 1st Minervia legion. Capital (1949–90); the Federal government remained in Bonn until moving to Berlin (1999).
Brandenburg	Branibor, Brennaburg	Originally a Slavic fortress, its name may mean 'Burnt fortress', from *brand*, burning or fire, and *burg*. At one time a *mark* and an electorate of the Holy Roman Empire. The base on which the Kingdom of Prussia was founded and with which it merged (1701). A province of Prussia (1815–1945), but lost its status as a state (1952) until restored (1990).
Bremerhaven	Wesermünde	Means the 'Port, *haven*, for Bremen'. Bremen is derived from the Old German *brem*, marshland. It is situated inland. The previous name meant 'Mouth of the (River) Weser'.
Brunswick	Braunschweig	Means 'Bruno's settlement' after the Saxon Duke Ludolf's son who founded it (861).
Chemnitz	Karl-Marx-Stadt	Named (1953–90) after Karl Marx (1818–83), political and economic theorist who inspired the creation of communism.
Coblenz	Confluentes	Named after its position at the confluence of the Rivers Moselle and Rhine.
Cologne	Colonia Claudia Ara Agrippinensis	The Roman name means the 'Colony of Agrippina the Younger', whose birthplace this was. She (15–59) was the great granddaughter of Caesar Augustus, daughter of Agrippina the Elder, mother of Nero and wife of the Emperor Claudius; so named at Agrippina's request (50) and shortened to Colonia. Garrison of the Roman 30th Ulpia Victrix legion. Gave its name to *eau de Cologne*, the brand 4711 coming from the number of the house in which Wilhelm Mühlens, its creator, lived (c.1794).

Constance	Konstanz	Named after the Roman Emperor Constantius Chlorus (c.250–306). The Swiss part of the town is called Kreuzlingen. The Council of Constance (1414–18) brought to an end the Great Schism of the Roman Catholic church (1378–1417) during which there were two, and for a short time three, rival popes.
Dresden	Dreždane	The original Slavonic name means 'Forest dwellers on the plain'. The Treaty of Dresden (1745) ended the Second Silesian War (1744–45) and confirmed Prussia's conquest of Silesia. Site of Napoleon's last major victory in Germany (Aug 1813). Almost entirely destroyed by Allied bombing (Feb 1945).
Duisburg	Castrum Deutonis, Diuspargum, Duisburg-Hamborn	Means 'Diu's fort' and may be connected with Tiu, the Anglo-Saxon name for the Norse god of war who gave his name to Tuesday.
Düsseldorf		Means 'Village on the Düssel', a tributary of the Rhine.
Eisenhüttenstadt		Formed (1961) from the amalgamation of Fürstenberg, Schönfliess and Stalinstadt, it means 'Iron-works town'. It was at this time that many cities and towns in the USSR which bore Stalin's name were renamed.
Flensburg		Means 'Fort on a mound'. The site of the surrender to the Allies of the German government under Admiral Karl Dönitz (May 1945).
Frankfurt	Franconofurt, Frankfurt am Main	Means the 'Ford where the Franks crossed the River Main'. Capital of Germany (1816–66). The Treaty of Frankfurt (1871) brought the Franco-Prussian War to an end and imposed harsh terms on France. Gave its name to the highly seasoned sausage, the frankfurter, which is also

known as the wiener or hot dog. Not to be confused with Frankfurt an der Oder, Frankfurt on the (River) Oder.

Place	Origin	Description
Göttingen	Gutingi	Named after the Goding family.
Halle		Derived from the Illyrian *hall*, salt, for which the city is distinguished.
Hamburg	Hammaburg	Developed (c.825) around a moated castle between two rivers, it means 'Fortress in the marshland (*hamma*)'. May have given its name to the hamburger, taken by 19th century German emigrants to the USA.
Hanover	Hannover	Means 'High bank', derived from *hoch*, high, and *ufer*, bank. It lies on the River Leine and Mittelland Canal. The House of Hanover provided six monarchs for the British crown. The first, George Louis (1660–1727), Elector of Hanover (1698–1727), came to the British throne as George I (1714–27); he was followed by four more kings and Queen Victoria (r. 1837–1901).
Heilbronn	Heiligbronn	Means 'Holy spring' in reference to a stream that appeared from under the altar of St Kilian's church.
Hesse	Franconia	A state named after a Frankish tribe, the Hatti or Chatti.
Homburg	Bad Homburg	The full name is Bad Homburg vor der Höhe which denotes a spa town at the foot of the (Taunus) mountains. Homburg means 'High fortress', a reference to the Hohenburg Fortress (12th century, but now in ruins). The town has given its name to the felt hat with a lengthways furrow in the crown which was first made here.

Karlsruhe		Means 'Karl's retreat'. Its development began when Karl Wilhelm, Margrave of Baden-Durlach, built a castle close to his hunting lodge (1715).
Kassel	Chassala	Derived from the Latin *castellum*, fortress.
Kiel	Kyle	Possibly derived from the Anglo-Saxon *kille*, a safe haven for ships. In Denmark (1773–1866). The Treaty of Kiel (1814) ended the war between Denmark and Sweden, and forced Denmark to cede Norway to Sweden.
Leipzig	Urbs Libzi	Originally a Slav settlement, the name is derived from the Slav *lipa*, lime tree. Site of the Battle of Leipzig (Oct 1813) (also called the Battle of the Nations) at which Austrian, Prussian, Russian and Swedish troops defeated Napoleon.
Lübeck	Liubice	Named after the Liubichi, a Slav tribe.
Ludwigsburg		Means the 'Castle of Ludwig', the city being developed (early 18th century) around the palace of Duke Eberhard Ludwig (Louis) of Württemberg.
Ludwigshafen	Rheinschanze	Originally a fortification on the west bank of the Rhine, it was renamed (1843) after Ludwig I (1786–1868), King of Bavaria (1825–48) and means 'Ludwig's harbour'.
Mainz	Mogontiacum	A Celtic settlement, originally named by the Romans after the Celtic god, Mogo. Garrison of the Roman 22nd Primigenia legion.
Mecklenburg-West Pomerania	Mecklenburg-Vorpommern	A state. Mecklenburg was the name of a dynasty which took its name from the family castle, Mikilinborg. Pomerania comes from the Polish *pomorze*, coastland.

Mönchengladbach			Developed around a monastery, it means 'Monks' tranquil stream' from *bach*, stream, and the Old German equivalent of *glatt*, smooth, and here meaning tranquil.
Munich	München	Zu den Munichen	Meant 'To the Monks' and now 'Home of the Monks'. Originally a tiny settlement of friars with a church. Henry the Lion, Duke of Bavaria, wanted to levy a toll over the River Isar. He destroyed the only bridge nearby (1157), built a new one at the settlement and allowed the monks to develop a market by the toll station. The Munich Agreement (Sep 1938), signed by Britain, France, Germany and Italy, enabled the Germans to annex the Sudetenland in Czechoslovakia (1938).
Münster		Mimigernaford	Situated on the River Aa, its original name meant 'Ford over the Aa'. Renamed (1068) after the Latin *monasterium*, monastery, a bishopric having been founded (804). The Treaty of Westphalia was signed here and at Osnabrück (1648).
Niedersachsen			A state. Means 'Lower Saxony'. Named after the Saxons who themselves are possibly named after the Old German *sahsa*, *dagger*, said to be their favourite weapon.
Nordrhein-Westfalen			A state. Means 'North Rhine-Westphalia'. Westphalia means 'Western plain-dweller' from the Old German *falaho*.
Nuremberg	Nürnberg	Nourenberc, Norimberg	Site of Nazi Party rallies and congresses (1920/30s), and the War Crimes Tribunal (1945–46). The 'Law of the Reich Citizen' (depriving Jews of German citizenship) and the 'Law for the Protection of German Blood and German Honour' (forbidding marriage or sexual relations between Jews and German citizens) was passed in the city (1935). Known as the Nuremberg Laws, they effectively expunged

Osnabrück		all Jews from German life. While *berg* means mountain the meaning of the first part of the name is unknown.
		Means the 'Bridge over the River Hase', the latter name coming from the Old German *asa*, current. One of the two sites where the Peace of Westphalia was signed (1648).
Potsdam	Poztupimi	Means 'Under the oaks' from the Slav *pod*, under, and *domb*. Site of the post-World War II Conference (Jul/Aug 1945) which discussed the administration of Germany and the future of Europe.
Regensburg	Radasbona, Casta Regina Regeneopurc, Ratisbon	All names mean 'Fort on the Regen river'.
Remagen	Riomagus	Means 'King's market'.
Rottweil	Arae Flaviae	The Roman name may have meant the 'Altar-like monuments of Flavia'. The present name may mean the 'Camp of a company (of troops)' from *rotte*, company or file, and *weiler*, camp or settlement. Gave its name to the Rottweiler, a working dog left by the departing Roman legions in Rottweil.
Saarbrücken		Means 'Bridge over the River Saar' and is the capital of Saarland, the state of Saar.
Saarlouis	Saarlautern	Founded and named (1680) by King Louis XIV of France (1638–1715) after himself. Occupied by both the French and the Germans a number of times before becoming German when the Saarland became a state of Germany (1957). Saarlautern (1936–45) after a majority of the population of

the Saarland voted in a plebiscite (1935) to return to Germany.

Saxony	Sachsen	Previously a kingdom, now a state, named after the Saxons. Two other states are named Saxony-Anhalt and Lower Saxony.
Schleswig-Holstein		A state. Schleswig is derived from the Old Norse *sle*, reed, and *vik*, harbour; Holstein is corrupted from *holt*, wood, and *sittan*, to be settled. Both were previously duchies acquired by Prussia (1864); the northern half of Schleswig was returned to Denmark after a plebiscite (1920).
Starnberg		*Star* means 'Starling'. It is in the area of Lake Starnberg that thousands of starlings gather each year before flying south to Africa. Starnberg lies at the northern tip of the lake.
Stuttgart	Stuotgarten	Originated (c.950) as a stud farm, *stutengarten*.
Thuringia	Thüringen	A state. Named after the Thuringi.
Trier	Augusta Treverorum, Trèves, Treveris	Named (c.15 BC) after the Treveri, a local Germanic tribe, by its founder, Caesar Augustus (63 BC-14 AD), the first Roman emperor (23 BC-14 AD). Ceded to France (1801–15).
Weimar	Wimare	Derived from the Old German *wisa*, meadow, and *mari*, lake. The place where the new National Assembly met (Feb 1919) to draw up a constitution for the new German Republic, henceforth known as the Weimar Republic (1919–33).
Wiesbaden	Aquae Mattiacae Wisibada	Known as a spa from Roman times, it means 'Meadow baths'.

Worms	Borbetomagus, Civitas Vangionum	The main town of the Vangiones, the original name means 'Borbeto's market'. It was changed from the Celtic by the Romans. Martin Luther defended his beliefs at the Diet (of the Holy Roman Empire) of Worms (Apr 1521); they were condemned at the Edict of Worms (May 1522) which declared Luther to be a heretic and an outlaw. Annexed by France (1797–1816).

[1] German nouns begin with a capital letter, but this convention has not been followed here.

GREECE

Greece is the only country in the world where the adjectives 'ancient' and 'modern' have to be applied to the country, the people and their language to differentiate between the past and the present. Greece's histiry comprises four phases: pagan ancient Greece, the Orthodox Christian Byzantine Empire, Ottoman rule and 170 years of independence.

The Minoan civilization, named after King Minos of Knossos, took root on the island of Crete around 2000 BC to be followed by that of the Mycenaeans on the mainland some 500 years later. For some 300 years after the Dorians, coming out of the mountains to the north, had colonized southern Greece about 1100 BC, little is known of the region. Not until about 800 BC (the first Olympic Games were held in 776 BC) did classical Greece (so-called because it was a time of outstanding artistic and intellectual creativity) and the *polis* or city-state appear. Greek colonies began to form in many parts of the Mediterranean and the Black Sea. Eventually two city-states came to dominate Greece: the Ionian city of Athens and Sparta, a Dorian city.

In 500 BC the Greek cities in Asia began their six-year rebellion, known as the Ionian Revolt, against the Persians. It failed, but Athens' support for it was used as a pretext for the Persians to launch an invasion of the Greek mainland in 490 BC. At the Battle of Marathon, however, they were routed and withdrew. Ten years later they were back. To meet them the Greeks formed the Hellenic League which included Sparta and its allies, but some Greek states joined the Persians. Despite a brave Spartan stand at Thermopylae, the Persians sacked Athens. At sea it was a different matter: the Persian fleet was defeated and forced to withdraw. In 479 BC the Persian army was defeated. Hopes of conquest were abandoned.

With the Persian threat removed Sparta left the League which Athens then gradually transformed into its own empire. Sparta's fear of Athenian expansion aided by the growth of its maritime power, however, led to the main Peloponnesian War (431–04 BC) between the two cities. Sparta emerged the victor. Nevertheless, in 371 BC Sparta was defeated by Thebes, a region north of Athens, and from then on ceased to be militarily significant.

In 359 BC Philip II came to the throne of Macedon and within two years he had organized his army and moved south. In 340 BC he declared war on Athens and two years later won a famous victory. But it was his son, Alexander the Great, who carved out a huge empire, uniting the Greeks and Persians and spreading Greek culture as far as India and North Africa. He founded 30 towns which bore his name. Following his death in 323 BC, his generals started squabbling and the empire began to disintegrate. The Romans took advantage of the situation. Corinth was sacked in 146 BC and Achaea (the northern part of Peloponnesus) and Macedonia became Roman provinces. Within a hundred years the Romans were in the ascendant throughout the Greek world. Yet Cleopatra, the last of the Greek Ptolemaic dynasty in Egypt, still yearned to rule the world. The showdown came at the naval Battle of Actium (off the west coast of Greece) in 31 BC when Octavian defeated Mark Antony, her husband. Cleopatra committed suicide, and peace and stability followed under the *Pax Romana*.

As Christianity began to spread during the third century AD so did the barbarian threat from the north; Athens was sacked in 267. The Roman Empire began to fragment. Too large to administer from a single centre, Rome, Emperor Constantine divided the empire in two and in 330 established his eastern capital at Byzantium – New Rome – and renamed it Constantinople. At this time eight Roman provinces roughly made up what is Greece today: six were part of what was known as Moesia which extended as far north as the Danube, Rhodope was part of Thrace and the islands were separate.

Despite the depredations of invasions by Avars, Bulgars, Huns and Slavs between the fourth and seventh centuries, Greek language and culture flourished within the Byzantine Empire; in due course the invaders became largely civilized and hellenized. During the ninth century some areas were reconquered although Crete was lost to the Saracens in 826. The empire was seriously weakened in 1054 when the Latin and Greek-speaking churches split. At much the same time Normans from Sicily invaded the Greek mainland, laying waste to Thebes and Corinth, while Venetians took control of the islands.

In 1204, after Constantinople had been sacked by the Fourth Crusade, its leaders reorganized much of the Byzantine Empire into new crusader states. The Latin Empire, comprising 'Romania' centred on Constantinople, the Kingdom of Thessaloniki, the Duchy of Athens and the Principality of Achaea, was founded. Venice gained control of Crete, Corfu and Euboea. Byzantine control was confined to the Despotate of Epirus. In 1261 Constantinople was recaptured by the Nicaean emperor, Michael Palaeologus. The Latin Empire was brought to an end and the Byzantine Empire re-established; but Achaea and the Duchy of Athens remained in Latin hands.

During the 14th century the Serbs moved south into northern Greece while the Ottoman Turks advanced from the east, taking Thessaloniki in 1387. Constantinople held out, but the Byzantine Empire finally fell when Sultan Mehmet II took the city in 1453. By this time most of peninsular Greece was in Ottoman hands. Although the islands were taken later (except for the Ionian Islands which were part of the Venetian Empire), mainland Greece remained part of the Ottoman Empire for the next 379 years; independence was only achieved in 1832.

The failure of the siege of Vienna in 1683 heralded the decline of Ottoman power in Europe. When the French Revolution broke out in 1789 the Greeks began to think of liberation. In 1821 scattered outbreaks of violence coalesced into a general revolt. After a protracted struggle, and with British and French help, Greece finally became free in 1828; formal recognition of independence followed four years later. A monarchy was established with Otto, the 17-year-old son of King Ludwig I of Bavaria, becoming king in 1833. Still a Roman Catholic in an Orthodox country and without an heir after 29 years on the throne, he was forced by the Army to abdicate in 1862. He was replaced by a Danish prince who was proclaimed George I, King of the Hellenes.

The drawing of the frontiers at independence left some two-thirds of the Greeks under Turkish rule: Greece's northern boundary lay between the Gulf of Arta in the west and the Gulf of Volos in the east. The desire to unite all Greeks in one homeland, known as the Great Idea (*Megali Idea*), grew. In 1864 Great Britain ceded the Ionian Islands (a British protectorate since 1815) and in 1881 Greece acquired Thessaly and part of Epirus from the Ottoman Empire. Greece and Macedonia, a region with an explosive ethnic mix but still under Ottoman rule, now shared a border. Greece participated in both Balkan Wars in 1912/13 (q.v.Macedonia). Following the second, during which Greece joined Romania and Serbia against Bulgaria, the rest of Epirus, part of Macedonia, western Thrace, Crete and the north-eastern Aegean islands were obtained. By this time Greece had more than doubled in size since independence. Nevertheless, the Greeks were upset by the establishment of an independent Albania, which included the Greek-speaking part of northern Epirus, in 1912.

Divided by the 'National Schism' (King Constantine I demanded neutrality, Prime Minister Venizélos support for the Allies) for three years, Greece finally entered the First World War on the Allied side in 1917. The Treaty of Sèvres in 1920 assigned all Thrace, the Aegean Islands and the province of Smyrna (now Izmir) on the Turkish mainland to Greece. The same year the Greek army in Smyrna, threatened by Turkish nationalists led by Kemal Atatürk, launched a major offensive towards Ankara; within two years it had been driven back to the coast and Smyrna sacked. In July 1923 at the Treaty of Lausanne the Greeks agreed to abandon all claims to territory in Asia Minor and ceded those parts of Thrace east of the River Évros (now Meriç) and the islands of Tenedos (now Bozca) and Imbros (Gökçe) to Turkey. In addition, a huge exchange of populations was agreed: Greece received 1.3 million Turkish-speaking Orthodox Christians in exchange for 400,000 Greek-speaking Muslims. The Treaty more or less established the boundaries of today's Greece and, by means of the population exchange, made it virtually ethnically homogeneous.

In April 1939 Italy invaded Albania. Believing Greece to be a push-over and wishing to match Hitler's victories, Mussolini ordered his troops into Greece in October 1940. Within weeks they had been driven out and the Greeks had gone on to occupy southern Albania (Northern Epirus). To protect his southern flank Hitler invaded Greece (and Yugoslavia) in April 1941 and within two weeks forced surrender. Crete was captured in May. By June the Germans were in control of all the key areas, the Bulgarians occupied Thrace and most of Macedonia, and the Italians had the rest. The Greeks, particularly the communists, resisted until the Germans withdrew in October 1944.

With liberation came civil war. The Communist National Liberation Front attempted to usurp power but, with American and British help, the Greek government eventually prevailed in 1949. Two years earlier, the Dodecanese Islands, seized by Italy in 1912, were handed over to Greece.

The 'colonels' coup' in 1967 introduced a military dictatorship and led to the expulsion of the king. But it was not until 1974 when the regime collapsed, that the monarchy was abolished and a republic proclaimed. Disagreements over Cyprus and the control of the airspace, territorial waters, islands and seabed of the Aegean have continued to bedevil relations with Turkey despite the fact that both countries are members of NATO.

Area:	131 957 sq km (50 949 sq miles)
Population:	10.6 million; Greeks, Turks
Language:	Greek
Religions:	Greek Orthodox, Muslim
Capital:	Athens
Administrative Districts:	13 regions
Date of Independence:	7 May 1832
Neighbouring Countries:	Former Yugoslav Republic of Macedonia, Bulgaria, Turkey, Albania

English-speaking Name	Local Name	Former Names	Notes
GREECE	Ellas	Hellas	The Hellenic Republic (Elliniki Dhimokratia) since Jul 1973. The Kingdom of Greece (1830–1924, 1935–74) the monarchy was not abolished until Dec 1974. The Greeks were called Hellenes after Hellen, eldest son of Deucalion and Pyrrha. Originally restricted to a section of the Thessalians and only later applied to the Greeks as a whole. Referred to as Graecia by the Romans. The phrase 'It's all Greek to me' comes from Shakespeare's play *Julius Caesar* to mean talk that is incomprehensible to the listener.
Aegean	Aiyaion		Sea and islands named after Aegeus, King of Athens. When his son, Theseus, returned, having been offered up to the Minotaur, he forgot to change the colour of his sails from black to white to indicate his survival. When Aegeus saw this he flung himself into the sea.
Alexandroúpolis		Dedeagaç	Founded (1860) by the Turks as Dedeagaç which means 'Tree of the Holy Man', being derived from a colony of dervishes. Ceded to Bulgaria (1913) but given to Greece by the Treaties of Neuilly (1919) and Sèvres (1920). It is not clear after which Alexander the city, *polis*, is now named: Tsar Alexander II whose Russian armies drove the Turks out (1878) or after the Greek King Alexander (1893–1920, r.1917–20).
Amphipolis		Ennea Hodoi	Originally a Thracian communications centre meaning 'Nine Roads', its present name means 'Surrounded city' because of its position in a loop of the River Strymon.

Athens	Athinai		According to legend, the gods of Olympus proclaimed that a city founded by Cecrops, a Phoenician, should be named after the god who could produce the most valuable legacy for mortals. Athena (in Latin, Minerva), goddess of war, produced an olive tree, the symbol of peace and prosperity, while Poseidon, god of the sea, produced a horse, a symbol of strength and endurance (although it is also said that he caused a spring of salt water to flow on the Acropolis). The gods chose Athena. Under Turkish domination (1456–1829). Capital since 1834.
Attica	Attiki		A region whose name means the 'Territory of Athens'; Athens is its chief city.
Corfu	Kerkira	Corcyra	An island. The name is an Italian corruption of the Greek *coryphai*, crests, referring to the twin mountain peaks. According to legend, called Scheria. The Corfu Declaration (20 Jul 1917) called for the establishment of a unified Yugoslav state, the Kingdom of the Serbs, Croats and Slovenes.
Corinth	Kórinthos		Probably derived from *kar*, point, a reference to its position on the isthmus of Corinth. St Paul wrote letters to the Corinthians. Although the Roman Emperor Nero (37–68, r.54–68) began to carve a canal through the isthmus (67), the Corinth Canal was not opened until 1893.
Cyclades	Kikládhes		A group of 39 islands which roughly encircle the sacred island of Delos. The name comes from *kyklos*, circle.
Dodecanese	Dhodhekánisos		Means '12 islands' although there are more. However, the name was designated by the Turks (16th century) who recognized only 12 as composing the group because they had voluntarily accepted Turkish rule. This lasted until the Italians took control (1912). Finally returned to Greece (1947).

Epirus	Ípiros		A region whose name comes from *epeiros*, mainland, to differentiate from the many islands. The Despotate of Epirus (1204–1337) was a Byzantine principality in north-west Greece and southern Albania, formed when the Crusaders took Constantinople (1204). Part of Epirus still lies in Albania.
Évvoia		Hellopia, Euboea, Egripo, Negraponte	An island whose name may be derived from *euboia*, rich in cattle; or it may be named after the nymph Evia. Hellopia was named after the Hellopes.
Ioánnina		Janina, Yanya	Believed to take its name from a local monastery dedicated to St John the Baptist. Centre of the Despot Ali Pasha's rule over mainland Greece and Albania. His revolt against the Sultan helped to initiate the Greek War of Independence (1821–28).
Ionian Islands	Iónioi Nisoi		Often called Eftanisa or Heptanesos, Seven Islands, there being seven principal islands.
Kavála		Neapolis, Christopolis	Means 'on horseback', a reference to the site as a relay station for changing horses. Originally Byzantine as Christopolis, held by the Turks (1387–1912) and occupied by the Bulgarians three times during the First and Second World Wars.
Lesbos	Lésvos	Pentapolis	An island. The word lesbian is derived from the island of Lesbos on account of the practices attributed to the poetess Sappho. Also called Mitilini after its capital. Its original name alluded to its five cities.
Marathon			Takes its name from *marathron*, fennel, an aromatic herb that grows here. Gave its name to the marathon race after a courier ran from Marathon to Athens to report the victory of the Athenians over the Persians (490 BC). After running about 25 miles he delivered his message but then fell dead. The

			distance for the modern marathon is 26 miles 385 yards – the distance from the start at Windsor Castle to the finish opposite the royal box in the stadium in London at the 1908 Olympics.
Megalopolis			Founded (c.370 BC) as a 'Great City' to oppose Sparta.
Metéora			From *meteoros*, meaning 'suspended in the air', a reference to the pillars of rock in the area, some of which are crowned with monasteries and hermitages.
Monemvasia		Minoa	Linked to the mainland by a causeway, the name means 'City with only one entrance' from *moni embasis*.
Nafpaktos		Lepanto	Name derived from the Greek *naus*, ship and *pegnumi*, I make fast. A Christian fleet from Italy, Spain and Malta defeated the Turks at the naval Battle of Lepanto (1571), thus demonstrating Turkish weakness.
Náfplion	Návplio(n)	Nauplia	Means 'Naval station' probably because it was the main port for Argos, the dominant city-state in the Peloponnese (7th century BC). Alternatively, some claim that it is named after Navplius, King of Euboea and a hero of the Trojan War. Capital (1829–34).
Nea Sparti		Sparta, Lacedaemon	Destroyed by the Visigoths and re-populated by the Byzantines and called Lacedaemon. The inhabitants of Sparta were known for their frugality, stern discipline and endurance. Thus the word spartan to describe one who can withstand pain and hardship without complaint.
Pella		Bounomos, Diocletianopolis	Capital of Macedonia at the height of its greatness (4th century BC). The Romans named it after Emperor Diocletian (245–316, r.284–305).

Peloponnese	Pelopónnisos	Morea	A region meaning 'Island of Pelops', *Pelops nisos*, son of the Lydian King Tantalus and grandson of Zeus in Greek mythology. Morea means 'The black mulberry'.
Philippi	Filippoi		Founded (356 BC) by, and named after, Philip II of Macedon (382–36 BC, r.359–36 BC) who took the site from the Thracians to protect the gold mines. Site of the decisive Battle of Philippi (42 BC) when Mark Antony and Octavian defeated Brutus and Cassius. St Paul wrote an Epistle to the Philippians.
Pythagóreio	Samos, Tigáni		The ancient name gave way to Tigáni which means 'Frying pan' because of the shape of its harbour. However, Pythagoras, philosopher and mathematician, was born here (c.580 BC) and his name has recently been considered more suitable.
Rhodes	Ródhos		The name may be derived from the Phoenician *erod*, snake, in recognition of the fact that the island used to be infested with snakes; or it may be derived from the Greek *rhodos*, rose. Site of one of the Seven Wonders of the Ancient World, the Colossus of Rhodes, a 32m-high bronze statue dedicated to the sun god Helios to commemorate the raising of a long siege (built c.292–280 BC; destroyed by an earthquake c.224 BC). Ceded to Greece (1947).
Santorini	Thera		An island in the Cyclades group. Named after the martyr St Irene (d.304).
Thessaloniki	Thessalonica, Salonika		Means 'Victory in Thessaly' from *Thessalia* and *nikë*, victory. So named after a sister of Alexander the Great, Thessaloníkë, by her husband Kassandros (358–297 BC) who founded it (315 BC). St Paul wrote two Epistles to the Thessalonians. Capital of the Roman province of Macedonia. Under Turkish rule (1430–1912).

Thera	Calliste, Santorini	The original name means 'Most beautiful'. Santorini is named after St Irene of Salonika, one of three sisters who were all martyred (304).
Thivai	Thebes, Ogygion	Known by some classical poets as Ogygion from King Ogygos. Seat of the legendary King Oedipus who unwittingly killed his father and married his mother.
Crete Kriti	Kaptaru, Candia	A region and an island. The birthplace of the Minoan civilization. Thereafter under the control of the Dorians, the Romans (67 BC–395 AD), the Byzantines (395–824), the Arabs (824–1204), the Venetians (1204–1669) and the Turks (1669–1898). United with Greece (1913). Battle of Crete won by the Germans (May 1941).
Heraklion Iráklio(n)	Herakleion, El Khandaq, Chandax, Candia, Megalo Kastro	El Khandaq, meaning 'Castle in the moat' in Arabic and referring to the great ditch which surrounded it, was corrupted to Chandax by the Byzantines and to Candia by the Venetians to whom the island was sold (1204). Under Turkish rule (1669–1897) as Megalo Kastro, Great Fort. Iraklion is derived from the ancient Roman port of Heracleum. Named after Hercules (Heracles) who successfully carried out the seventh of his 'labours' here: the capture of the fire-breathing mad bull that had been terrorizing Crete. Capital (1971).

HUNGARY

The Magyars, a Finno-Ugric people, not Slavs, erupted out of the steppes to seize Bulgarian territory north of the Danube and an area west of it, Pannonia, during the ninth century. Ever since, Hungary has been subject to change, contracting, expanding and being partitioned.

In the first century AD the Romans incorporated Pannonia into their empire as a buffer against the Ostrogothic and Vandal incursions from the east. A century later the province of Dacia was formed in the region of Transylvania (q.v. Romania). After the Roman departure in the fourth century the area was occupied by a number of warring tribes, the most prominent being the Avars who survived until the end of the eighth century before being crushed by Charlemagne. Driven westwards from the area north of the Black Sea by the Pechenegs, the Magyars were living on the western edge of the steppes when they were invited by the Carolingian emperor Arnulf to help him suppress the Moravians (q.v.The Czech Republic). Under their elected chieftain, Árpád, the seven Magyar tribes crossed the Carpathian Mountains in 896.

They destroyed the Moravian Empire in 906 and took Pannonia the next year where Árpád entrenched himself. For the next fifty years the Magyars terrorized central Europe until defeated by the Holy Roman Emperor Otto I in 955. The tribes were now faced with a choice: either to settle and adjust to European life or face extinction like other nomadic peoples. Árpád's great-grandson, Géza, chose the former, establishing his authority over the tribes and beginning the conversion to Christianity. His son Stephen (István), later canonized, consolidated his control and asked the pope to recognize him as King of Hungary. Stephen was crowned on Christmas Day, 1000 and the Kingdom of Hungary, independent of the Holy Roman Empire, was founded.

After Stephen's death in 1038, Hungary acquired new territories: Transylvania (populated by a majority of Germans who called themselves Saxons), Slavonia, Croatia and a frontier zone near Belgrade. The Mongol Tatar invasion of 1241, however, devastated Hungary and half the population was killed. It might have been worse, but news of the death of the Great Khan arrived in 1242, inducing the Tatars to withdraw. In 1301 the Árpád dynasty came to an end. Until the monarchy was abolished in 1918 Hungary was to be ruled by foreign kings with only two exceptions, of which one was disputed.

The expansionist Ottoman Empire began to pose a threat to the country at the end of the 14th century. In 1446 János Hunyadi, an outstanding military leader, was appointed the young king's regent. Raising an army, he thwarted Turkish ambitions. However, the Turks could not consider the conquest of Hungary until they had taken Belgrade, a pivotal fortress guarding the southern border. By raising the siege of Belgrade in 1456, Hunyadi inflicted a crushing defeat on the Turks which ensured Hungarian security for another half century. This was enjoyed by the only Hungarian king to rule over all Hungary, Matthias Corvinus (Mátyás Hunyadi, reigned 1458–90 and called Corvinus because of the raven on his coat-of-arms).

In 1521 the Turks under the sultan, Süleiman the Magnificent, took Belgrade and five years later launched a large-scale offensive against Hungary. At the Battle of Mohács in 1526 the weak and poorly-armed Hungarian army was decisively defeated and the 19-year old king killed. Surprisingly, the Turks withdrew. Two men now claimed the throne: Ferdinand, Habsburg ruler of Austria, who was supported by most of the nobles, and János Zápolya, governor of Transylvania. When Ferdinand prevailed over Zápolya in a civil war, Zápolya appealed to Süleiman. Ferdinand's occupation of the western and northern third of the country was confirmed on payment of tribute, while Zápolya took the rest as a vassal of the sultan. After Zápolya died in 1541 Süleiman occupied Buda and annexed central and southern Hungary, thus partitioning the country; Ferdinand was allowed to hold his territory in return for continued tributes.

From Vienna the Habsburgs controlled Royal Hungary – the counties along the Austrian border, modern Slovakia and parts of north-western Croatia. When Ferdinand's successor, Maximilian II, renewed hostilities in 1566 Süleiman responded: Transylvania became an autonomous vassal principality ruled by Zápolya's son.

The partition between the Habsburg and Ottoman Empires lasted for more than 150 years. Vienna's view of Royal Hungary as an unimportant buffer against the Turks and its wish not to antagonize them conflicted with the Hungarian desire to end Ottoman rule. The Hungarians became anti-Habsburg and xenophobic. Relegated to the status of a colony, Hungarian discontent grew. Although attempts to rid themselves of Habsburg rule failed, the Hungarians did force concessions from their masters. This encouraged the Turks to send a large army into Hungary that went on to reach the gates of Vienna in 1683. It got no further and, after more fighting during which the Turks were driven south of the Danube, the sultan was forced to relinquish virtually all his Hungarian and Croatian possessions, including Transylvania which became an independent imperial principality, at the Treaty of Karlowitz (now Sremski Karlovci, Yugoslavia) in 1699. Already by 1687, with these successes, the Habsburgs had regained the favour of the Hungarian nobles who had declared them to be the hereditary rulers of Hungary.

As the Habsburgs gained control over the Hungarian crown lands, they caused outrage by the suppression of the Protestants and the re-distribution of land, previously held by the Turks, to foreigners. In 1703 a national rebellion – a war of independence – broke out which took eight years for the Habsburgs to put down. Thereafter a more enlightened view was taken in Vienna. In 1713 the Habsburg Emperor, Charles VI, proclaimed the Pragmatic Sanction: the indivisibility of the empire and the rights of inheritance of the Habsburg's female offspring. One aim of this decree was to ensure that his daughter, Maria Theresa, should inherit his territories. The emperor or empress was to rule Hungary as a king or queen subject to Hungary's constitution. Ten years later the Hungarian Diet approved the Sanction, thereby agreeing to become a hereditary monarchy under the Habsburgs for as long as their dynasty lasted.

For the Hungarians the danger now lay in the possibility that their country might be reduced to the status of a province. Maria Theresa's son, Joseph II, proceeded to try to do just this by pursuing a policy of Germanization of the language and the business of government. By this time the ethnic composition of Hungary had become diverse. Substantial numbers of Croats, Germans, Romanians, Serbs and Slovaks had been introduced, so that by the end of the 18th century the Magyars comprised only about 40 per cent of the population. Most of these nationalities harboured grievances against the Magyars.

By the time a revolutionary spirit began to sweep Europe in 1848 the Magyars were ready. The following year reformers, led by Lajos Kossuth, declared Hungary to be a Magyar nation-state, which included Transylvania, independent of Habsburg rule. However, with Russian assistance, imperial troops overwhelmed the Hungarian rebel forces and reasserted Habsburg control. Nevertheless, undermined by internal resistance and defeat by the Prussians in 1866 (q.v.Austria), the Habsburgs were forced to make concessions to the Hungarians which resulted in the Compromise (*Ausgleich)* of 1867 and the creation of the Dual Monarchy of Austro-Hungary. Although Francis Joseph, already Habsburg emperor and King of Bohemia, was crowned King of Hungary, the country (which comprised modern Hungary, Slovakia, Transylvania, the Banat and Croatia) enjoyed greater internal independence than at any time since 1526. Foreign affairs and defence were conducted jointly.

Austria-Hungary's defeat at the end of the First World War resulted in the dismemberment of Hungary and the proclamation of an independent republic. At the Treaty of Trianon in June 1920 Hungary was forced to cede more than two thirds (189 000 sq km) of its 'historic' lands: Burgenland to Austria, Carpatho-Ukraine and Slovakia to Czechoslovakia, Transylvania and part of the Banat to Romania, and another part of the Banat, Croatia-Slavonia and Vojvodina to Yugoslavia; Poland and Italy also received small parcels of land. These transfers, particularly that of Transylvania, were considered a national humiliation and, furthermore, they consigned more than three million Hungarians to living outside Hungary; the new country, however, became ethnically homogeneous, although the population was reduced from 18 million to 7.6 million.

Hitler's rise to power encouraged demands for a revision of the Treaty. By means of the Vienna Awards between 1938 and 1940 Hungary recovered parts of southern Slovakia, Carpatho-Ukraine and northern Transylvania; when Germany invaded Yugoslavia in 1941 Hungary annexed Vojvodina.

Having signed the Tripartite Pact in 1940 and thus been drawn to the side of Nazi Germany in the Second World War, Hungary shared in the defeat. The Red Army entered the country in September 1944 and by 4 April 1945 had driven the Germans out. The 1947 Paris Treaty restored the Trianon frontiers but for some small adjustments: strategically important was the acquisition by the Soviet Union of 110 km of common border with Hungary. The country became a Soviet satellite and member of the Warsaw Pact. In 1956, however, the Hungarians rebelled, announced their withdrawal from the Pact and proclaimed their neutrality. The revolution was brutally suppressed by the Soviet Army which maintained troops in the country until 1991.

Communist rule lasted until 1989 when the communist party voluntarily gave up its monopoly of power. Relations with Hungary's neighbours were not improved in 1990 when the prime minister declared himself the 'prime minister of all Hungarians', an assertion that was interpreted as a claim of Hungarian sovereignty over territory in adjacent countries.

Area:	93 030 sq km (35 919 sq miles)
Population:	9.81 million; Magyars, Gypsies, Germans, Slovaks
Language:	Hungarian
Religions:	Roman Catholic, Protestant
Capital:	Budapest
Administrative Districts:	19 counties and Budapest
Date of Independence:	16 November 1918
Neighbouring Countries:	Slovakia, Ukraine, Romania, Yugoslavia, Croatia, Slovenia, Austria

The Ottoman Empire in Europe 1683

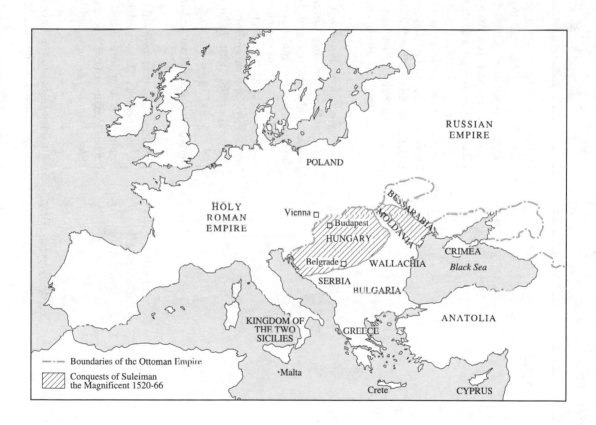

English-speaking Name	Local Name	Former Names	Notes
Hungary	Magyarország		Republic of Hungary (Magyar Köztársaság) (1918–49) and since 1989; the Hungarian People's Republic (1949–89); the Kingdom of Hungary (1000–1918). Hungary is derived from *On Ogur*, meaning 'Ten Arrows', the name of a group of tribes (seven Magyar and three Kavar) living along the north shore of the Black Sea before they moved to modern Hungary. In due course, the name of the strongest tribe, the Megyers, came to be used as a generic name for the whole group. The theory that the violence of the marauding Magyars reminded their victims of Attila and his Huns, thus attracting the name Hungarians, is not tenable.
Budapest		Aquincum, Ofen	The city acquired its present name (1873) when the towns of Buda, Pest, meaning 'furnace' or 'kiln' (German *ofen*), and Óbuda (Old Buda) merged. Buda may be named after Attila the Hun's brother, Buda; alternatively, it may be named after the first constable of the new fortress (11th century). Destroyed by the Tatars (1241), occupied by the Turks (1541–1686) and sacked by the Romanians (1919). Buda was the royal capital (1361–1873); capital since 1873.
Csongrád			Derived from the Slavonic *Czernigrad*, Black Castle.
Dunaújváros		Intercisa, Dunapentele Sztálinváros	The original Roman town fell into decline to become the village of Dunapentele. This was developed (1950) into the town of Sztálinváros after Joseph Stalin (1879–1953), the Soviet leader. Renamed (1961) to mean 'Danube New Town' from *Duna*, Danube, *új*, new, and *város*, town.
Esztergom		Strigonium, Gran, Etzelburg	All names testify to the town's prominence as a grain market. Capital of the early Árpád kings (until 1361).

Györ	Arrabona, Raab Janik-Kala	Name possibly derived from the Avar *gyürü*, circular fortress. The Roman name denotes a fort on the River Raba. The Turkish name means 'Burnt-out town'.
Gyula		Named after a Magyar tribal chieftain (10th century).
Hajdúság		A region around Debrecen taking its name from the *Hajduk*, partisan fighters against the Turks (or bandits, according to the Turks). Several towns in the area are pre-fixed Hajdú-.
Kecskemét		Name derived from *kecske*, goat.
Mosonmagyaróvár	Wieselburg	Means the 'Magyar fortress on the (River) Moson', *vár* meaning 'fortress' or 'castle'.
Nyiregyháza		Means 'Church among the birch trees', from *Nyir*, birch, and *egyház*, church.
Pécs	Sopianae, Fünfkirchen	Means 'Five churches' in German. Pécs means 'cave'.
Sopron	Scarabantia, Ödenburg	Although in Burgenland, which was awarded to Austria at the end of the First World War, the people of Sopron voted to be part of Hungary (1921).
Szeged		Possibly derived from *szeg*, angle, a reference to its position on a bend of the River Tisza.
Székesfehérvár	Alba Regia, Stuhlweissenburg	The name comes from *fehér*, white, and *vár*. Founded by Prince Géza, whose son made it his royal seat, *szék*. Briefly capital of the Hungarian kingdom (10th century).

Szentendre		Means 'St Andrew'.
Szombathely	Savaria, Steinamanger	Means 'Saturday market' from *szombat*, Saturday, and *hely*, place.
Tiszaújváros	Leninváros	Having been named 'Lenin Town', it was then renamed after the Tisza family as 'Tisza New Town'. Both Kálmán Tisza (1830–1902) and his son, Count Stephen (István) (1861–1918) were premiers of Hungary.
Veszprém		Named after the Polish prince Bezbriem (9th century).
Višegrád		Means 'High castle'. In seeking admittance to European institutions, such as NATO and the European Union, after the collapse of the Soviet Union, COMECON and the Warsaw Pact, Czechoslovakia, Hungary and Poland held a summit conference here (1991); as a result the three countries became known as the Višegrád Three.

ICELAND

Before Iceland was gradually settled between 870 and 930 by Vikings from Norway, it was populated only by some Irish monks who wished to live in isolation. When the Vikings arrived the monks left. Nevertheless, at the turn of the millennium, missionaries from Norway won the Icelanders over to Christianity. This did not stop local chieftains feuding. Not until 1262–64 were they forced to swear allegiance to the Norwegian king and forbidden to fight each other. Peace was established and the chieftains were permitted a significant degree of autonomy.

When Norway and Denmark were united under King Olav in 1380, Denmark assumed control. Until the mid-17th century this was quite even-handed. Absolute monarchy was imposed in 1662 and power of government transferred to Copenhagen. Thereafter, Iceland's fortunes declined.

Not until the middle of the 19th century did the first steps towards independence take place under the leadership of Jón Sigurdsson. In 1874, 1000 years after the notional colonization of the island, King Christian IX of Denmark granted the Icelanders their own constitution and self-government in domestic affairs. Despite this, the Icelanders felt that this constitution fell short of their desires. The executive power in internal affairs was vested in the governor, who was responsible to the minister of Icelandic affairs in Copenhagen and not to the Althing (Parliament). The Danes, on the other hand, considered Iceland to be an integral part of the Danish kingdom. Only in 1904 did Iceland get home rule with a national government being set up in Reykjavik.

Home rule was not sufficient for the Icelanders. The dissolution of the union between Norway and Sweden in 1905 gave impetus to Icelandic demands for complete independence. In December 1918 the Act of Union, which could be abrogated after 25 years, provided for Iceland to become an independent state under the Danish crown; the Danes would be responsible only for foreign affairs.

The Germans invaded Denmark in April 1940 and the union with Iceland was broken. A month later British forces arrived on the island to drive them out and protect sea lanes in the North Atlantic. In 1941 the Americans assumed responsibility for Iceland's defence. In June 1944 the Althing decided to terminate the Act of Union and sever all ties with Denmark. A republic was declared and this was approved later by plebiscite.

Area:	103 000 sq km (39 768 sq miles)
Population:	282 000; Icelanders
Language:	Icelandic
Religions:	Evangelical Lutheranism, Roman Catholic, Protestant
Capital:	Reykjavik
Administrative Districts:	8 regions
Date of Independence:	17 June 1944

English-speaking Name	Local Name	Former Names	Notes
ICELAND	Island	Snæland	Republic of Iceland (Lydhveldidh Island) since Jun 1944. Originally named Snow Land (c.850) and then allegedly renamed Iceland from *ís*, ice, by the first Norse settlers in the ninth century to deter visitors coming to an island which had in fact comparatively moderate annual temperatures (because of the Gulf Stream) and was much warmer than would be expected for a country so far north.
Akranes			Means 'Field peninsula'.
Akureyri			Means 'Meadow sand spit'.
Ísafjördur		Eyri	The original name meant 'Sand spit'.
Keflavik			Means 'Bay of sticks' from *kefli*, *stick*, and *vik*, bay, supposedly a reference to the debris found in the water.
Neskaupstadur		Nordfjördur	Meant 'Northern fjord'.
Reykjavik		Ingolfsholdi	Originally named after Ingolfr Arnarson, an early Norse adventurer to Iceland (c.870) and first permanent settler on the island. Reykjavik means 'Smoky bay', a reference to the steam given off as a result of volcanic activity. Capital in all but name since 1786 and of the Republic since 1944.
Siglufjördur		Pormodsseyri	Originally named after Pormodur Rammi, the first settler, with *eyri*, sand spit, added.

IRELAND

Sometime during the second half of the first millennium BC the Celts came to Ireland where for centuries they remained undisturbed; the Romans never extended their conquest of Britain to Ireland. In the fifth century AD St Patrick arrived to convert the Irish to Christianity and at about this time their communities or petty kingdoms (*tuatha*), numbering perhaps about a hundred, became subordinate to five provincial kings. Known as the Five Fifths, these groups of *tuatha* were Connaught, Leinster, Meath, Munster and Ulster. Not until the end of the tenth century did a king (of kings) of all Ireland emerge.

In the meantime, in 795, the Vikings began their invasions, intent on plunder and pillage. Gradually these raids advanced further inland. In time military camps were established and then, during the course of the next century, Norse settlements sprang up to promote trade and provide the nucleus of such cities as Dublin, Limerick, Waterford and Wexford. Intermarriage became common between the Norse (who called themselves Ostmen) and the Irish. Any further thoughts of Viking penetration were halted after the victory of High King Brian Boru at the Battle of Clontarf in 1014 when he defeated the Scandinavian allies of the King of Dublin. Brian was killed, however, and his incipient dynasty was cut short. The struggle for dominance between the provincial kings was renewed.

Shortly after he became King of England in 1154, Henry II was given the right to rule Ireland and take the papal title 'Lord of Ireland' by the English Pope Adrian IV. When the King of Leinster, Dermot MacMurrough, was driven out of his kingdom in 1166 he turned to Henry for help. Henry agreed to several Norman knights from Wales going to his assistance, among them the Earl of Pembroke (better known as Strongbow) who was given hereditary rights. In 1167 Dermot was restored. When he died four years later Strongbow became king. Henry now perceived this development – the possible creation of an independent Norman kingdom – as a threat.

To preclude any such thing, Henry went to Ireland, arriving near Waterford at the head of a large army in October 1171. He granted Leinster to Strongbow and other territory to some of his Norman knights in return for homage; the Irish kings, including the High King, Rory O'Connor, submitted to Henry's overlordship which was enshrined in the Treaty of Windsor in 1175. Although he remained High King of Ireland except for Leinster, Meath and Waterford, O'Connor assumed the title of King of Connaught. His power dwindled to the extent that he was forced to abdicate in 1183.

In 1177 Henry transferred his rights as Lord of Ireland to his youngest son, later to become King John. By now the emphasis had shifted from acquiring the submission of the Irish to colonizing the island. English law was introduced. The Irish reacted in the mid-13th century to try to limit the extent of the Anglo-Norman conquest. Within a century practically all of Ulster had been lost and the Irish living outside English control were regarded as enemies. In 1366 the Statutes of Kilkenny were brought to the Irish Parliament by Lionel, one of Edward III's sons and Earl of Ulster, in an attempt to prevent further Irish encroachment; they were also directed against those Englishmen who had assumed Irish habits and spoke Irish, and against intermarriage. Many of the statutes proved to be largely ineffective. Nevertheless, as the Anglo-Norman (or, by now, the Anglo-Irish) aristocracy became absorbed into Gaelic culture (some even changed their names), the authority of the English government progressively eroded until, by the end of the 15th century, only the four eastern counties of Louth, Meath, Kildare and Dublin remained loyal to the English crown. A fortified earthen rampart, known as the Pale[1], was built to enclose them, although in fact half of Meath and half of Kildare were excluded. As lords deputy for the English and viceroy, the Earls of Kildare became the effective rulers of Ireland.

In 1534 Henry VIII ended the primacy of the 9th Earl of Kildare and thereafter the Tudor monarchs restored the authority of the English crown throughout Ireland. But this only stored up trouble for the future. During this century large numbers of English and Scottish Protestants began to settle in the north. Speaking a different language and with a different religion, they did not merge with the inhabitants. In 1641 Irish Catholics in Ulster rebelled and the conflict between Catholics and Protestants spread throughout the island, raging on until Oliver Cromwell brought it to an end in 1652 in a campaign of great brutality. As conquered territory, Irish land was confiscated by an Act of Settlement in 1652 and awarded to officers of the Army and local Protestants. Only those Irish landowners who could prove their loyalty to Cromwell's cause were allowed to retain their property; most land still in Catholic hands lay west of the River Shannon.

Having been forced to flee to France in 1688, the Catholic James II of England moved on to Ireland. His successor as King of England was William III; to confront his rival he took a large army to Ireland. Most Protestants rushed to his support. In 1690 he defeated James at the Battle of the Boyne. More territory was surrendered and Protestant ascendancy was confirmed. Dispossessed and without political power, the Catholics were bitter, but Catholicism was not extinguished.

During the 18th century, while the lot of the Catholics gradually improved, the Irish Protestants began to realize that their commercial interests did not always coincide with those of the English crown. They began to demand autonomy, or at least greater freedom, to run their own affairs. In 1782 the Irish Parliament was given the right to initiate legislation. However, the outbreak of the French Revolution encouraged those Catholics and Protestants keen to break the connection with England to rebel in 1798. The rebellion failed; its outcome was not what the rebels wanted. On 1 January 1801 Ireland became part of Britain when the United Kingdom of Great Britain and Ireland was created. Hopes of independence disappeared and a Catholic majority became subject to rule by a Protestant minority.

Perhaps up to a million people died during the potato famine of 1846–51 and another million emigrated to the United States. In 1856 the Irish Republican Brotherhood, also known as the Fenians, was founded. A secret revolutionary society, it believed that the only way to achieve independence was by armed action. However, an attempted uprising in 1867 failed. Nevertheless, their activities, in the shape of campaigns for land reform and home rule, convinced the British Prime Minister, Gladstone, of the need for reforms.

Home rule, however, was anathema to most Protestant 'loyalists', and when the third Home Rule Bill was introduced in the House of Commons in 1912, they established the Ulster Volunteer Force to resist it, by force if necessary. To oppose it, the nationalists organised their own militia, the Irish Volunteers, in Dublin in 1913. Ireland was on the brink of civil war when the First World War broke out.

Many Irish Volunteers joined the British Army on the understanding that the Home Rule Bill would be enacted at the end of the war. Many others, however, prepared to take advantage of Britain's difficulties. On Easter Monday, 24 April 1916, an armed uprising, led by the Fenians, took place and the Irish Republic was proclaimed. Within a week the leaders had been arrested, but their subsequent execution aroused further support for the Republican cause. The Irish public now switched their votes to the opposition party, Sinn Féin (We Alone). In the election of December 1918 Sinn Féin won 70 per cent of the seats. When its MPs met in January 1919 they reaffirmed the independence of Ireland. At the same time they inaugurated their provisional government, the Irish Assembly (Dáil Éirann), in opposition to the British administration. It created a defence force, the Irish Republican Army (IRA), which soon began to attack British troops and police. Law and order broke down and a guerrilla war ensued.

In 1920 the British Parliament passed the Government of Ireland Act which created separate subordinate parliaments with limited self-governing powers, one for six of Ulster's nine counties for the Unionists, and another for the remaining three in Ulster, which had large nationalist majorities, and the 23 other counties in the rest of Ireland. Southern Ireland was to remain a part of the British Empire. While the nationalists abhorred the idea of a divided island, republican Sinn Féin wanted to leave Britain and its empire without delay. To this end the IRA increased the tempo of its operations. Nevertheless, a truce was agreed in July 1921 and this was followed by the Anglo-Irish Treaty in December. Southern Ireland was to become a dominion of the British Commonwealth, enjoying the same degree of independence as the other dominions, and call itself the Irish Free State; the separate existence of the six north-eastern counties in Ulster was safeguarded. In December 1922 the Free State was formally inaugurated. The Republicans were outraged, but were forced to abandon their armed resistance in May 1923. The existing boundary between North and South was agreed in December 1925.

A new constitution declaring Ireland's full independence by renouncing British sovereignty was ratified in 1937, the position of Governor General (who represented the king in Ireland) was abolished and the Irish Free State became Éire. Ireland remained neutral during the Second World War. The External Relations Act of 1936 was repealed by means of the Republic of Ireland Act 1948. This came into force in April 1949 when Ireland became a republic and left the British Commonwealth. Ireland's new status was recognized by Britain with the proviso that only the Parliament of Northern Ireland could agree to any change in its status.

1. The word 'pale' comes from the Latin *palus*, stake, to indicate a specific area with clear boundaries. In Ireland it marked the area under English control. A Jewish Pale of Settlement was created in Russia after an influx of Jews as a result of the partitions of Poland in the 18th century. This is the origin of the expression 'beyond the pale'.

Area:	70 283 sq km (27 136 sq miles)
Population:	3.71 million; Celtic, English
Languages:	English, Irish (Gaelic)
Religions:	Roman Catholic, Anglican
Capital:	Dublin
Administrative Districts:	4 provinces
Date of Independence:	6 December 1922
Neighbouring Countries:	UK

English-speaking Name	Local Name	Former Name	Notes
IRELAND	Éire	Hibernia, Ériu, Scotia, Irish Free State	Four fifths of the island is occupied by the Republic of Ireland (Poblacht na hÉirann), the remainder by Northern Ireland which is part of the United Kingdom. A corruption of Iar-en-land, the land in the west; sometimes called Erin, short for Iar-innis, west island. The Latin Hibernia is a corruption of the Old Celtic Iverna, possibly itself derived from the Greek Ierne. The Irish Free State (Saorstát Éireann) (1922); Éire (1937).
Cavan	An Cabhán		Means the 'Hollow place'. One of the three counties of the old province of Ulster.
Clonmel	Cluain Meala		Means a 'Meadow of honey', a reference to the fertility of the Suir Valley.
Connaught			Also spelled Connacht. A province – and one of the five ancient kingdoms of Ireland – named after the Connachta tribe.
Cork	Corcaigh		Means 'Marshy place' from *corcach*. Developed from a church built on an island in the River Lee.
Donegal	Dún na nGall	Tyrconnell	Means 'Fort of the foreigners'; that is, Vikings, who built a fort here. From *dún*, fort, and *gall*, foreigner.
Drogheda	Droichead Átha		Means 'Bridge of the ford'; that is, the bridge built in place of the earlier ford after the Vikings had built settlements (911) on both sides of the River Boyne. Sacked (1649) by Oliver Cromwell (1599–1658), then commander-in-chief of the army and lord lieutenant.

Dublin	Baile Átha Cliath, Dubh Linn	Eblana	Means 'Dark Pool' from the Irish *dubh*, black, and *linn*, pool and named for the dark waters of the River Liffey on which a trading post was first built. The Irish name means 'Town of the Ford of the Hurdle'. Capital (1921).
Dún Laoghaire		Kingstown	Named after Laoghaire, High King of Ireland (5th century BC) who built a fortress here.
Galway	Gaillimh		Means 'stony' from the Irish *gall*, stone, having originated as a crossing place over the River Corrib.
Kerry	Ciarraí		The name comes from Ciar, the son of Fergus, King of Ulster.
Kildare	Cill Dara		Means 'Church of the oak' from *cill*, church, and *doire*, oak grove.
Kilkenney	Cill Chainnigh		Named after St Canice who founded a church here (6th century). The Statute of Kilkenney (1366) made it high treason for an Anglo-Norman to marry an Irish woman.
Killarney	Cill Airne		Means 'Church of the sloe'.
Leinster	Laigin		A province meaning the 'Land of the Laigin', one of the earliest tribes to settle in Ireland, from *tír*, land.
Leitrim	Liatroim		A county meaning 'Grey ridge'.
Limerick	Luimneach		Means 'Open land' from *lom*, bare, or possibly 'Rich land' from the Norse *Laemrich*. Possibly gave its name to a form of humorous verse from the chorus of a soldier's song, Will you come up to Limerick?

Mayo	Maigh Eo	A county meaning 'Plain of the yew trees'.
Meath	An Mhi	A county meaning 'The middle', a reference to its central location as one of the five ancient kingdoms.
Monaghan	Muineachán	Means 'Place of thickets'.
Munster	Muma	A province meaning 'Land of the Mumu'. One of the five ancient kingdoms of Ireland.
Offaly	Uibh Fhailai	A county meaning the 'Descendants of Failghe', an obscure chieftain.
Roscommon	Ros Comáin	A county meaning 'Comán's wood'. St Comán established a monastery here (7th century).
Sligo	Sligeach	Means 'River of shells' from the bed of the River Garavogue.
Tipperary	Tiobraid Árann	Means 'House of the Well of Ara'. Famous for its connection with the song 'Tipperary' which was sung by British troops during the First World War.
Waterford	Vadrefjord	Means 'Wether inlet' with wether meaning an uncastrated ram; these were loaded onto ships here for export. Ford is from the Old Norse *fiorthr*, fjord or inlet. The city was the most important Viking settlement in Ireland (8th century). Site of the defeat of the High King of Leinster by the Anglo-Normans (1170). The victor married the king's daughter, thus cementing the first Norman/Irish alliance.
Wexford	Loch Garman	Menapia, Waesfjord	Lying at the mouth of the River Slaney, it means 'Inlet by the sandbank' and is derived from the Old Norse

			Waesfjorthr. Named Menapia by Ptolemy in his map (2nd century). The Slaney was originally the Garma, headland. Sacked (1649) by Cromwell's troops, and captured and occupied (1690).
Wicklow	Cill Mhantáin	Wykingalo	Named after St Mantan. The English name is derived from the Old Norse for Viking meadow, *Vikingr-aló*.

ITALY

Given its location in the Mediterranean, it is no wonder that Italy attracted invaders from the very earliest times, the Alps providing no real barrier. Competition, often violent, for land and domination have characterized its history. Between the fall of the Western Roman Empire in 476 and the unification of the Italian peninsula in 1870, Italy largely comprised a collection of major and minor independent states.

The eighth century BC saw the Greeks and Etruscans inaugurating city-states. Both civilisations lasted about six centuries. The Greeks penetrated as far north as the Bay of Naples, establishing Hellenic colonies along the shores of southern Italy and Sicily in what came to be called Magna Graecia (Greater Greece). The native Etruscans reached northwards into the Po Valley and southwards to Latium and Naples, giving to a Latin settlement called Rome the first hint of forthcoming grandeur when the Etruscan king of Tarquinii conquered it in about 550 BC.

In 509 BC the Latins of Rome overthrew their tyrannical Etruscan king and founded the Roman Republic. Thereafter, despite some setbacks, Roman power waxed as Etruscan power waned. Celtic expansion across the Alps into northern Italy, however, blocked the Roman progression, but by 265 BC the whole of Italy south of Cisalpine Gaul ('Gaul this side of the Alps' to the Romans) was under Roman control.

At this time Rome's main rival was the North African city-state of Carthage which controlled the western Mediterranean. The first two Punic Wars (264–41 BC and 218–02 BC) resulted in defeat for Carthage and the loss of Corsica, Sardinia and Sicily. The final Punic War (149–46 BC) saw the destruction of Carthage, which became a Roman province with the name of Africa, and Roman dominance of the western Mediterranean.

By 133 BC Roman power stretched from the Atlantic to Asia Minor and Roman citizenship had been extended to the whole of modern Italy. But the ambitions of successful generals and those that supported them induced rivalry and open confrontation, culminating in conflict. By the time of Julius Caesar's assassination in 44 BC, the Roman Republic had effectively ceased to exist.

Four years after defeating Anthony at the Battle of Actium in 31 BC, Octavian, Caesar's adopted son and heir, received the title of Augustus Caesar and began to wield imperial powers. The Empire flourished, its civilization being exported well beyond its borders. But from its zenith in 180 AD the power of Rome started to decline as a succession of weak, corrupt and unpopular emperors followed one after the other; none was able to defend Rome's vast frontiers against marauding barbarians. The Empire began to retract.

Meanwhile, Christianity gained a foothold in the peninsula. By the end of the second century the authority of the Bishop of Rome had begun to emerge in the city where St Peter had lived and been martyred. A year after becoming emperor in 312, Constantine, the first to become a Christian, issued the Edict of Milan which gave complete freedom to all religions. The transition from persecution to pre-eminence for Christians was concluded in 395 when Emperor Theodosius decreed that Christianity should be the official religion of the Empire.

In 330 Constantine moved his capital to Byzantium, renaming it Constantinople (now Istanbul) after himself; it was sometimes, however, called New Rome. The division of the

Empire was formalized in 395 with the death of Theodosius and in 402 Emperor Honorius moved the capital of the western half to Ravenna which he thought he could defend more easily than Rome. The Western Empire finally toppled under the onslaught of invading barbarians in 476 and Odoacer became the first barbarian King of Italy, although subservient to the Eastern emperor in Constantinople. When Theoderic, King of the Ostrogoths, removed Odoacer he ushered in over 60 years of Germanic rule in Italy.

More invasions followed: by 540 Belisarius, Emperor Justinian's greatest general, had conquered Italy as far north as its capital Ravenna and in 554 Italy was made a province of the Byzantine (Eastern) Empire. The Lombards, another Germanic tribe, made an even greater impact. Invading in 568, they established a kingdom in northern Italy, which included Tuscany, despite opposition from the Franks and Byzantines. By about 700 Italy comprised the Kingdom of Lombardy, the two virtually independent Lombard duchies of Benevento and Spoleto, and the territory still belonging to the Byzantine Empire – the Exarchate of Ravenna, the Duchy of Rome, the Duchy of Naples, Calabria, Sicily and Sardinia.

As the Lombards expanded their kingdom and captured Ravenna in 751, the people of central Italy turned to the pope, rather than to Byzantium, to provide protection. He turned to the King of the Franks, Pepin III, for help. Pepin won back Ravenna and gave it to the pope. Byzantium was dismayed. When, however, the Lombards threatened Rome in 773, one of Pepin's sons, Charlemagne, invaded Italy and conquered Lombardy. He confirmed the pope's temporal control of a wide belt of territory slanting diagonally across central Italy; it was to become known as the Papal States. Between 756 and 1870 its possession gave to the pope considerable temporal power. In 800 Pope Leo III crowned Charlemagne Emperor of the Romans; he called himself King of the Franks and Lombards, thus demonstrating the separation between his Frankish Empire and Italy.

From 827 Muslim Arabs – Saracens (possibly derived from the Arabic *sharc,* east) – wrested Sicily and Sardinia from Byzantium and raided the mainland. For the rest of the century Italy was riven by strife. The Carolingian Empire and the Lombard Kingdom began to disintegrate. But in Germany, in 919, the Saxon dynasty had been founded and in 962 Pope John XII invited the German king, Otto I, to intervene. The leading figure in Europe, he wished to have a title that acknowledged his supremacy. His wish was granted and the same year he assumed the title of Holy Roman Emperor – the master of Germany, Italy, Burgundy and of Rome. With Otto's coronation the Kingdom of Italy was now tied to that of Germany: whoever the German nobles elected as King of Germany automatically became King of Italy as well.

Otto was not slow in exercising his imperial power: he deposed the pope and appointed a new one the following year. Ultimately, this was to lead to a power struggle between Church (the papacy) and State (the Holy Roman Empire) at the end of the 11th century which became known as the Investiture Controversy.

As Norman knights, travelling and pillaging their way to and from the Holy Land on pilgrimages at the beginning of the 11th century, rode through southern Italy they saw rich land ripe for conquest. Within 100 years they had driven the Arabs out of Sicily, gained control of southern Italy and in 1130 created the 'Kingdom of Sicily' which stretched as far north as the Papal States. Norman rule did not last long. It was supplanted by that of the Hohenstaufens when Henry VI, the Holy Roman Emperor, was crowned as king in 1194; he claimed the kingdom through his wife, Constance, daughter of King Roger II.

During the 13th century relations between the Hohenstaufens and the papacy deteriorated. When Frederick II died in 1250 he was succeeded by his son and then by his illegitimate son, Manfred, in 1254. This dismayed the papacy which turned to Charles of Anjou, brother of Louis IX of France. By 1266, when he defeated Manfred, Charles was in nominal control of all Italy; certainly in Sicily and Naples, but the Papal States remained and papal influence in northern Italy was strong.

A popular uprising against the French in Palermo in 1282, known as the Sicilian Vespers, was supported by the King of Aragon who was married to Manfred's daughter. He was offered the throne and thus the Kingdom of Sicily was split: the Angevins in Naples with no hereditary claim but supported by the papacy, and the Aragonese in Sicily founded on a Hohenstaufen claim. The two kingdoms were only rejoined under Alfonso V of Aragon in 1442.

When the French Pope, John XXII, moved the papacy to Avignon in 1309 Rome's importance dwindled and the Papal States began to disintegrate. Their disarray caused the papacy, paradoxically, both to remain in Avignon until 1377 and to encourage it to return to Rome. This period was known as the Babylonian Captivity.

On the eve of invasion in 1494 by Charles VIII of France, Italy was little more than a collection of city-states: the Kingdom of the Two Sicilies (Sicily and Naples which, since 1458, had had separate rulers, although both from the House of Aragon), the Papal States, the Venetian Republic (which included possessions along the east coast of the Adriatic Sea and Cyprus), the city-states of Milan and Florence, Corsica and a few other minor territories. Disunited and weak, Italy was incapable of resisting invasion and foreign domination.

Charles was invited into Italy by the uncle of the Duke of Milan in the hope that he would receive French support in his quest to become duke. Charles was keen to oblige to further his claim to the Kingdom of Naples, not long before ruled by the Angevins. Charles invaded in October 1494 and within five months was in possession of Naples. But, with over-extended lines of communication, Charles was forced to withdraw and the French garrison left in Naples to surrender. The French presence lasted less than a year.

Between 1494 and the Congress of Vienna in 1815 the peninsula was engulfed in turmoil. As the Venetian Empire continued to expand, Austrians, French and Spaniards competed for power and land. The French were ejected from Milan at the Battle of Pavia in 1525 by the Spanish, who then sacked Rome two years later. Between 1559 and 1700, when the Spanish Habsburg line became extinct, the Spaniards were the predominant power: Milan, Naples, Sicily and Sardinia were all Spanish possessions.

At the end of the War of Spanish Succession in 1713 Spanish Habsburg rule was swapped for Austrian Habsburg rule. In 1720 Charles, son of Philip V of Spain, received Sicily in exchange for Sardinia which went to Piedmont to become the Kingdom of Sardinia. Austrian rule in Naples was not a success and in 1734 Charles was crowned King of Naples and Sicily, founding the Neapolitan branch of the House of Bourbon[1] and declaring the kingdom to be independent. When Charles became King of Spain in 1759 he left Naples-Sicily to his son Ferdinand. His wife was an Austrian and as the French Revolution got under way she began to realign the kingdom with Austria and Britain.

In March 1796 Napoleon's army swept into Italy. Piedmont was defeated and Austrian-occupied Milan entered. By the middle of 1797 most of northern Italy had succumbed and the Cisalpine Republic was born. Venetian independence was ended the same year. The Papal States were invaded in 1798 and the Roman Republic declared. French troops entered Naples in 1799 and the Parthenopean Republic created. But in 1798 Napoleon had left for Egypt and the following year the French were forced out of all Italy save Genoa. In 1800 Napoleon was back. The Cisalpine Republic was resurrected to become the Italian Republic and in 1805 the Kingdom of Italy; in 1806 the French took Naples and Napoleon's brother, Joseph, became King of the Two Sicilies until 1808 when he was succeeded by General Murat. In 1809 Napoleon abolished the temporal power of the pope, having annexed the Papal States the year before. He was master now of all northern and central Italy, but not Sardinia or Sicily.

After the fall of Napoleon in 1815 the Congress of Vienna terminated these arrangements and the peninsula returned to being a conglomeration of seven major and minor states: Lombardy-Venetia (a Habsburg province ruled by Francis I of Austria), Tuscany, Parma, and Modena (all ruled by members of the House of Habsburg), the Kingdom of the Two Sicilies (the Spanish Bourbon Ferdinand I), the Kingdom of Piedmont-Sardinia (the Italian Victor Emmanuel I of Savoy) and the Papal States (the Italian Pope Pius VII). Italy was defined as "merely a geographical expression" by the Austrian chancellor, Metternich, and he did his best to keep it that way.

The first attempts to rid Italy of foreign rule began in 1820 and proved to be failures. The movement towards Italian unification, known as the *Il Risorgimento* (The Resurrection), began in Piedmont. Both France and Piedmont were opposed to Austrian hegemony of northern Italy. In January 1859 Count Cavour, the Piedmontese prime minister, sealed an alliance with Napoleon III. This was enough to provoke the Austrians into declaring war in April. French and Piedmontese troops were successful on the battlefield. Nice and Savoy were ceded to France while Lombardy, Tuscany and Emilia joined Piedmont. Despite their defeat, the Austrians retained Venetia and still dominated Italy.

It was clear that it would be easier to free Italy by starting in the south where the Sicilians were ready to rebel. Giuseppe Garibaldi, an experienced guerrilla leader, and The Thousand, without official backing from Piedmont, conquered Sicily in August 1860, crossed the Strait of Messina and marched on Naples which they reached the next month. Piedmontese troops struck from the north. Although incomplete (the pope, with the support of French troops, was still in control of Rome and the Austrians in Venetia), the Kingdom of Italy was proclaimed in March 1861 with Victor Emmanuel II of Piedmont as king.

An agreement in 1866 between Prussia and Italy committed Italy to declaring war on Austria as soon as Prussia started hostilities in return for the acquisition of Venetia. By the Peace of Vienna in October Venetia was ceded to Italy.

There remained the problem of the Papal States and Rome. The French were determined to keep the city out of Italian hands, while the Italians were intent on making it their capital. The French garrison was withdrawn from Rome when the Franco-Prussian war broke out in 1870. Italian troops entered the city shortly afterwards. Unification was now a reality, although San Marino remained independent. In 1929 Mussolini and Pope Pius XI signed the Lateran Pact whereby the Vatican was recognized as a sovereign state and it recognized the Kingdom of Italy.

Starting off as neutral, Italy joined the Allies in 1915, having been promised some Austrian-held territory by the Treaty of London. As one of the victors in 1918, it gained this Alpine territory (Alto Adige, or South Tirol, and Trentino) at the Treaty of St Germain in September 1919, as well as Istria, parts of Dalmatia and Austria's Adriatic islands. That same year the Fascist Party was founded by Benito Mussolini and by October 1922 he had been asked by the king to form a government.

In the Second World War Italy was at first one of the Axis Powers. Land won in Europe between 1939 and 1942 was lost when the Allies invaded in 1943. Italy surrendered and declared war on Germany in October. Between 1945 and 1947 Italy lost Istria and Zadar to Yugoslavia, some small areas to France, the Dodecanese Islands to Greece, and the colonies of Eritrea and Libya which had been acquired in 1890 and 1912 respectively. In June 1946 the Italians voted to replace the monarchy with a republic.

1. The House of Bourbon is one of the foremost ruling houses of Europe. It was founded by Louis I when Bourbon was raised to a Duchy for him in 1327. Bourbons ruled in France, Naples and Sicily, and in Spain where the present king is a Bourbon.

Area:	301 225 sq km (116 303 sq miles)
Population:	57.46 million; Italians
Languages:	Italian, French, German
Religion:	Roman Catholic
Capital:	Rome
Administrative Districts:	20 regions
Date of Independence:	Unification: 17 March 1861
Neighbouring Countries:	Switzerland, Austria, Slovenia, San Marino, Vatican City, France

The Kingdom of Italy 1861

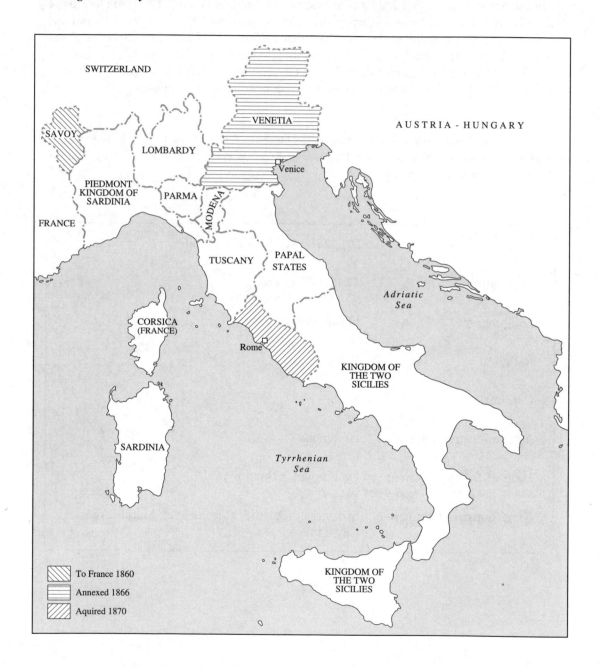

SWITZERLAND

SAVOY

VENETIA

AUSTRIA - HUNGARY

LOMBARDY

Venice

PIEDMONT
KINGDOM OF
SARDINIA

PARMA

MODENA

FRANCE

TUSCANY

PAPAL
STATES

CORSICA
(FRANCE)

Rome

*Adriatic
Sea*

KINGDOM OF
THE TWO
SICILIES

SARDINIA

*Tyrrhenian
Sea*

KINGDOM OF
THE TWO
SICILIES

To France 1860

Annexed 1866

Aquired 1870

English-speaking Name	Local Name	Former Names	Notes
ITALY	Italia		The Italian Republic (Repubblica Italiana) since 1946. Previously the Kingdom of Italy (1851–1946). Traditionally thought to be named after the Vitali tribe whose name may have some connection with the Latin *vitulus*, calf, or *vitaloi*, sons of the bull. Another theory is that the name is derived from *divi-telia*, land of the day or land of the light.
Abruzzi		Aprutium	A region, possibly named after an ancient tribe, the Praetutii, or from the Latin *abruptus*, steep, a reference to the Apennine Mountain range.
Ancona			Name derived from the Greek *angkon*, elbow, referring to the shape of the coastline.
Aosta		Augusta Praetoria	Founded (24 BC) by Emperor Augustus (63 BC-14 AD, r.31 BC-14 AD) and named after him. The present name is a shortening of the original name.
Apulia	Puglia		A region in the south-east with a long coastline whose name may be connected with *ap*, water.
Basilicata		Lucania	A region in the south whose name is derived from *basilikós*, the name given to the local Byzantine administrator or governor.
Benevento		Malies, Maleventum, Beneventum	In Sanniti, the language of the Sannites, Malventum meant 'Town in the mountains'. Thinking that it sounded like the Latin *male ventum*, the Romans misinterpreted it to mean 'Ill wind'. They therefore changed it to Beneventum, meaning 'Fair wind'. Site of the final victory (275 BC) of the Romans over Pyrrhus (319–272 BC), King of Epirus.

Bergamo	Bergomum	Situated in the foothills of the Alps, the name (196 BC) is derived from the German *berg*, mountain.
Bologna	Felsina, Bononia	Founded by the Etruscans but named by the Romans after the Gallic Boii tribe whose capital it had been (350–190 BC).
Brindisi	Brentesion, Brundisium	Named originally after the deer (in Illyrian, *brento* or *bretto*) which roamed the area and then changed (266 BC) to 'stag's head' because of the antler-shaped harbour.
Calabria	Ager Bruttius	A region whose name is derived from a pre-Indo-European root word *cal*, rock. For 1000 years up to the end of the 7th century Calabria referred to the south-eastern 'heel' of Italy rather than, as now, to the 'toe'.
Campania		A region whose name is derived from the Latin *campus*, plain, to describe the area round Naples.
Campobasso		Although the name literally means 'Low field', here it probably means the 'Field of Bassus', a personal name.
Capri		An island whose name may be derived from the Etruscan *capra*, land of tombs, the Latin, *capreae*, wild goats, or the Greek *kapros*, wild boar.
Carrara		Derived from the Latin *quadraria*, quarry, which gives its name to the famous Carrara marble, much favoured by Michelangelo (1475–1564), sculptor and painter.
Como	Comum	Associated with the Celtic word for valley (now *coomb* in English).

Cremona		Probably named after the Cenomani tribe. But it may be derived from the pre-Latin *carra*, stone, which later became corrupted to carm and then crem.
Emilia-Romagna	Romagna	A region whose name is derived from the Via Aemilia, a Roman road which joined Rimini and Piacenza. When the region was given to the papacy (756) by the Frankish King Pepin III (714–68, r.751–68) it was called Romagna (until 1948).
Faenza	Faventia	Derived from its Roman name which means 'favourable' in the sense that it is located in a very fertile area. Gave its name to faience, glazed pottery.
Ferrara	Forum Alieni	The present name may have a connection with a forge or smithy from the Latin *ferrarius*, pertaining to iron. It might instead come from *ferraria* to mean 'Land where farro (a type of cereal used to prepare soups) is cultivated'.
Florence	Colonia Florentia / Firenze	Means 'Flowering colony' or 'City of flowers', from the medieval *fiorenza*, perhaps because the city was built on a flowery meadow; or the name may be derived from the Latin *fluentia*, flowing, referring to the River Arno. Originally a colony of the Etruscan city of Fiesole (Faesulae). Capital of the Kingdom of Italy (1864–70).
Friuli-Venezia Giulia	Forum Julii	A region, the first part of which is a contraction of the Roman name, 'Market place of Julius (Caesar)'.
Genoa	Genua / Genova	The name may be derived from the Latin *janua*, door, or the Indo-European root word *gen*, curve. Gave its name to jeans (from the French name for the city, Gênes), twilled cotton cloth trousers worn by Genoese sailors.

Gorgonzola	Argenza		With surplus milk on their hands, the people of Argenza began to make a cheese they called Gorgonzola. In due course, the town came to be known as Gorgonzola.
Herculaneum	Resina	Ercolano	Traditionally associated with the name Hercules (Heracles), the son of Zeus and Alcmena in Greek mythology, who was possessed of superhuman physical strength. The ancient city was destroyed in the eruption of Vesuvius (79).
L'Aquila	Aquila degli Abruzzi		Founded (1240) on the order of Frederick II (1194–1250, Holy Roman Emperor 1220–50) who gave it an imperial eagle, *aquila*, for its emblem from which the town took its name. The previous name means 'Eagle of the Abruzzi (Mountains)' (1863–1939).
Latium		Lazio	A region taken from the Latin *latus*, broad, i.e. the plain, to describe the flat area in west-central Italy. Settled by the Latini whose language was thereafter widely adopted.
Liguria			A region named after the Ligurians.
Lombardy	Langobardus	Lombardia	A region named after the Germanic Lombards or Langobards, 'Long beards'. They established a kingdom in Italy (568–774).
Manfredonia			Founded (1256) and named after King Manfred of Sicily (1232–66, r.1258–66), the illegitimate son of Emperor Frederick II.
Mantua		Mantova	Name may be derived from Mantus, the Etruscan god of the underworld. Napoleon completed his conquest of northern Italy after an eight month siege (Jun 1796-Feb 1797).

Marches, The	Marche	A region whose name means 'Boundary'. In the early Middle Ages the region comprised three border provinces which acted as a buffer for Rome against invasion.
Milan	Mediolanum	Means 'Middle of the plain'. Founded (3rd century BC) by the Gauls on the banks of the River Po in the centre of an economically and strategically important region. Emperor Constantine I the Great (c.274–337, r.306–37) published his Edict (313) making Christianity the official religion of the Roman Empire. Destroyed (1162) by the Holy Roman Emperor Frederick I 'Barbarossa' (c.1123–90, r.1152–90). Made capital of the Cisalpine Republic by Napoleon (1797) and of the Italian Republic (1802).
Naples	Partenope, Neapolis	According to legend, the city was originally named after the siren Partenope. Founded (about 600 BC), Neapolis means 'New town'. Capital of the Kingdom of Naples and of the Two Sicilies (variously 1266–1860).
Orvieto	Urbs Vetus	Derived from the Roman name meaning 'Ancient town'. It was founded by the Etruscans.
Padua	Patavium	Possibly named after the local pine forests (in Gallic, *padi*, pine).
Parma		Named after the River Parma or possibly comes from the Celtic *parma*, round shield, which might have described the shape of the original settlement. Gave its name to the ham, violets and Parmesan (in Italian, *Parmigiano*), a hard cheese.
Pavia	Ticinum	Named after the Roman Papiria tribe. At the Battle of Pavia (1525) the French king, Francis I (1494–1547, r.1515–47), was defeated and captured by the Holy Roman Emperor

		Charles V (1500–58, emperor 1519–56). This Habsburg victory confirmed Spanish authority in Italy.
Pesaro	Pisaurum	At the mouth of the River Foglia, the ancient name for which was Isaurus, the name means 'Close to the Isaurus'.
Pescara	Aternum, Piscaria	Name derived from *pesce*, fish, to mean 'Full of fish'.
Piacenza	Placentia	As the Carthaginian general, Hannibal (247 BC–c.183 BC), advanced southwards, the Roman general, Publius Cornelius Scipio (? – 211 BC), prepared to defend a newly founded colony a little beyond the confluence of the Rivers Po and Trebbia. The armies met on Trebbia (Dec 218 BC); the Romans were defeated. The survivors of Scipio's army wintered in the colony where they found the conditions 'pleasing'. The colony was thus named Placentia.
Piedmont	Piemonte	A region whose name is derived from the old Italian *pie di monte*, foot of the mountain, i.e. the Alps. Part of the Kingdom of Sardinia (1720–1861).
Pozzuoli	Dicaearchia, Puteoli	The name comes from the Latin *putere*, to stink, which could refer to the smell of the sulphurous fumes from nearby Solfatara, 'sulphur place'. Dicaearchia means 'City of Justice'.
Ravenna		Derived from *rava*, a stony slope subject to landslides, with an Etruscan suffix, *enna*. Capital of the Western Roman Empire (402–76), then of the Ostrogothic kingdom (476–540) and of Byzantine Italy (540–751).
Reggio di Calabria	Rhegium, Regium Julium	Means the 'Royal town of Julius Caesar in Calabria'. Razed by an earthquake (1908).

Rome   Roma   According to legend, Romulus and his twin brother Remus resolved to found a new settlement on the hills above the river where their lives had been saved by a she-wolf (753 BC). They quarrelled and in the ensuing fight Romulus killed Remus; Romulus then began to increase the population by offering asylum to refugees and providing wives by arranging the rape of the Sabine women. However, Romulus and Remus were mythical figures. The name may actually come from Etruscan or Greek – possibly from the Greek *rhome*, strong. A strong theory is that the name comes from *Rumo*, one of the ancient names for the River Tiber on which the city lies. Capital of the Roman Republic (509–44 BC), then the Roman Empire (31 BC–402 AD). The Roman Emperor Commodus (161–192, r.180–92) had it called Colonia Commodiana, Colony of Commodus, until he was strangled. Sacked by the Visigoths (410), the Vandals (455), the Ostrogoths (546), the Normans (1084) and by the Holy Roman Emperor Charles V (1527). Became the centre of Christianity (end of the 2nd century). Under absolute Papal rule (1420–1870). Joined the Kingdom of Italy (1870) when King Victor Emmanuel II (1820–78, r.1849–78), seized Rome despite resistance by the Papal Guard and became the first king of a united Italy. Capital (1870). Occupied by the Germans (Sep 1943) when Italy surrendered; liberated by the Allies (Jun 1944). Gave its name to the Romance group of European languages as a result of the Roman occupation of Western Europe and Romania. The English word is derived from the Old French *romanz* to mean the speech of the people, that is the vernacular as opposed to classical latin. A romance was originally a novel written in the vernacular before the meaning changed to indicate the type of work rather than the language.

San Gimignano   Named after the Saint Bishop of Modena, Gimignano (?-397), who saved the town from the barbarians.

Senigallia	Sena Gallica Sinigaglia	Named (289 BC) after the Senonian Gauls.
Siena	Sena Julia	May be named after the Senonian Gauls who invaded Etruria (391). The colours of yellowish-brown raw sienna and reddish-brown burnt sienna derive from the colour of the earth of Siena. This ferruginous earth is used as a pigment in paint.
Taranto	Tarentum	May be named after a village called Taras, itself named after the River Tara. Or the name may come from *darandos*, oak, in recognition of the numerous trees of this kind previously in the region. Gave its name to the tarantula, originally the wolf-spider found in south-east Europe.
Tivoli	Tibur	May be derived from the Sabine word *teba*, hills. Famous as a resort, it gave its name to the Tivoli Gardens in Copenhagen.
Trentino-Alto Adige	Venetia-Tridentina	See Trento below. Alto means 'high' and Adige is the name of a river after which this region is named. It passed to Italy (1919); renamed (1947).
Trento	Tridentum, Trent/Trient	The original name came from the Latin *tres*, three, and *dens*, tooth, a reference to a local mountain with three peaks. Site of the Council of Trent (1545–63) which considered the restructuring of the Catholic Church and launched the Counter-Reformation. Became a part of Italy (1918).
Trieste	Tergeste, Triest	May be derived from the root *terg*, the Illyrian for market. Main port of the Austro-Hungarian Empire. Occupied by Italy (1918) following the secret Treaty of London (Apr 1915) by which Great Britain, France and Russia agreed to give the city to Italy at the conclusion of the First World War. Seized by Germany (1943), liberated by the Yugoslavs

Turin	Torino	(1945), but became a Free Territory under Anglo-American military administration (Zone A which included the city and port) and Yugoslav military administration (Zone B to the south). Zone B and part of Zone A became Yugoslav with the remainder of Zone A, which included the city, going to Italy (1954).
	Taurisia, Julia Taurinorum, Augusta Taurinorum	Named after the Taurini, a Ligurian tribe which might have taken its name from an ancient root word, *tauro*, mountains; thus Taurini would mean 'People of the mountains'. Or it may come from *taurus*, bull, which was used as an emblem. Annexed to France (1536–62). Became capital of the Kingdom of Sardinia (1720). Capital of Italy (1861–64).
Tuscany	Toscana	Tuscia	A region which takes its name from an Etruscan tribe, its original inhabitants. Created the Kingdom of Etruria (1801) by Napoleon for Louis of Bourbon-Parma, annexed to the French Empire (1808) and became a republic (1849).
Umbria		A region named after the pre-Etruscan Umbri tribe.
Venice	Venezia	Venetia	Nicknamed 'La Serenissima', 'The Most Serene'. Named after the Veneti tribes who withdrew into the lagoon under the onslaught of barbarian invasions (5th century). Forming a federation, they elected their first Doge (726). Having secured maritime ascendancy, the Venetians began to look inland. With the fall of Constantinople to the Turks (1453) and the consequent struggle between them, and the opening of new trade routes, the power of the Venetian Republic went into a gradual decline until it abolished its own constitution (1797). With the Treaty of Campo Formio (Oct 1797), which marked the end of Napoleon's first successful campaign in Italy, Venice came under Austrian domination

Ventimiglia		Albintimulium	as compensation for other losses. Joined the Kingdom of Italy (1866).
Verona		Vernomago	Literally means 'Twenty miles' but it is in fact a corruption of Albintimulium from the name of a Ligurian tribe.
Vicenza		Vicetia/Vicentia	Means 'Field', *mago*, of elder trees, *verno'*. Possibly derived from the Etruscan *Veru*, the name of a person.
			May be derived from the Latin *vicus, a* district or quarter in a city.
Sardinia	Sardegna	Sardan	An island and region. Probably named after the Iberian Sardi tribe. Occupied by the Greeks, Phoenicians, Carthaginians, Romans, Vandals, Pisans, Genoese, Spanish, Austrians and Piedmontese. The Kingdom of Sardinia was ceded to the French-oriented House of Savoy (1720) and joined to Piedmont. Annexed by France (late 18th century until the defeat of Napoleon). Victor Emmanuel II of Sardinia proclaimed King of Italy (1861). May have given its name to the sardine and the adjective 'sardonic' from the *Herba Sardonia,* an arid herb which affects the facial muscles, producing a bitter grin.
Cagliari		Cardllis, Caralis	Capital.
Iglesias		Argentaria	Named Argentaria, meaning 'Place of silver', because of the silver mines, opened (13th century). Then from the Latin *ecclesia, a* Christian congregation or church.
Sicily	Sicilia	Trinacria	An island and region named after the Sicels/Siculi (before the 8th century BC). Greek colonization (8th century BC) followed by Roman domination (210 BC), then Byzantine

Agrigento	Akragas, Agrigentum, Girgenti		(535), Arab (invaded 827), Norman (1060), Swabian (1194), French Angevin (1265), Spanish Aragonese (1282), Savoy (1713), Austrian (1720), Spanish Bourbon (1734). Garibaldi began his drive to unify Italy at Marsala (May 1860). Became part of Italy (1861). Occupied by the Germans during the Second World War until liberated by the Allies (Jul 1943). Founded (c.581 BC) by the Greeks and meaning 'summit'. It is situated on a hill. Assumed present name (1927).
Alcamo	Manzil Alqamah		The Saracen name means 'Resort of lotus fruit'.
Marsala	Lilybaeon, Marsah el Allah		The Arab name means "Port of God", but it may be Marsah Ali, 'Port of Ali'. Gives its name to the dark sherry-type wine.
Messina	Zankle, Messana		Zankle means 'sickle' to describe the shape of the harbour. Renamed Messana by Anaxilas, the tyrant of Rhegium on the other side of the Strait, after his birthplace Messenia in Greece. Destroyed by an earthquake (1908).
Palermo	Ziz, Panormus, Balarma		Means a 'Safe haven for all boats' from the Greek pan, all, and ormus, chain of boats. Capital of Sicily (1072).
Syracuse		Siracusa	Derived from suraka, the Phoenician name for a nearby marsh. Became a maritime power and defeated the Athenian fleet (413 BC).
Taormina	Tauromenium		Takes its name from Mount Tauro which may have a connection with 'mediterranean'. Capital of Byzantine Sicily until destroyed by the Muizziyah Arabs (902).

Trapani	Drepanon	Derived from the Greek *drepanon*, sickle.
Vulcano	Thermessa, Terasia, Hiera	One of the Aeolian Islands with three volcanoes. Named after the fire god, Vulcan.

LATVIA

For more than 2000 years the Baltic tribes which settled on the eastern shores of the Baltic Sea were left undisturbed. But in the tenth century they were overrun by the Vikings from the west, while Slavs made incursions from the east. Before the end of the 12th century, as Viking power waned, German merchants had visited the area and reported that it was ripe for colonization. In 1200, in crusading spirit, Bishop Albert arrived from Bremen and founded the city of Riga a year later. With the help of a small military Order, the Sword Brothers, which he created in 1202, he proceeded to subdue neighbouring Livonia (now southern Estonia and northern and eastern Latvia), Courland and Zemgale and convert them to Christianity. The German Livonian Confederation was formed. The Order of the Teutonic Knights, which had absorbed the Sword Brothers in 1237, completed the conquest of the areas populated by the Latvian tribes. In 1282 Riga joined the Hanseatic League, a trading association active in northern Europe between the 13th and 15th centuries.

Although acquiring Estonia in 1346, the Teutonic Order suffered a severe setback in 1410 when its knights were decisively defeated at the Battle of Grünwald (Tannenberg) by Polish-Lithuanian forces. This initiated a steady decline in their fortunes. The Livonian War (1558–83) began when Tsar Ivan IV invaded Estonia and Livonia. One of his ambitions (q.v.Estonia) was to become the King of Poland: when Sigismund II died the Jagiełłonian dynasty would be extinguished – as it was in 1572. The Livonian Knights were too weak to resist, disbanded their Order and partitioned Livonia in 1561. The part north of the Daugava (Western Dvina) River (Vidzeme and Latgale) went to Lithuania; Courland and Zemgale to the south were combined into an autonomous duchy and became a Lithuanian fief, although ruled by the last Master of the Order as hereditary Duke of Courland. In fury, Ivan launched another invasion of Livonia, but was decisively defeated by Poland's elected king, the Transylvanian prince Stephen Bátory.

Sweden, already in control of Estonia, entered Livonia in 1621 and by the Peace Treaty of Altmark in 1629, took formal possession. Only Latgale in the south-east remained in Lithuanian hands. In 1700 the Great Northern War (q.v. Estonia) broke out. It came to an end in 1721 at the Treaty of Nystad when most of Livonia, Vidzeme, came under Russian rule. Latgale was annexed at the first partition of Poland (1772) and at the third (1795) Courland also fell into Russian hands; the Russian Empire at last included the whole of Latvia, and Russia and Germany had a common border. Within Latvia the peasants and workers lived under the rule of German landowners and German culture flourished in the towns.

The German military occupation of Latvia began in May 1915, but it did not spread beyond the River Daugava. Almost immediately after the revolution in Russia in March 1917 nationalist demands to create a combined administrative region from Courland, Latgale and southern Livonia to be called Latvia were made. Following the Russian Revolution and Lenin's seizure of power in November 1917, the Germans, posing as liberators, occupied almost the entire region as Russian regiments deserted *en masse*. At the Treaty of Brest-Litovsk in March 1918 Soviet Russia ceded the country to Germany.

Latvia protested and continued the struggle against the Germans. When the armistice was signed in November 1918 Latvia declared its independence.

Aiming to restore Russian control and spread communism, the Soviet government denounced the Treaty of Brest-Litovsk and sent its troops into the country, supported by some Latvian infantry regiments which welcomed the Bolsheviks. By the end of the year almost the whole of Latvia was in communist hands and the Soviet Republic of Latvia was proclaimed in Moscow in December.

The violent behaviour of the Bolsheviks soon led to the undoing of the regime. In May 1919 Riga was recaptured by the Germans and by July the Red Army had been forced to withdraw from most of the country. By the end of the year the final remnants had been ejected from Latgale by Latvian troops, the Germans also having been forced out in December. That they had remained for so long after the German capitulation was due to the fact that the Western Allies had wanted them to provide the first line of defence against the Bolshevik invaders. A Latvian-Russian Peace Treaty 'for eternity' was signed in Riga in August 1920, the Soviet government abandoning all claims to Latvia. Latvia's eastern border was drawn along ethnic lines so that it acquired some slices of Soviet territory. Latgale, separated from Livonia in 1629 and part of the Russian province of Vitebsk, was reunited with the rest of Latvia.

For the next 22 years only, Latvia enjoyed independence. The non-aggression pact signed by Germany and the Soviet Union in August 1939 included a secret protocol providing for the annexation of Latvia (and Estonia) by the USSR. In October a Treaty of Mutual Assistance was forced upon the Latvians whereby they had to accept the stationing of Soviet troops on their soil. In June 1940 Latvia was occupied by Soviet troops and within a few days a pro-Soviet government was in power. In August Latvia was admitted into the Union as the Latvian Soviet Socialist Republic.

A year later German troops invaded, receiving a short-lived welcome. Latvia became a province of Ostland which also included Byelorussia, Estonia and Lithuania. Not until July 1944 did Soviet troops re-enter Latvia. In 1945 a very small area around Pytalovo (Jaunlatgale), acquired in 1920, was transferred back to the Soviet Union. As in Estonia and Lithuania, Sovietization and considerable Soviet immigration began. It was not until the emergence of Mikhail Gorbachev as General Secretary of the Soviet Communist Party in 1985 and his policy of *glasnost* (openness) that the Soviet grip began to relax.

In July 1989 the Latvian Supreme Soviet declared sovereignty and economic independence. In May 1990 Latvia's incorporation into the Soviet Union in 1940 was declared as unlawful; a transitional period to full independence was announced, but this was annulled a few days later by President Gorbachev. Violence broke out in January 1991 when Soviet special riot police stormed government buildings in Riga. Such events merely served to encourage the Latvians to break completely free of the Soviet Union. On the day when the attempted *coup d'état* in Moscow was seen to have failed, 21 August 1991, Latvia declared full independence.

The withdrawal of Russian troops began in early 1992 and was completed, but for the manning of a radar station, in August 1994. As with Estonia and Lithuania, Russia considers Latvia to be within its natural sphere of influence and this is reinforced by the fact that nearly half the population is non-Latvian; a third is ethnic Russian. The possibility exists of the Latvians becoming a minority in their own country and this makes them deeply distrustful of the Russians. The Latvians maintain that the 1920

Latvian-Russian Peace Treaty of Riga is still valid and they insist that their independence dates from 1918 and not 1991.

Area:	64 589 sq km (24 938 sq miles)
Population:	2.4 million; Latvians, Russians, Belarusians, Ukrainians, Poles, Lithuanians
Languages:	Latvian (Lettish), Russian, Lithuanian
Religions:	Protestant/Lutheran, Roman Catholic, Orthodox
Capital:	Riga
Administrative Districts:	26 districts and 7 cities
Dates of Independence:	18 November 1918/21 August 1991
Neighbouring Countries:	Estonia, Russia, Belarus, Lithuania

English-speaking name	Local Name	Former Names	Notes
LATVIA	Latvija	Livonia/Livland	The Republic of Latvia (Latvijas Republika) from 1918–40 and since Aug 1991. The Latvian Soviet Socialist Republic (1940–41, 1944–91). Derived from *latvis*, forest clearer. Livland is the German translation of the Latin Livonia. Embraces what was southern Livonia and Courland. Only after the War of Liberation (1920) did the three Baltic provinces of Estonia, Courland and Livonia become the two states of Estonia and Latvia. The Latvians are also called Letts.
Courland	Kurzeme	Curonia	A region meaning 'Land of the Kursi (in English, Cours)'. Created as a duchy out of the territory of southern Livonia (1561) and included Semigallia. Passed to Russia at the third partition of Poland (1795)
Daugavpils		Dünaburg, Dvinsk, Latgale, Borisoglebsk	Situated on the Western Daugava River (in English the Dvina), Daugavpils means 'Castle on the western Dvina'. Its original name (1278) was the German Dünaburg, meaning 'Fort on the Dvina', and during the German occupation (1941–44) this name was resurrected. Dvinsk (to 1920), Latgale (1920–41). The Russian name honours two half brothers, Boris and Gleb, who were murdered by their brother (1015) and later revered as saints.
Jekabpils		Jakobstadt	Means 'Jacob's (James) castle'.
Jelgava		Mitau, Mitava	Founded (1226) as a castle by the Sword Brothers. Became capital of the Dukes of Courland (1561).
Jurmala			Means the 'Sea shore'.

Latgallia	Latgale, Letgallen	A region named after the Latgallian (or Lettigallian) tribe. Sometimes known as 'Polish Livonia' (1629–1795).
Liepaya	Libau, Libava	Name derived from the Lettish *liepa*, limetree.
Livonia	Vidzeme, Livland	The Latin name (German, Livland) for an area north of Lithuania, incorporating Latvia and the southern part of modern Estonia, named after the Livs, a Finno-Ugric people. The name was also used to describe the territory ruled by the Teutonic Order. The northern part was populated by Estonians, the southern part by Latvians. Only in 1918 were these two peoples able to found their own separate states. Passed to Sweden (1622). Formally absorbed into the Russian Empire at the Treaty of Nystad (1721).
Riga		Name may be derived from the Latvian *rija*, barn or warehouse. The Vikings constructed such buildings to store their warlike materials for further attacks inland. German merchants may then have changed the *j* to the hard *g*. Or it may come from the Old Latvian *ringa*, curve, in reference to the curve of the River Daugava on which Riga lies. Or it may be named after the Ridzeme River which was filled in over 100 years ago. Founded by Bishop Albert I (1201) who formed the Brothers of the Sword. Under Russian rule (1710–1918, 1944–91). Treaties of Riga (1920 and 1921) established the Russo-Polish border. Capital (1918).
Semigallia	Zemgale	A region named after the Semigallian tribe.
Tukums	Tuckum	Derived from the Liv word *tukku maegi* which means a 'Group of hills'.

| Valdemarpils | | Named after the Danish King Valdemar II (1170–1241, r.1202–41) and meaning 'Valdemar's castle'. |
| Ventspils | Venta, Windau | Means 'Castle on the Venta', at the mouth of which river the town lies. |

LITHUANIA

The Lithuanians began to settle along the Baltic coast from the River Vistula to well beyond the Daugava (Western Dvina) River long before the birth of Christ. They were largely undisturbed until the Order of the Teutonic Knights, a military crusading organisation, began its drive to convert the pagan Lithuanians to Christianity early in the 13th century. It failed, largely due to the difficulty of penetrating the forests and marshlands in which the Lithuanians lived. German pressure, however, encouraged the deeply divided Lithuanian tribes to unite under Mindaugas, one of their chieftains. From then on until 1917 Lithuania was to follow a completely different path to that of its northern neighbours, Estonia and Latvia. In 1236 Mindaugas defeated the crusading Sword Brothers (q.v.Estonia). But it was not until 1253 that he was crowned king with the approval of Pope Innocent IV, having adopted Christianity two years earlier. Before he was murdered in 1263, however, he had reverted to paganism; Lithuania was to remain pagan until the end of the 14th century, the last country in Europe to convert. Repeated attempts by the Teutonic Knights to conquer Lithuania from the west and north failed. Indeed, under Gediminas, who ruled from 1316–41, Lithuania began to extend its frontiers to the east and south. Filling the vacuum left by the Tatars (the Golden Horde), he incorporated vast areas of Ukraine and Byelorussia into Lithuania and strongly resisted the territorial ambitions of the Teutonic Knights in the west. Two of his seven sons followed in his footsteps: Kestutis opposed the knights in the west, while Algirdas became Great Prince and conquered more territory in the east, thus increasing the Slavic element of the population. In 1362 he defeated a powerful Tatar army, arguably saving the West from Tatar domination, and gained Kiev.

Algirdas was succeeded by his eldest son, the Grand Duke Jogaila. Jogaila was soon at loggerheads with his uncle. In 1381 Kestutis deposed him and assumed the title of Great Prince. But the following year Jogaila was back, capturing and killing Kestutis and imprisoning his son, Vytautas. The threat from the Teutonic Knights now began to increase and, to add to Jogaila's problems, Vytautas escaped and turned to the knights for support. Jogaila needed an ally: he could turn either to Muscovy (Moscow), which would entail conversion to Orthodox Christianity, or Poland, which would mean accepting Roman Catholicism. He chose the latter, concluding a pact with the Poles in 1385 in which he agreed to marry the 12-year-old Polish Crown Princess Jadwiga, become King of Poland and abandon paganism. The following year this personal union was achieved and Jogaila took the Polish name Vladislav II Jagiełło. Although King of Poland, he was also Grand Duke of Lithuania. So that the Grand Duchy did not become a part of Poland and lose its autonomy, Jagiełło patched up his differences with his cousin Vytautas, who was allowed to assume control.

Under Vytautas, Lithuania continued to expand, becoming the most powerful state in eastern Europe. By 1392 its borders stretched from the Baltic Sea to the Black Sea. In 1410 Polish-Lithuanian forces united to annihilate the Teutonic Knights at the Battle of Grünwald (Tannenberg); the German threat faded and the territory of Samogitia was acquired in 1422.

After the death of Vytautas, now 'the Great', in 1430, Lithuanian power began to decline in the face of growing threats from both the Ottomans and Muscovites. Territory was lost to both. In 1561, however, Lithuania gained northern Livonia and the nominal suzerainty of the Duchy of Courland from the Teutonic Knights (q.v.Latvia). By this time it had become clear that Sigismund II Augustus (Grand Duke of Lithuania 1544–72 and King of Poland 1548–72), still childless, would be the last Jagiełłonian king and Lithuania would lose the personal link with Poland. Under pressure from the east and south, Lithuania might cease to exist without Polish help. The Poles were also concerned about the Russian threat and the lack of an heir, and demanded union of the two countries. The Union of Lublin in 1569 created the Polish-Lithuanian Commonwealth. It was to have a joint king, but both countries were to remain separate states with their own armies. Lithuania ceded its Ukrainian provinces to Poland in return for protection. Polonization of all but the peasants was inevitable and Poland soon became the dominant partner.

From the second half of the 17th century the fortunes of the Commonwealth began to ebb. The first two partitions of Poland in 1772 and 1793 resulted in the loss of Lithuania's East Slav lands to Russia which, at the third partition two years later, annexed practically all of what was left of Lithuania; the remainder was acquired in 1815.

Russian rule lasted 120 years. A glimmer of hope flared briefly in 1812 when Napoleon passed through on his way to Moscow, but thereafter a policy of strict Russification was followed. This stopped only in 1915 when German troops occupied the country after the outbreak of the First World War. Their rule was benign. Cut off from Russia for three years but still under German control, Lithuania declared independence in February 1918. On the basis of a December 1917 declaration by the Lithuanians that they would ally themselves with the German Reich, the kaiser recognized this independence a month later. German troops were withdrawn after the armistice in November 1918.

The Bolshevik Red Army invaded almost immediately and by early January 1919 had captured the capital, Vilnius, and a communist puppet government was installed. However, the Lithuanian National Army launched an offensive in May and by August had ejected all Russian troops from the country. At much the same time, however, German and White Russian (anti-Bolshevik) troops had invaded northern Lithuania, but they too were removed before the end of the year.

At the close of the First World War international borders in the region were not clearly demarcated. The situation was further complicated by Polish claims to Vilnius. In April 1919 the Polish Army drove the Red Army out of the city and refused to let the Lithuanians in. In July 1920 a peace treaty was signed between Lithuania and Soviet Russia which recognized Vilnius as Lithuanian. However, it was the Red Army which occupied the city during the Russo-Polish war of 1920. When the Russians withdrew the Lithuanians rushed in to replace them. In October the Poles suddenly crossed the armistice line demarcating the Polish-Lithuanian border which had been agreed only two days earlier. They overran the southern part of the country and re-captured Vilnius which was annexed in 1922. Because the Poles retained the capital and about a third of Lithuanian territory, diplomatic relations were severed and not restored until 1938. As some sort of compensation Lithuania occupied the former German port of Memel (now Klaipeda), which it had never owned, in 1923. The capital was transferred to Kaunas.

The secret protocols of the non-aggression pact signed by Germany and the Soviet Union in August 1939 provided for Lithuania to be assigned to Germany. However, a month later a second treaty between the two countries reassigned Lithuania to the Soviet Union. In October the Lithuanians were forced to sign a so-called Treaty of Mutual Assistance, which allowed Soviet troops to establish several military bases in the country; the Soviet Union then, cynically, took the opportunity to return Vilnius and the surrounding territory, which it had captured from the Poles in September 1939, to Lithuania. Nevertheless, relations declined to the point when, in June 1940, the Soviet Union demanded a change of government. The next day more Red Army units burst into Lithuania.

A puppet government was installed and in August Lithuania entered the Union as the Lithuanian Soviet Socialist Republic. This was overturned when German troops began their attack on the Soviet Union on 22 June 1941. By a decree of 17 July 1941 Hitler created 'Ostland' which consisted of Estonia, Latvia, Lithuania and Byelorussia. The second Soviet invasion of Lithuania occurred in July 1944 as the Red Army drove westwards. The Lithuanian Soviet Socialist Republic was restored. Harsh Soviet rule, although met with armed resistance until 1952, followed in its wake.

Liberalization only began with Mikhail Gorbachev's accession to power in Moscow in 1985. Very quickly Lithuanian nationalism re-emerged and anti-Soviet feeling began to swell. In August 1988 the Nazi-Soviet Treaty was denounced and the Soviet Union was accused of occupying the country illegally. Lithuanian independence was declared in March 1991 as a continuation of its earlier independence, but when the Soviet Union imposed an economic embargo the Lithuanians agreed to suspend further moves towards full independence for six months.

Hostility and tension increased as Soviet riot police engaged in intimidation and murder. When an attempt at a *coup d'état* in Moscow in August 1991 failed, success for the Lithuanians was in sight. Western countries began to recognize Lithuanian independence and in September 1991 the Soviet Union followed suit. Within two years all Russian troops had been withdrawn. Besides regarding Lithuania as in its sphere of influence, Russia has a particular interest in the country because it is one of those situated between the main part of the Russian Federation and the detached *oblast* (province) of Kaliningrad. Unlike Estonia and Latvia, Lithuania does not have a significant ethnic Russian minority.

Area:	65 200 sq km (25 174 sq miles)
Population:	3.69 million; Lithuanians, Russians, Poles, Belarusians, Ukrainians
Languages:	Lithuanian, Russian, Polish
Religions:	Roman Catholic, Russian Orthodox
Capital:	Vilnius
Administrative Districts:	10 provinces
Dates of Independence:	16 February 1918/6 September 1991
Neighbouring Countries:	Latvia, Belarus, Poland, Russia

The Polish-Lithuanian Commonwealth 1569

English-speaking Name	Local Name	Former Names	Notes
LITHUANIA	Lietuva		The Republic of Lithuania (Lietuvos Respublika) from 1918–40 and since 1991. The Lithuanian Soviet Socialist Republic (1940–41 and 1944–91). The origin of the name is not known. It was first mentioned in the *Annales Quedlinburgenses* in 1009. The connection between modern Lithuania, and the historical Grand Duchy of Lithuania and the Polish-Lithuanian Commonwealth is slight.
Druskininkai			Means 'A man or people who work with salt' from *druskas*, salt.
Kaunas		Kowno, Kovno	Passed to Russia at the third partition of Poland (1795). Destroyed by Napoleon (1812). Provisional capital (1920–1940) while Vilnius was occupied by the Poles.
Klaipeda		Memelburg, Memel	Destroying the original settlement, the Teutonic Knights (1252) then built a new fortress which they called Memelburg, meaning 'Fortress on the (River) Memel' (in Lithuanian, the Nemunas, and known in English as the Neman), and later just Memel. Part of East Prussia until becoming international territory as a result of the Treaty of Versailles (1919) and placed under French administration by the League of Nations. The Lithuanians drove the French out (1923) and annexed the area. Renamed Klaipeda (1925) until severed from Lithuania by Hitler (Mar 1939) when it became Memel again. Ceded to the USSR (1945–91) and renamed Klaipeda (1945). The meaning is unknown, but the second half, *peda*, means 'territory' in Lithuanian.
Marijampole		Starapole, Kapsukas	Previously named (1955–90) after Vincas Kapsukas-Mickevicius who headed the short-lived, Moscow-created 'Lithuanian Soviet Government' imposed on

Šiauliai	Lithuania as the German Army withdrew at the end of 1918. Returned to its original name which would appear to mean the 'City of the Blessed Virgin Mary'.
Šiauliai	Schaulen	Possibly the site of the battle (1236) where Mindaugas defeated the Sword Brothers. This defeat led to their amalgamation with the Teutonic Knights.
Taurage	Tauroggen	The Convention of Tauroggen was signed (Dec 1812) between Russia and Prussia to form an anti-French alliance.
Vilnius	Vilna, Wilno	Named after the River Vilnia (Lithuanian *vilnis*, wave) on which it stands. According to legend, while hunting in the area, the Lithuanian ruler Gediminas, had a dream about an iron wolf which howled continuously. The next morning Gediminas asked a mystic in his entourage the meaning of his dream. It was that Gediminas should build a city here which would become known throughout the world. Gediminas did so and his city became capital of the former Grand Duchy of Lithuania (1323). Destroyed by the Teutonic Knights (1377). Under Russian rule (1655–60, 1795–1918), Polish rule (1920–39), German rule (1941–44) and Soviet rule (1944–91). Restored as capital (1940).

LUXEMBOURG

As part of the Low Countries, the historical development of Luxembourg is closely linked with that of The Netherlands and Belgium. For more detail see the sections on both these countries.

Inhabited by Belgic tribes at the time of the Roman conquest in 53 BC, the area that constitutes modern Luxembourg was invaded by the Franks some time after 400. Part of Charlemagne's empire and then Lotharingia after the Treaty of Verdun in 843, Luxembourg became independent in 963 when Count Siegfried of the Ardennes chose to develop a small Roman castle on the Alzette River into a fortress and to make it his capital. He and his successors gradually enlarged the territory by conquest, inheritance and marriage so that by about 1060 its ruler was calling himself Count of Luxembourg. Its importance was recognized when Count Henry IV was crowned Holy Roman Emperor as Henry VII in 1312. In 1354 Luxembourg became a duchy. By the end of the 14th century the duchy extended as far north as Malmédy, to Metz in the south and Sedan in the west.

From then on, however, the duchy went into decline, neglected by its rulers and losing its significance. Following Burgundian penetration of the Low Countries, Philip III the Good, Duke of Burgundy, acquired Luxembourg in 1443. The marriage of Mary, Duchess of Burgundy, and Maximilian of Habsburg in 1477 united the Houses of Burgundy and Habsburg. When Mary died in 1482 the duchy became fully Habsburg. When Habsburg possessions were divided following the abdication of Charles V in 1555, Luxembourg became part of the Spanish Netherlands.

During the Revolt of the Netherlands (1568–1609) the duchy remained true to Catholicism and Spain. It succeeded in keeping out of the Thirty Years' War (1618–48) until 1635. Thereafter, war brought devastation and famine which did not end in 1648 with the Peace of Westphalia. Its involvement in the Franco-Spanish struggle ended only in 1659 when Spain, punch-drunk and on the brink of financial collapse, conceded defeat at the Treaty of the Pyrenees. Swiftly developing its military strength to become the most powerful state in Europe, France invaded Luxembourg in 1679. By 1683 the French were in total control and, in their opinion, had achieved 're-union'. French domination, however, lasted only until 1697 when the duchy was restored to Spain.

The Treaty of Utrecht in 1713–14 brought the War of the Spanish Succession to an end and resulted in the Spanish Netherlands (Belgium and Luxembourg) being handed over to the Austrian branch of the Habsburgs and becoming known as the Austrian Netherlands. In 1789 the French Revolution broke out and three years later French troops entered the duchy only to withdraw in 1793. They returned the next year and Luxembourg was formally annexed in 1795. French rule endured only until the fall of Napoleon in 1814.

In January 1815, during the Congress of Vienna, Luxembourg was raised in status to a grand duchy. The Congress also decided that its future lay as part of Prussia. William I of the Netherlands was outraged. To soothe him Prussia agreed that he could regard it as a personal possession and take the title grand duke. But William had to agree that it would be free to join the German Confederation and that a Prussian military unit would be based in the city of Luxembourg; the garrison's strength was to be 6000 at a time when the city's population itself was only 10 000. Not all of the grand duchy, however, was to go to William. Prussia received those parts that lay east of the Rivers Our, Sûre and Moselle. William's part was subjected to the new Dutch constitution where this did not conflict with the requirements of the Confederation; there was no political link between the Netherlands and the grand duchy. It was a bizarre arrangement.

William regarded the grand duchy as no more than conquered territory and so, when the Belgians rebelled against him in 1830, so did the people of Luxembourg who placed themselves under Belgian authority. Luxembourg was declared to be part of Belgium when that country proclaimed its independence. William rejected this. The Great Powers (Austria, Britain, France, Prussia and Russia) stepped in the next year and decreed that the French-speaking part of the grand duchy should become a province of Belgium, still called Luxembourg; William was allowed to keep the rest, also called Luxembourg, although it had to remain part of the German Confederation. Again, William rejected this and the settlement was not implemented until he relented in 1839. Luxembourg's present borders stem from this time.

In 1866 the German Confederation was dissolved and replaced by the North German Confederation. Bismarck did not include Luxembourg in the new confederation. Unclear as to the grand duchy's new status, William III let it be known that he would be willing to sell it to Napoleon III of France who was keen to seal the deal. It fell through when Bismarck voiced his disapproval. The following year the Great Powers declared Luxembourg an independent state, guaranteed its neutrality and arranged for the withdrawal of the Prussians.

Luxembourg required its head of state to be a male descendant of the House of Nassau[1]. When William III died in 1890 without a male heir the grand duchy passed to another branch, the current ruling House of Nassau-Weilburg. The link with The Netherlands was thus broken. Luxembourg's neutrality was violated by the Germans during the First World War and the grand duchy was again occupied by German troops in the Second World War until liberated in September 1944. Its policy of neutrality was abandoned when it joined NATO in 1949.

1. Nassau is in Germany. The first Count of Nassau assumed the title in the 12th century. One branch of the House acquired the Dutch principality of Orange in the 16th century and thereafter produced Princes of Orange-Nassau. In 1806 William VI of Orange lost his German possessions to Napoleon. As compensation he was appointed Grand Duke of Luxembourg by the Congress of Vienna in 1815 and succeeded to the Dutch throne as William I the same year.

Area:	2 586 sq km (998 sq miles)
Population:	430 000; Luxembourgers, Portuguese, Italians, French, Belgians, Germans
Languages:	Letzeburgesch, French, German
Religion:	Roman Catholic
Capital:	Luxembourg
Administrative Districts:	3 districts
Date of Independence:	1867
Neighbouring Countries:	Belgium, Germany, France

English-speaking Name	Local Name	Former Names	Notes
LUXEMBOURG			Grand Duchy of Luxembourg (Grousherzogdem Lëtzebuerg (Letzeburgesch), Grand-Duché de Luxembourg (French), Grossherzogtum Luxemburg (German)) since 1815. Takes its name from a Roman castle.
Luxembourg		Lucilinburhuc	Means 'Little fortress' from a Roman castle around which the city developed.

MACEDONIA

Macedonia is a land-locked geographical area at the heart of the Balkan Peninsula, populated by a wide, and somewhat fragile, ethnic mix. It has never been a single state in its own right, but those who call themselves Macedonians claim they are a distinct nation. This is denied by the Bulgars, Greeks and Serbs. Today ancient Macedonia spreads over three countries: the former Yugoslav republic of Macedonia, Bulgaria and Greece. Rivalry between the Serbs, Bulgars, Greeks and Turks has characterized the area for centuries and at one time or another Macedonia has featured in the dreams of a 'Greater Serbia', a 'Greater Bulgaria' and a 'Greater Greece'.

The ancient Greek-speaking Kingdom of Macedon came to the fore under Philip II, whose son, Alexander III the Great, took it to the peak of its power. When he died in 323 BC his huge empire disintegrated. After a shattering defeat by the Romans at the Battle of Pynda in 168 BC, Macedon lost its independence and in 148 BC it became a province of the Roman Empire.

The Slavs first arrived in the sixth century, the Bulgars gradually encroaching on the lands of the East Roman, or Byzantine, Empire. By the start of the 11th century Byzantine authority had been reimposed and a Greek renaissance began. As the Serbian Empire blossomed in the late 13th and 14th centuries Macedonia came under Serbian control. This was not to last long. With the defeat of the Serbian Army at the Battle of Kosovo Polje in 1389, Turkish Ottoman possession of most of Macedonia was consolidated. Thereafter, Turks mixed with Albanians, Slavs and Greeks, and Orthodox Christians with Muslims, but the next 520 years were the most stable in Macedonia's history.

The Treaty of San Stefano in March 1878 brought an end to the series of wars between Russia and the Ottoman Empire. The Treaty proposed that, in accordance with what was thought to be the national character of the region, now Slav-speaking, practically all Macedonia should be awarded to Bulgaria – thus creating a 'Greater Bulgaria' – under Turkish suzerainty; this proposal was speedily overturned, to the fury of the Bulgars, by the Congress of Berlin which decreed that Macedonia should remain in Turkish hands.

As Turkish power waned in the closing decades of the 19th century, Serbia, Bulgaria and Greece all harboured territorial aspirations at the expense of Turkey and all were keen to advance their historical claims to Macedonia. Following an Albanian insurrection against the Turks and the Albanian occupation of Skopje (now the Macedonian capital) in 1912, a Balkan League was formed, uniting Bulgaria, Greece, Serbia and Montenegro for the first time in 500 years. Its express intention was to evict the Turks from Europe. Anti-Turkish violence grew until in October 1912 Montenegro declared war on Turkey. The sultan responded by declaring war on the League. Within seven weeks, although without a coordinated plan, the League defeated the Turks, and Macedonia was occupied by the Serbs and Montenegrins. Fighting flared up again in February 1913, but the Treaty of London in May 1913 brought hostilities formally to an end. Both Bulgaria and Greece joined Serbia in possessing parts of Macedonia.

The spoils of victory sowed the seeds of disaster. Serbia, Greece and Bulgaria could not agree on the partition of Macedonia and threatened war if their claims were not met. The Bulgarians, determined not to permit Serb territorial supremacy in Macedonia, invaded Greek and Serb-held territory in Macedonia at the end of June 1913. Montenegro and Romania supported the Serbs. The Second Balkan War ended a month later with the total defeat of Bulgaria. At the Treaty of Bucharest in August 1913 Serbia was awarded substantial areas (38 per cent) of northern and central Macedonia, to be known as South Serbia (or Vardar Macedonia), while Greece received the southern regions (Aegean Macedonia, including the port of Thessaloniki, – 52 per cent) which approximated to historical Macedonia. Bulgaria received 10 per cent of the territory, known as Pirin Macedonia. With a few minor changes, these borders exist today. Inevitably, Vardar and Pirin Macedonia contained a substantial number of Greeks, while thousands of Bulgarians remained on Greek soil.

The First World War gave the Bulgarians a new opportunity to right perceived wrongs and by 1915 the Bulgarian Army was in control of Vardar Macedonia. Recaptured in 1917, it was included as part of Serbia in the new Kingdom of the Serbs, Croats and Slovenes when it was proclaimed on 1 December 1918. In 1929 when the Kingdom was reorganized and renamed the Kingdom of Yugoslavia, Vardar Macedonia became the Vadarska *banovina*; it included a part of what is modern Kosovo.

On the capitulation of Yugoslavia in April 1941, Albania, under Italian tutelage since April 1939, was enlarged to include the western part of Vardar Macedonia while the rest of it was occupied by the Bulgarians. A policy of Bulgarization forfeited any pro-Bulgarian sympathy among the Yugoslav Macedonians. At the close of the war in 1945 the territory was returned to Yugoslavia. When the 1946 federal constitution was promulgated in January 1946, however, Macedonia (South Serbia) was removed from Serbia and became a republic in its own right. It was to have the name Macedonia and the people were to be called Macedonians. It was an exercise in nation-building. The aim was to impose a new nationality on a geographical area rather than create a new ethnic identity, and distinguish it from the Serbs, and Bulgarian and Greek Macedonians. The possibility that Yugoslav Macedonia might later serve as a focus for the creation of a Macedonian nation-state for all Macedonians alarmed the Bulgars and Greeks.

The disintegration of Yugoslavia in 1991 led the Macedonians, fearful of Serb dominance, to hold a referendum on independence in September. The majority was in favour and Macedonia applied to the European Community for recognition. Greece refused approval on the grounds that Macedonia was only a geographical term and was an Ancient Greek name; furthermore, three regions of Greece already included Macedonia in their title. The Greeks were concerned that using the name might encourage the new state to lay claim some time in the future to Greek Macedonia. Consequently, in April 1993, Macedonia was formally admitted into the UN as the 'Former Yugoslav Republic of Macedonia (FYROM)'.

Macedonia remains vulnerable to events in the Balkans which could destabilize the uneasy balance between the Albanian minority, some 25% of the population, and the Slav majority.

Area:	25 713 sq km (9928 sq miles)
Population:	2.23 million; Macedonians, Albanians, Turks, Serbs, Gypsies
Languages:	Macedonian, Serbo-Croat
Religions:	Serb Orthodox, Muslim (Sunni)
Capital:	Skopje
Administrative Districts:	125 communes
Date of Independence:	20 November 1992
Neighbouring Countries:	Yugoslavia, Bulgaria, Greece, Albania

English-speaking Name	Local Name	Former Names	Notes
MACEDONIA	Makedonija		Republic of Macedonia (Republika Makedonija) since Nov 1991, but a member of the United Nations under the name 'Former Yugoslav Republic of Macedonia' (FYROM). A republic within Yugoslavia (1946). Named after the Greek tribe of Makedones. This name may be derived from the fact that the early tribes came from the rugged Pindos mountains and took their names from the physical features of the landscape; from the Greek root *mak*, tall or high. Gives its name to *macédoine*, a French word for mixed vegetables or fruit, an allusion to the ethnic mix in Macedonia. Macedonia, together with Kosovo and south Serbia, comprised Dardania (1st millennium BC).
Bitola		Heraclea Lyncestis, Pelagonia, Monastir, Bitolj	Name taken from the Latin *monasterium*, monastery, after falling into Turkish hands (1382). Capital of Macedonia (19th century).
Ohrid		Lychnidos	Named after the cliff, *hrid* in Serbo-Croat, on which it stands. Associated with Saints Methodius (826–85) and his brother, Cyril (827–69), Apostles of the Slavs, who created the Cyrillic script. Capital of the Slav Macedonian state (and of Bulgaria) under Emperor Samuilo (10th and 11th centuries). Incorporated into the medieval Serbian state (c.1334).
Skopje	Skoplje	Skupi, Prima Justiniana, Üsküb	The native town of Justinian I (c.482–565), Byzantine emperor (527–65), who rebuilt it (535) after an earthquake (518) and renamed it after himself 'First (Town) of Justinian'. Part of medieval Serbia (1282–1392). Capital of Stephen Dušan of Serbia who was crowned Emperor of the Serbs, Albanians, Bulgarians and Greeks (1331). Renamed and ruled by the Turks (1392–1912). Captured by the

	Austrians and, to prevent a cholera epidemic, burnt to the ground (1689). Incorporated into Serbia (1913). Occupied by the Bulgarians during both World Wars. Destroyed by an earthquake (1963). Capital (1945).
Struga	Means 'Fishing channel' – channels through which the River Crni Drim flows out of Lake Ohrid.
Strumica Tiberiopolis	Originally meant the 'City of Tiberius', (42 BC-37 AD), Roman Emperor (14–37). Renamed after the River Strumica.
Tetovo Veles Kalkandelen Bilazor/Bylazora, Kuprili, Titov Veles	The Turkish Kalkandelen meant 'helmet piercer'. The Slav Veles means the 'God of horned animals'. The Turkish Kuprili means 'bridge', so named because the town spread on both sides of the River Vardar with a bridge spanning the river. The Macedonian town named (1947–92) after President Tito (1892–1980).

MOLDOVA

Most of Moldova lies between the Prut and Dniester rivers in the north-eastern corner of the Balkans, an area historically called Bessarabia (which stretched down to the River Danube and the Black Sea) and almost completely inhabited by Romanians. Lying on a major route into the Balkans, it was frequently disputed between the Russian and Ottoman Empires. It is not to be confused with a region of the same name in Romania.

Roman rule prevailed in much of the area between 118 and about 270 when the province was given up to the Visigoths. For the next thousand years and more a 'dark age' enveloped Moldova (then Moldavia) and little is known of it until it began to emerge as a political entity in the 15th century. By the time he died in 1504, Prince Stephen IV the Great had extended Moldovan rule over a region stretching from the Carpathians to the Black Sea, although the Turks had begun to penetrate southern Bessarabia. Moldovan power, however, did not long survive his death. In 1512 Moldova became a Christian vassal state of the Ottoman Empire and remained so for the next 300 years.

Spasmodic uprisings by the local nobles and an unsuccessful invasion by the Russian Tsar Peter the Great in 1711 led to Moldova losing its autonomous status and coming under direct Turkish rule. The Turkish administrators were known as Phanariots since they came from the Greek Phanar district of Constantinople (now Istanbul, Turkey). Greek became the official language.

The Russo-Turkish wars of 1768–74 and 1806–12 ended with the Treaty of Bucharest under the terms of which Russia took control of Bessarabia, approximately half of historic Moldavia. That part west of the Prut remained subject to Ottoman rule (and joined Wallachia in 1859 to form Romania). In 1828 Bessarabia's autonomy was revoked and a policy of Russification was instituted: in 1854 Russian became the official language and, to dilute Bessarabia's Romanian character, other ethnic groups such as Jews, Bulgarians and Gagauz (Turkic-speaking Christians) were encouraged to settle in the region.

The Bolshevik Revolution in 1917 opened a window of opportunity for the Bessarabians who declared a Bessarabian Democratic Moldovan Republic the following year. The Bolshevik threat and potential Ukrainian schemes for territorial acquisitions persuaded the Bessarabians to seek Romanian protection. Shortly after, they voted for union with Romania and to sever all links with Soviet Russia. Bessarabia's new status was recognized by the Treaty of Paris (also known as the Treaty of Trianon, 1920). The Soviet Union refused to recognize the new situation and in 1924 established the Moldavian Autonomous Soviet Socialist Republic (ASSR) in a small strip of land on the left (east) bank of the River Dniester in Ukraine: Transdniestria.

The secret protocols in the Nazi-Soviet pact of non-aggression signed in August 1939 acknowledged the Soviet Union's wish to regain Bessarabia. In June 1940 the Red Army entered Bessarabia and in August it became a part of the USSR. Parts of it were united with the Moldavian ASSR to form the Moldavian Soviet Socialist Republic (SSR) which then became a republic within the Soviet Union. Land in the north and south was given to Ukraine and Moldavia became land-locked. Having entered the war as Germany's ally, Romania occupied Bessarabia once again in July 1941 and extended the eastern border of Transdniestria to the Bug river. The Moldavian SSR was resurrected in 1944 as Soviet armies advanced westwards and drove the Romanians out. A new wave of Russification began: particularly the imposition of the Cyrillic script on the 'Moldavian' language (actually a dialect of Romanian) and the reintroduction of Russian as the official language. Stalin's aim was to create a Moldavian nation quite separate from the Romanians. Nevertheless, it proved impossible to suppress entirely Moldavian nationalism.

In June 1990 sovereignty within the Soviet Union was proclaimed and the name of the Republic changed to Moldova from the Russian Moldavia. Fearful of Moldovan nationalism, the Gagauz in the south declared a separate 'Gagauz Republic' in August and the following month Transdniestria (to all intents a Russian colony within Moldova) announced its secession from Moldova and the creation of the 'Transdniestrian Moldavian Republic'. Although with a considerable Moldovan population, Transdniestria, which comprises only 15 per cent of Moldovan territory, has never been considered a part of traditional Moldova.

On 27 August 1991, following the failed *coup d'état* in Moscow, Moldova declared its independence from Moscow, but made no attempt to seek reunification with Romania. Four months later fighting broke out between the Transdniestrians and Moldovans. Soviet/Russian troops intervened – on the side of the separatists. Eventually, in July 1992, the fighting was stopped and a security zone was established along the Dniester river, patrolled by Russian, Moldovan and Transdniestrian peace-keeping troops.

The status of the breakaway republic, not recognized internationally, remains to be resolved. Although the ceasefire has held, negotiations between the Moldovans and the Transdniestrian separatists have so far failed to produce a long-term settlement. Russian troops remain in Transdniestria which is regarded as an area of special strategic interest, although 40 per cent of the population is Moldovan and only 25 per cent Russian. The 'Gagauz Republic' (Gagauz-Yeri, the Gagauz Land) ceased to exist in 1995 when it became a special autonomous district within Moldova.

Area:	33 700 sq km (13 000 sq miles)
Population:	4.46 million; Moldovans, Russians, Ukrainians, Gagauz, Bulgarians
Languages:	Moldovan, Russian, Gagauz
Religions:	Orthodox, Catholic, Protestant, Jewish
Capital:	Kishinev
Administrative Districts:	10 cities and 14 rural districts
Date of Independence:	27 August 1991
Neighbouring Countries:	Ukraine, Romania

English-speaking Name	Local Name	Former Names	Notes
MOLDOVA		Bogdania, Bessarabia, Moldavia	The Republic of Moldova (Republica Moldova) since Aug 1991. Previously the Moldavian Soviet Socialist Republic within the USSR (1940). According to legend, the present name is derived from the favourite hunting hound, Molda, of Dragoş, a Transylvanian prince. While out hunting with her master Molda was drowned and Dragoş named the river Molda (1359). In time this became Moldovo; the people living by it became known as Moldovans and the principality as Moldova. Originally, when it achieved independence (1349), this region between the Carpathian Mountains and the River Dniester was named after Prince Bogdan. To justify the creation of the Moldavian Soviet Socialist Republic Stalin coined the name 'Moldavian' to describe the language and nationality of the ethnic Romanians in Bessarabia which he claimed were different from those of the ethnic Romanians in Romania.
Bessarabia	Basarabia		A region, and former principality, lying between the Prut and Dniester rivers and named after the Basarab family which had ruled parts of Wallachia (14th century), not Moldova-Bessarabia.
Kishinëv	Chişinau		May derive its name from the Turkic word for winter, in modern Turkish, *kiş*. Adopted Russian name of Kishinëv when ceded by Turkey to Russia (1812) until included within Romania (1918–40) when it became Chişinau once more. Restored to Kishinëv (1940–1991) and then changed back again to Chişirau when Moldova achieved independence (1991). Capital (1940).
Tighina		Bendery	Founded by the Genoese (12th century), occupied by the Turks (16th century), won and lost by Russia (late 18th

century) and finally incorporated into Russia (1818). Within Romania (1918–40). Reverted to original name (1991).

Tiraspol

Means 'City on the (River) Dniester'. Founded as a Greek colony known as Tyras, the Greek name for the Dniester, and developed into an independent city-state; *pol* is from the Greek *polis*, city. Re-founded (1792) by Count Alexander Suvorov (1729–1800), military commander, as a fortress built on the site of the ancient Moldovan settlement of Staraya Sukleia which had been burned by the Turks (1787). Named Tiraspol (1795). Capital of the Moldavian Autonomous Soviet Socialist Republic (1929–40). Occupied by the Germans (1941–44). Capital of Transdniestria.

Transdniestria
Transnistria

A region between the River Dniester (Nistru to the Moldovans and Romanians) and Moldova's eastern border with a name meaning the 'Land beyond the Dniester'. By the Transdniestrians it is known by its Russian name Pridnestrovye, the land on the Dniester (in Russian Dnestr). Also currently known by its majority Slav inhabitants as the Transdniestrian Moldavian Republic.

THE NETHERLANDS

U ntil 1579 the Low Countries – Belgium, Luxembourg and the Netherlands (and, indeed, parts of northern France) – shared a common history and this is described below. Thereafter, this section concentrates solely on the Netherlands.

As the Romans penetrated the Low Countries during the first century BC they came across a variety of Celtic and Frankish tribes. The Romans remained south of the River Rhine, founding the imperial province of Gallia Belgica, while the Frisians settled to the north and the Batavi along the river. Roman strength began to wane in the third century in the face of renewed Frankish invasions and the occupation ended at the beginning of the fifth century.

Pushing the Romans southwards, the Franks stopped at a line roughly corresponding to the present dividing line in Belgium between the Flemings and the Walloons. Those that remained north of the line (Flanders) retained their Germanic languages which in time evolved into Dutch and the almost identical Flemish. Some Franks continued southwards (Wallonia) in the wake of the departing Germans and adopted the language of the Romanized Gauls which became French. This language frontier exists today with the north and west of Belgium inhabited by Dutch-speaking Flemings and the south and east by French-speaking Walloons.

During the Merovingian dynasty (476–751) the Franks expanded their empire, notably into Frisian territory to the north. The sea and the rivers are particularly influential in the Low Countries and the Frisians turned their attention to waterborne trade. The Carolingian dynasty followed, Charlemagne further expanding the Frankish empire and making the entire area of the Low Countries a part of it. After his death in 814, during the reign of Louis I the Pious of Aquitaine, this empire began to decline. When Louis died in 840, fighting broke out between his sons and this led to the Treaty of Verdun in 843. The empire was divided into three parts: Francia Occidentalis (France), Francia Orientalis (Germany) and a buffer state to separate them. This was called Francia Media (a huge territory stretching from the North Sea to the Mediterranean which included Belgium) or the Middle Kingdom. Charles II the Bald, a son of Louis, received Flanders (now spread across Belgium, France and the Netherlands) while the rest of the Middle Kingdom – and the major part of the Low Countries – went to an elder son, Lothair; the northern part was later named after him as Lotharingia (Lorraine).

The break-up of the Carolingian empire was accompanied by Viking and Magyar attacks which encouraged the development of more or less independent territorial principalities between the tenth and 15th centuries. However, they were rivals and thus had to contend with attempts by each other to extend their power and political influence. Nevertheless, Flanders, in particular, prospered, although the count held his land as a vassal of the King of France. Following the example of the Frisians and keen to exploit the geography of the land, these states began to develop trade and industry.

In 1369 Margaret, heiress of the Count of Flanders, married Philip II the Bold, Duke of Burgundy, thus starting Burgundian penetration of the Low Countries. By means of marriage, inheritance and war, succeeding Dukes of Burgundy hoped to create an empire by adding to Burgundy their expanding possessions in the Low Countries. When Charles the Bold (also called the Rash) provoked war with France he was killed in battle in 1477. He was succeeded by his daughter Mary who in the same year married Maximilian of Austria, son and heir to the Holy Roman Emperor. The Houses of Burgundy and Habsburg thus were united and when Mary died five years later the Low Countries became a full Habsburg possession. In 1494, the year following his election as Emperor, Maximilian handed over the Low Countries to his son Philip I the Handsome.

In 1504 Philip's wife, Joanna the Mad of Aragon and later Castile, inherited the Spanish crown. Their son Charles, born in 1500 in Belgium, became Prince of the Netherlands at the age of 15, King of Spain at 16, King of Germany at 19 and, as Charles V, Holy Roman Emperor in 1520. By now the Low Countries were more unified than they had ever been before, taking pride in the fact that their ruler was home-grown and the greatest emperor since Charlemagne. To Charles, however fond he was of his birth-place, the Low Countries could only be regarded as a sideshow in his struggle for European hegemony; little more, indeed, than a source of revenue for his interminable wars against France. Absent for most of the time, he appointed his paternal aunt Margaret of Austria and then his sister, Mary of Hungary, as regents.

In 1555 Charles abdicated, his son Philip II inheriting Spain and the Low Countries which by now were often referred to as the Seventeen Netherlands. Already joint sover-eign of England as the husband of Mary I, the haughty Philip was a staunch Spaniard. His Catholic bigotry and ruthless persecution of heretics encouraged the spread of Calvinism. His attempts to reduce the independence of the 17 provinces exacerbated the situation and discontent grew. In 1567 Philip sent the Spanish Duke of Alba to the Low Countries to restore order. His harsh measures and new taxes merely served to inflame passions and harden resistance.

What came to be known as the Revolt of the Netherlands (1568–1609), led by Prince William I of Orange[1], began in the more developed Southern Netherlands (ten provinces). It failed. But in the Northern Netherlands (seven provinces), despite the efforts of Alba and his successor, it took root, particularly in Holland and Zeeland where William became the stadtholder (governor). Structural, religious, economic and social differences between the North and the South eventually led to the division of the country. In January 1579 Artois and Hainaut formed the Union of Arras, committed to Catholicism and pledging allegiance to the Spanish king. The Calvinist northern provinces, in retaliation, signed the Union of Utrecht with a view to establishing a military league to fight a war of independence. In effect, a new and separate state emerged: the United Provinces of the Netherlands or the 'Dutch Republic', which before long included five southern cities including Antwerp. Internationally it was known as the States General[2].

In 1581 the States General declared that Philip had forfeited his sovereignty over the Low Countries, but it was not until 1587 that the northern provinces abandoned their efforts to obtain foreign protection and themselves declared their own sovereignty. Dutch commercial activities began to spread into all parts of Europe and beyond, particularly into Asia.

Except for a truce between 1609 and 1621, almost continuous warfare raged from 1579 to 1648 when Spain finally recognized Dutch independence at the Peace of Westphalia at the conclusion of the Thirty Years' War; thus ended the Eighty Years' War for Dutch independence. The Dutch Republic was simply a loose federation with no head of state or central institutions besides the States General. Holland, the most powerful province, and four others refused to appoint a successor to Prince William II of Orange when he died in 1650.

Conflict with France and England which began in 1672 revealed a need for a single leader to save the country. After three provinces had been quickly overrun by French troops Prince William III was appointed stadtholder of the United Provinces. He recovered the territory taken by France and in 1677 married his cousin Mary Stuart, daughter of the English Duke of York (later King James II). The English crown was offered jointly to William and Mary in 1689 as a result of the rebellion against the catholic James and his flight to France; he was considered to have abdicated. Mary died in 1694, William then serving as King of England and stadtholder of the United Provinces of the Netherlands until his death, childless, in 1702.

Again, Holland led the way in denying the need for a stadtholder, the Dutch Republic remaining 'leaderless' for the next 45 years. Some individual provinces, however, appointed the stadtholder of Friesland, Prince William of Orange-Nassau, as their stadtholder. In 1747, in the face of yet another French invasion, it was realized that a single figure was necessary to harmonize the endeavours of such a decentralized republic. William became the hereditary stadtholder of all the provinces as William IV.

Following the French Revolution in 1789, the French declared war on Britain and the Dutch Republic in 1793. In January 1795 French forces entered the Republic and removed the government; William V fled to England. Unlike Belgium, the French decided not to annex the country, but rather make it a protectorate under the title of the Batavian Republic. Eleven years later Napoleon decided to create a kingdom out of the Republic. The country was renamed the Kingdom of Holland and Napoleon's younger brother Louis became king. Louis, however, fell foul of Napoleon when he put the interests of his new country before those of France. He was removed from his throne in 1810 and Holland was incorporated into the French Empire to complete the blockade of Britain. The Dutch did not protest.

French control began to ebb after the retreat from Moscow and defeat at the Battle of Leipzig in 1813. William V having died in 1806, his son was invited by the provisional Dutch government to return from exile and become sovereign prince of the Republic. In 1814 a constitutional monarchy was established. At the Congress of Vienna in 1815 it was decreed that he should become King William I of the United Kingdom of the Netherlands which included Belgium and Luxembourg. However, although William was keen to promote the concept of the Low Countries once again, the feeling of common interests and culture amongst the populations was lacking. William's heavy-handed treatment of the Belgians and the decline in their social standards caused resentment in the South. This grew into resistance which developed into open revolt in 1830. Alarmed, the Great Powers recognized Belgium's independence at the London Conference in January 1831. William refused to accept the situation, believing that Belgium was not viable by itself and that it would soon return to him. Dutch troops invaded Belgium in August and withdrew only when the French came to the rescue of the Belgians. Only in 1839 was a final agreement achieved between the Belgians and the Dutch; Belgium and The Netherlands were recognized as independent kingdoms. Disillusioned, William abdicated the next year in favour of his son.

During the First World War The Netherlands mobilized but remained neutral, and tried to remain so when the Second World War broke out in 1939. But when the Germans invaded in May 1940 the government and royal family withdrew to England and a Dutch resistance movement was founded. Hitler's orders to the *Reichskommissar* of The Netherlands was to incorporate the country into his 'New Europe' as *Westland,* Land of the West. At the end of the war the policy of neutrality was abandoned and the Dutch joined NATO in 1949 as a founder member.

1. The House of Orange does not have Dutch origins. It stems from a principality in south-eastern France. Through marriage it became allied to the German House of Nassau. In 1544 the Count of Nassau became the Prince of Orange, inheriting additional land in Brabant (now Belgium). Thereafter the House of Orange-Nassau, founded by William I, produced a succession of stadtholders. In 1815 the Prince of Orange became King of the United Netherlands. The Netherlands is a hereditary monarchy. William's great-grandson became William III of England. He embodied the Irish Orange Order, established to preserve Protestant supremacy.

2. Instituted in the 15th century to assist in government by a foreign ruler, the States General comprised deputies of the provincial assemblies. The Union of Utrecht modified its responsibilities so that it concerned itself primarily with the foreign policy and military affairs of the United Provinces. However, a unanimous vote from the delegates of the seven sovereign provinces, which comprised the States General, was necessary for any important national decision.

Area:	33 936 sq km (13 103 sq miles)
Population:	15.87 million; Dutch, Moroccans, Turks
Language:	Dutch
Religions:	Roman Catholic, Dutch Reformed Church, Calvinist, Muslim
Capital:	Amsterdam (Seat of Government: The Hague)
Administrative Districts:	12 provinces
Date of Independence:	1648
Neighbouring Countries:	Germany, Belgium

English-speaking Name	Local Name	Former Names	Notes
THE NETHERLANDS	Die Nederlanden	United Provinces of the Netherlands, Dutch Republic/ States General, Batavian Republic, Kingdom of Holland	The Kingdom of the Netherlands (Koninkrijk der Nederlanden) since 1814; the Kingdom of Holland (1806–1810); the Bataviar Republic (1795–1806) from the Germanic Batavi tribe which was brought under Roman control; the United Provinces of the Netherlands or the Dutch Republic or the States General (1579–1795). The name Holland is quite often used by English-speakers to describe what the Dutch call The Netherlands, despite the fact that North and South Holland comprise only two of 12 provinces. (However, it is Holland that enters the Football World Cup!) Netherlands or 'Low lands' is the literal English translation of Nederlanden (itself a translation from the Latin *inferior terra*). This area is more commonly known as the Low Countries and refers to the Netherlands, Belgium and Luxembourg. However, the term 'Netherlands' came to refer only to the United Provinces during the Twelve Years' Truce (1609–21). (It is used in this book as a singular noun). The adjective Dutch, previously applied by the English to all German-speakers (Deutsch), was used to distinguish the people of the United Provinces of the Netherlands from the people of the Southern (Spanish and Austrian) Netherlands (Belgium) and thus to the present inhabitants of the modern Netherlands. A number of English terms, usually derogatory, using the name Dutch originate from Anglo-Dutch rivalry (17th and 18th centuries). They include 'Dutch courage', false courage derived from alcohol, because the Dutch were considered heavy drinkers; to speak 'double Dutch', is to speak incomprehensibly where Dutch means 'foreign' and the 'double' indicates excessively, thus reflecting a disagreeable English disdain for those not speaking English. A Dutch treat, when everybody pays for themselves, is clearly not a treat; and to 'go Dutch', when everybody agrees to share expenses equally, is another form of Dutch treat.

Place	Other names	Description
Amersfoort		Means 'Ford on the Amer' River, now called the Eem.
Amsterdam	Amsteldam	Founded (1270) when a dam was built between the dykes on both sides of the River Amstel. Nominal capital of the Kingdom of the Netherlands (1806) but not the seat of government.
Apeldoorn		Means 'Apple tree'.
Arnhem	Arenacum	A corruption of the Latin *arena*, sand, and the German *heim*, home, to indicate a settlement on the banks of the River Rhine. Site of the gallant failure of the British 1st Airborne Division to capture the bridge over the river during Operation Market Garden (Sep 1944).
Breda		Means 'Broad river' from the Old Dutch *brede*, broad, and *a*, river. The Treaty of Breda (1667) between England, France, the United Provinces and Denmark brought an end to the second Anglo-Dutch War (1665–67) and gave the colonies of New York (New Amsterdam) and New Jersey to England.
Delft		Takes its name from the Old Dutch *delf*, canal. It lies on the canalized Schie River. Gives its name to the tin-glazed earthenware, delftware.
Dordrecht	Dort, Dordt	Name derived from the Old Dutch *drecht*, channel. It lies at the junction of four rivers.
Eindhoven		Derived from the Old Dutch *eind*, end, and *hoven*, property, to mean the village at the end of a larger settlement.
Elburg		Name derived from *elle*, ridge, on which the original inhabitants were forced to build a settlement due to flooding in the area.

Place name	Original name	Description
Flevoland		A province, the name for which is taken from the Roman Lacus Flavo which referred to the Zuider Zee, the Southern Sea.
Flushing	Vlissingen	The English name stems from the local name and means 'flowing'. It is situated at the mouth of the western Scheldt estuary.
Friesland		A province named after the Frisians.
Gelderland		A province whose name is taken from the sandy soil of the hills in the south-east: *gelwa*, yellow, and *haru*, mountain, to which '*land* has been added.
Groningen	Villa Crucninga	A province whose name is derived from the Old German *grüni*, green.
Holland, North and South		Provinces. Holland may mean simply 'hollow land', a reference to the low-lying terrain, or 'scrub land' from the Germanic *hulta*, scrub, and *land*.
Lelystad		Named after Dr Cornelis Lely (1854–1929), the engineer who proposed plans (1891) to drain the Zuider Zee; they finally led to the Zuider Zee Reclamation Act (1918).
Limburg		A province named after the House of Limburg which takes its name from *lindo*, lime tree, and *burg*, castle. Split (1839) between Belgium and the Netherlands.
Maastricht	Trajectum ad Mosam	Means 'Ford over the River Maas (Meuse)'. Treaty of Maastricht (1991) established a European Union with Union citizenship for every person having the nationality of a member state.

Place	Alternative names	Description
Nijmegen	Novio Magus	A corruption of the Roman name which meant 'New market'. The bridges over the River Waal were captured by the US Army's 82nd Airborne Division during Operation market Garden (Sep 1944), but the planned link-up with the British at Arnhem failed.
Overijssel		A province with a name meaning 'Beyond, that is, north, of the River IJssel'.
Rotterdam		Site of a 13th century dam on the River Rotte. Much of the city and port was bombed to destruction by the Germans (May 1940).
's Hertogenbosch	Den Bosch, Bois-le-Duc	Means 'The Duke's wood' for Henry I, Duke of Brabant (12th century).
The Hague	's Gravenhage, Den Haag	Means 'The Count's hedge', a settlement developed round the original hunting lodge built (1248) in a woodland area called Haghe, 'hedge', for Count William II; it became the principal residence of the Counts of Holland. The Hague Convention (1907) laid down the law governing the conduct of international warfare. Seat of government and administrative capital.
Utrecht	Trajectum ad Rhenum Ultrajectum, Ouda Trecht	Means 'Ford on the Rhine' from the Latin *trajectus*, river crossing, with Ouda Trecht, meaning 'Old Ford'. The Union of Utrecht (1579) established the Dutch Republic whereby the seven northern provinces united against Spain. The Treaties of Utrecht (1713/14) between France and other European powers concluded the War of the Spanish Succession (1701–14) and re-arranged affairs in Europe and beyond.
Vreeswijk		Means 'Land of the Frisians'.

Zeebrugge

Means 'Bruges-on-Sea', it being the port for the city of Bruges ten miles to the south.

Zeeland

A province meaning 'Sea land', a reference to its continuing struggle against the encroachments of the North Sea.

Zutphen

Zuidveen, Zutphania Turria

Means 'Southern peat bog'.

NORWAY

Warm Atlantic currents, sheltered fjords and an abundance of fish encouraged settlement along the long western Norwegian coast. However, little is known of the Norwegian tribes and their independent kingdoms until the advent of the Viking Age which began in about 800. The Vikings – emanating from the bay, *vik*, now known as the Skagerrak – made their presence felt, raiding and settling in what is now Britain, France, Iceland, Ireland, Greenland and the Faeroe Islands. By 900 a Viking chieftain, Harald I Fairhair, had managed by peaceful agreement and battle to subdue all opponents and proclaim himself the first King of a unified Norway.

In 1015 one of Harald's descendants, Olav II Haraldsson, who had been in England, succeeded to the throne. Opposition to his efforts to increase his power and convert all his people to Christianity grew; it looked to Canute of Denmark (King of England from 1016 and of Denmark from 1018) for support. When Olav was killed in battle in 1030 Canute became King of Norway. When he died in 1035 the Harald dynasty reclaimed the throne and by the end of the 11th century had established Norway's independence. An attempt to subdue England, however, had failed when Harald III Hardraada was defeated by King Harold at the Battle of Stamford Bridge near York in 1066.

After Greenland and Iceland had established personal unions with the Norwegian king in 1261 and 1262, and with possession of the Scottish Isles and the Faeroe Islands, Norwegian power was at its greatest. In 1266 the Hebrides and the Isle of Man were ceded to Scotland in exchange for the Orkney and Shetland Islands.

The connection with Sweden began in 1319 when King Haakon V was succeeded by his grandson, Magnus VII Eriksson, the three year old son of a Swedish duke. This personal union lasted until 1355 when Magnus abdicated in favour of his younger son who became Haakon VI. He married Margaret, daughter of the King of Denmark (q.v.), and their son Olav first became King of Denmark in 1375 and then King of Norway in 1380 when Haakon died. With this dynastic union Norway lost its independence. When Olav died in 1387 his mother became regent of both countries. In 1388 she formally accepted the rule of Sweden. Having been nominated as her heir, her nephew, Erik of Pomerania, was crowned King of Denmark, Sweden and Norway in 1397 at Kalmar (in Sweden), thus heralding the start of the Kalmar Union.

After Erik was deposed from the Norwegian throne in 1442 the country was ruled by Danish kings until 1814. Foreigners were placed in positions of authority and Norway's political influence was undistinguished. The Orkney and Shetland Islands were pawned to the Scottish king in 1468 and never regained. In 1536, the Kalmar Union having foundered in 1523, Norway was proclaimed a Danish province. Danish was the official language. Norway's hereditary monarchy, however, was still recognized and, due to the remoteness of much of the country, a considerable degree of Norwegian independence was inevitable. As a result of war between Denmark and Sweden, Norway lost some territory in the east to Sweden at the Peace of Roskilde in 1658; the present border stems from that Peace. In 1661 the Twin Kingdoms of Denmark-Norway were created whereby every citizen was an equal subject of the king.

Denmark, and therefore Norway, entered the Napoleonic Wars on the side of France in 1807 after an English fleet had bombarded Copenhagen and captured the Danish fleet. In response to the continental blockade of England, a counter-blockade was imposed on Norway. With British command of the sea Norway was isolated from Denmark and suffered serious economic hardship. When Napoleon was defeated at the Battle of Leipzig in 1813, the Swedes, who opposed the French, saw their opportunity and attacked Denmark. By the Treaty of Kiel in January 1814, Denmark was forced to surrender Norway (but not its dependencies of the Faeroes, Greenland and Iceland) to Sweden.

Six years of separation from Denmark had encouraged the Norwegians to think that they could go their own way. Led by Prince Christian Frederick, governor of Norway and heir to the throne of the Twin Kingdoms, they refused to recognize the Treaty of Kiel and declared independence. Christian Frederick became king. The arrival of Swedish troops two months later in July 1814 combined with foreign pressure, however, forced the Norwegians to accept a personal union and the abdication of Christian Frederick. In November the union became fact and the Swedish king was elected King of Norway. The pill was sweetened by allowing the Norwegians to have their own constitution and thus be a distinct and self-governing state, although inferior to Sweden.

Integration of the two countries did not prove to be a great success. The Norwegians disliked being ruled by a Swedish king who had the right to appoint their government; furthermore, the Swedes conducted Norway's foreign policy. When the king refused to sanction a bill in 1905 on the creation of a Norwegian consular service the government resigned. With the king unable to form a new government, the Storting (parliament) declared that he was no longer able to function as Norwegian king and that the union with Sweden was dissolved. Nevertheless, a referendum on the future of the union was arranged. The result was an overwhelming vote to dissolve it. This unilateral declaration did not receive immediate Swedish assent, but, after negotiation, the dissolution became legal under the terms of the Treaty of Karlstad in September 1905. The Danish Prince Charles, a grandson of the reigning king, Oscar II, and a son-in-law of Edward VII of Britain, was elected in a second referendum as king, taking the name Haakon VII.

Norway remained neutral during the First World War and declared its neutrality at the outbreak of the Second in 1939. However, the Germans invaded in 1940 and quickly defeated the Norwegian Army. German occupation then continued until the end of the war.

Area:	324 220 sq km (125 182 sq miles)
Population:	4.41 million; Norwegians, Lapps (Sami)
Languages:	Norwegian (Bokmél and Nynorsk), Lapp
Religions:	Evangelical Lutheran, Roman Catholic, Protestant
Capital:	Oslo
Administrative Districts:	19 counties
Date of Independence:	7 June 1905
Neighbouring Countries:	Russia, Finland, Sweden

English-speaking Name	Local Name	Former Names	Notes
NORWAY	Norge		The Kingdom of Norway (Kongeriket Norge) since the ninth century and a constitutional monarchy since 1905. Settled by Germanic tribes, it means 'The way north' or 'The northern way'.
Bergen		Björgvin	Means 'Mountain pasture' from *björg*, mountain, and *vin*, pasture or meadow. Capital (12th and 13th centuries).
Finnmark			A county whose name means 'Borderland of the Finns'.
Fredrikstad			Founded and named (1567) after Frederick II (1534–1588), King of Denmark and Norway (1559–88). Razed by the Swedes (1570).
Halden		Fredrikshald	Founded (1661) and named (1665) after Frederick III (1609–70), King of Denmark and Norway (1648–70) as Fredrikshald (1665–1928). Besieged by the Swedes (1718), the Swedish king, Charles XII, being killed.
Hammerfest			With its harbour nestling at the foot of a cliff face, it takes its name from *hammer*, steep cliff, and *feste*, mooring place. Destroyed by the British (1809).
Kristiansand			Founded (1641) and named after King Christian IV (1577–1648), King of Denmark and Norway (1588–1648). *Sand* simply means sand.
Kristiansund			Named after Christian IV. *Sund*, inlet, is equivalent to the English 'sound'.

Lillehammer		Means 'Little hammer'.
Narvik	Victoriahavn	Means 'Narrow bay' from the fact that it lies at the end of a peninsula between two fjords. The original name (1887–98), Victoria's Port, was in honour of the Swedish Crown Princess. Seized by German troops (Apr 1940) but recaptured by an Anglo-French expeditionary force (May 1940) – Germany's only defeat in the first three years of the Second World War.
Oslo	Ansloga, Christiania, Kristiania	After destruction by fire (1624), rebuilt further to the west and renamed Christiania (1624–1877), after King Christian IV. Changed to Kristiania (1877–1925) before becoming Oslo, the modern version of its original name. It lies at the head of a fjord and the *os* may therefore mean 'mouth'. Capital (c.1300–1397, and from 1814). The Oslo Declaration of Principles signed (Sep 1993) between Israel and the PLO established limited Palestinian self-rule in the Gaza Strip and Jericho.
Tromsø		The administrative centre of Troms County, it is situated on the island of Tromsøy, which gives it its name from the Danish *ø*, island (and in Norwegian *øy*).
Trondheim	Kaupangr Nidaros, Trondhjem	May takes its name from the original name of the fjord, Throndr, which may itself be derived from the Old Norwegian *thorr*, thunder; or it could mean 'Home of the Throne', the city being the traditional site for royal coronations. Nidaros (1015–16th century) means 'Mouth of the River Nid(elva)'; Trondhjem (16th century-1930); Nidaros (1930–31); finally the new spelling of Trondheim. Norway's first capital (from c.1000).
Svalbard	Grumant, Spitsbergen	An archipelago of nine principal islands whose name means 'Cold, *sval*, coast, *bård*,'. Competing claims to possession

because of rich mineral deposits were resolved by treaty in Norway's favour (1920). Mineral rights, however, were granted to a number of countries. The Dutch explorer, Willem Barents (c.1550–97), discovered (1596) the archipelago which he named Spitsbergen. One of the islands is named after him. Over 60 per cent of the population is Russian.

Longyearbyen

Named (1906) after John Longyear, an American who founded the Arctic Coal Company on Spitsbergen.

Spitsbergen

The main island within Svalbard. Means 'Needle mountains' from the Dutch *spits*, points, and *bergen*, mountains.

POLAND

For over a millennium Poland's history has been punctuated by foreign aggression and war. For centuries it had the misfortune to be surrounded by militaristic Prussians, imperialistic Russians and acquisitive Austrians. Later, its fears centred on Germany and Russia, particularly when these two countries were on good terms. Because Poland has few natural barriers it has always been susceptible to invasion.

Poland's origins lie in the unification of various West Slav tribes in the tenth century. In 965 a Polanian tribal chieftain, Mieszko I, a member of the Piast dynasty which controlled what came to be known as Great Poland, emerged into history by marrying the sister of the Bohemian Duke Boleslav II to form an alliance against the threat from the German king and Holy Roman Emperor, Otto I the Great. Realizing that the Germans could use the excuse of Christianizing the pagan Poles to win control of Polish territory, Mieszko was baptised the following year and Polish conversion to Christianity followed. 966 is recognized as the year in which Poland was founded. Until his death in 992 Mieszko expanded his territory to include the Baltic coast, Silesia and Little Poland (an area round Cracow in southern Poland) which quickly assumed political leadership.

In the early years of the 11th century conquests to east, south and west enlarged Polish territory to incorporate non-Polish tribal areas. In 1024 Poland became a kingdom when Mieszko's eldest son, Boleslav I the Brave, was crowned. The struggle for territory continued, but by 1135 the German Emperor Lothair had extended his frontiers in the east to the Oder and Neisse rivers. At the death of Boleslav III in 1138 the kingdom was divided among his five sons. Inevitably, this resulted in disunity and conflict. For the next 200 years Poland was fragmented and peripheral territories were lost. In 1225 the Teutonic Knights, a German military Order of crusaders, were invited by Duke Konrad I of Mazovia to come and help beat off the pagan Prussians to the north. Successful, the Order carved out a papal fiefdom along the Baltic coast, resettling Germans in many Polish towns.

In 1241 the Mongols invaded, devastating much of the country during succeeding years. Nevertheless, the Piast dynasty, although weakened, continued to rule Poland, or parts of it (Silesia was lost in 1348), until the death of Casimir III the Great in 1370. With no male heir, he was succeeded by his nephew, Louis I of Hungary, with the approval of the aristocracy. This established the principle whereby the aristocracy gained the right to choose the king. This later turned out to be a disaster: its interests were elevated above the need for state unity and central authority was undermined.

Two years after Louis' death in 1382 his 11 year-old daughter, Jadwiga, was chosen to succeed him. Although Grand Duke Jagiełło of Lithuania (q.v.) was a heathen, the Polish aristocracy preferred a new dynastic association with that country. The succession crisis was solved when, as Crown Princess, Jadwiga married Jagiełło, who converted to Catholicism and became King of Poland as Vladislav II Jagiełło. Although remaining separate states, a Polish-Lithuanian alliance was forged. Lithuania was three times larger than Poland and possessed what is today Belarus, Ukraine and large sections of western Russia. The personal union with Lithuania was to last 187 years.

The continuing threat posed by the Teutonic Knights encouraged Polish-Lithuanian forces to combine and in 1410 they overwhelmed the Knights at the Battle of Grünwald (Tannenberg). Jagiełłonian power reached its height when Vladislav, son of the Polish King Casimir IV, was elected both King of Bohemia (1471) and King of Hungary (1492), although they remained separate kingdoms. The union was broken in 1526 when the Turks destroyed Hungarian independence at the Battle of Mohács.

When it became obvious that the still childless King Sigismund II Augustus was not going to produce an heir and thus the personal link between Poland and Lithuania would be broken, a Polish-Lithuanian Commonwealth was created by the Union of Lublin in 1569[1]; it also included Livonia (q.v.Latvia) and parts of Prussia. Poland and Lithuania remained distinct states with their own laws and administrations, though jointly governed by an elected king and a joint *Sejm* (Diet). Lithuania ceded its Ukrainian provinces, Volhynia and Podlasie to Poland in return for protection. The Jagiełłonian dynasty, however, ended three years later when Sigismund died and thereafter Polish kings were selected from the aristocracy and the royal houses of Europe. By now, the lower nobility were also eligible to elect the king whose powers were strictly limited. For the next 223 years, until 1795, Poland was ruled by 11 elected kings, of which seven were foreigners, three being Swedes. As a result the monarchy and Poland itself entered a prolonged period of weakness; the Commonwealth began to fragment.

The 17th century was characterized by warfare, no period being more disastrous for Poland than that known as the 'Deluge'. In 1654 the Russians and Cossacks invaded, capturing the eastern part of the country. The Swedish king, Charles X Gustav, joined in. Foreign armies tramped across Poland, leaving a wasteland in their wake. Between 1655 and 1660 practically the whole of the country was under Swedish domination. Only in 1673 did John Sobieski defeat the Ottoman Turks and, on the strength of his victory, he was chosen as king the following year. He was, however, unable to regain lands in the south-east lost to the Turks, Kiev, and parts of Ukraine east of the River Dnepr lost to Muscovy at the Truce of Andrusovo in 1667.

More fighting followed the accession in 1697 of Augustus II the Strong, the Elector of Saxony, who was chosen from 18 candidates. It was he, allied to Peter I the Great of Russia and Frederick IV of Denmark, who began the Great Northern War (1700–21) (q.v.Estonia) by attacking Sweden with the aim of seizing its Baltic possessions. Again, Poland, a geographical giant yet politically and militarily insignificant, provided a battleground for foreign armies – Russian, Saxon and Swedish – and again it was devastated. Half of its population was not Polish, but Cossack, German, Jewish, Lithuanian and Russian. It remained intact only because no other country had any interest in dismembering it. However, after its victory over the Swedes at Poltava (in Ukraine) in 1709, Russia began to increase its influence over Polish affairs to the extent that Poland all but became a protectorate.

In 1772, for reasons of territorial aggrandizement and increased security, more than a quarter of Poland was annexed by Austria (Galicia), Prussia (Polish Pomerania and other land in the north-west) and Russia (north-east Poland). A second partition followed in 1793 when Prussia and Russia seized over half the remaining territory. The former received Great Poland and parts of Mazovia while the Russians took what remained of Byelorussia and Ukraine. A year later, a national insurrection, led by Tadeusz Kosciuszko, was crushed and in 1795 Poland was erased from the map when the three countries divided what remained amongst themselves. Russia had acquired the most territory from the three partitions. For the next 123 years – until 1918 – Poland did not exist. But the Poles with their powerful feelings of national identity did.

Following the Treaty of Tilsit (now Sovetsk) in 1807, Napoleon created the Duchy of Warsaw out of some Polish-populated areas of Prussia seized during the partitions. It lasted only until 1812 when his armies were hurled out of Russia. In 1815, after Napoleon's defeat at the Battle of Waterloo, the Congress of Vienna established the Congress Kingdom of Poland within the Russian empire; supposedly autonomous, it had the Russian tsar as its king. Russification and Germanization by the Prussians in the partitioned lands followed, although the Austrians permitted a degree of autonomy in Galicia. Nevertheless, uprisings took place in 1830, 1846, 1848 and 1863; all failed.

In August 1914 war broke out between the occupying powers: Austria and Germany/Prussia on the one hand and Russia and its western allies on the other. This was a catastrophe for the Poles over whose land the three powers fought. Conscripted into the three armies, Poles found themselves fighting against Poles. However, the Bolshevik Revolution in Russia in October 1917 and the subsequent Civil War, the collapse of the Austro-Hungarian Empire in October 1918 and the armistice with Germany the following month, created a political vacuum which permitted the Poles, under Marshal Jozef Pilsudski and with the agreement of the Allies, to reconstitute their country and declare sovereignty. Attention had already been drawn to the 'Polish problem' by Woodrow Wilson, the American president, in his 'Fourteen Points' set out in January 1918. One of these was the creation of an independent Poland with access to the sea. The Second Polish Republic was founded on 11 November 1918.

The Treaty of Versailles in June 1919 awarded Poland the western part of Prussia, thus giving access to the Baltic Sea; at the same time a German minority was acquired. It meant also that East Prussia was separated from the rest of Germany and a 'Polish Corridor' was created. Poland's claims to Gdansk (then Danzig) were not approved and it became a 'Free City' although still German-populated. At the Treaty of St Germain three months later, Poland received the Austrian provinces of Western and Eastern Galicia. Poland's western border was delineated as a result of various plebiscites which brought in part of Upper Silesia; the rest of Silesia remained German.

To the east Poland was intent on territorial expansion, demanding a return to the 1772 border with Russia. Lenin refused. The Poles swarmed into Ukraine and captured Kiev in June 1920 before being driven back to Warsaw. In August, however, Pilsudski defeated the Red Army. The eastern border, reaching some 250 kilometres beyond the line proposed by the British Foreign Secretary Lord Curzon, later known as the Curzon Line (which used the River Bug), was finally established at the Treaty of Riga in March 1921. Land amounting to some 135,000 sq km of what is now Ukraine, Belarus and Lithuania was acquired. The existence of this new independent Poland was anathema to Germany and Russia and gave them a common enemy.

On 23 August 1939 Germany and the Soviet Union signed a non-aggression pact. More important in Stalin's view, however, was a secret protocol which defined the partition of Eastern Europe, including Poland, between the two countries. On 1 September 1939 Germany invaded Poland from the west and on 17 September the Soviet Union invaded from the east, advancing as far as the Curzon Line. By November Poland had been partitioned for the fourth time. This status quo only lasted until 22 June 1941 when the

Germans unexpectedly attacked the Soviet Union, pushing the Russians out of eastern Poland. For the next three years Poland was subjected to Nazi rule, the intention being once more to remove Poland from the map and Germanize those Poles still living in what was called the General Government.

The Soviet victory at Stalingrad in February 1943 heralded a turning point in the war on the eastern front. Thereafter, Soviet troops began their inexorable advance westwards. They flooded into Poland in July 1944 and by March 1945 had re-conquered Poland and entered Germany. Liberation by the Red Army did not restore Polish independence.

At the Yalta conference in February 1945 it was agreed that Poland would remain under Soviet domination and that the Curzon Line would be recognized as the Soviet/Polish border. A few minor adjustments were made later. At the Potsdam conference in August 1945 it was agreed that Poland's western borders would follow the line of the Oder and Neisse rivers. The effect of the agreements at these two conferences was to move the whole of Poland some 200 kilometres westwards and to give it the German provinces of Pomerania and Silesia and a part of the former East Prussia. The Soviet Union acquired some 180,000 sq km of Polish land and Poland regained about 100,000 sq km of territory after centuries of German occupation. Poles and Germans were evicted from land they had lived on for centuries. It was called population transfer; now it would be called ethnic cleansing.

Stalin's aim now was to Sovietize Poland. Soviet troops did not withdraw after the war. Not only did their presence demonstrate Soviet hegemony and underpin communist power, but they also provided a second echelon for any potential attack on Western Europe. On a day-to-day basis they guarded the lines of communication to the five Soviet armies based in East Germany. Polish defiance of Soviet communism, however, led to grave crises in 1956, 1970 and 1981. The rise of Solidarity, the first independent trade union in the Soviet *bloc*, in 1980 and the more liberal policies being pursued by Mikhail Gorbachev in the Soviet Union led to the demise of the communist regime at the end of 1989. Following the disintegration of the Soviet Union and an agreement to withdraw all Russian forces from Eastern Europe, all but a few minor units had left Poland by the end of 1993.

1. Since 1499 the combined kingdoms had been called a 'republic' – the First Polish Republic.

Area:	312 677 sq km (120 725 sq miles)
Population:	38.73 million; Poles, Ukrainians
Language:	Polish
Religion:	Roman Catholic
Capital:	Warsaw
Administrative Districts:	49 voivodships
Date of Independence:	11 November 1918 (Second Republic)
Neighbouring Countries:	Russia, Lithuania, Belarus, Ukraine, Slovakia, Czech Republic, Germany

The Partitions of Poland 1772, 1793, 1795

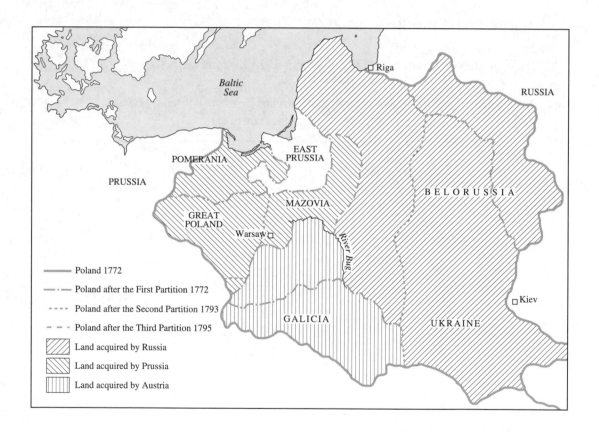

Poland 1772

Poland after the First Partition 1772

Poland after the Second Partition 1793

Poland after the Third Partition 1795

Land acquired by Russia

Land acquired by Prussia

Land acquired by Austria

English-speaking Name	Local Name	Former Names	Notes
POLAND	Polska		Republic of Poland (Rzeczpospolita Polska) since Dec 1989. People's Republic of Poland (1947–89). Second Polish Republic (1918–1939, although it is argued that the presence of a Polish government-in-exile in the UK from 1940–45 continued its existence). Congress Kingdom of Poland (known as Congress Poland) (1815–64) within the Russian Empire; the Russian sector of Poland was thereafter called the Western Region of the Russian Empire. Kingdom of Poland (1024–1795, although from 1569 called a Royal Republic). Named after the Polanie, the 'People of the fields or plain' from *pole*, field, who settled on the banks of the River Warta in the open plain between the Oder and Vistula rivers. Their tribal chief and progenitor of a line of princes, the legendary Piast (c.870), united the scattered groups into one unit which he called Polska. The Poles used to call themselves Polaks and Polish immigrants to the USA are sometimes called Polacks.
Augustów			Named after King Sigismund II Augustus (1520–1572), the last Jagiellonian king (1548–72). By means of the Union of Lublin (1569), he united Livonia (now Latvia and southern Estonia) and the Duchy of Lithuania legally with Poland which became a Royal Republic.
Białystok			Means 'White river from *biały*, white, and *stok*, river.
Bierutów		Brückenberg	Renamed after Bolesław Bierut (1892–1956), president (1945–52) and prime minister (1952–54). Bierut was a pseudonym; his real name was Krasnodebski.
Brzeg		Brieg	Lying on the River Oder, it means 'Bank' or 'riverside'.

Bydgoszcz	Bromberg	Situated near the confluence of the Brda and Vistula rivers, the name is derived from the Indo-European root word *bredahe*, swamp.
Cracow	Kraków	Named after a Polish knight, Krak or Krakus, who built a castle on a hill overlooking the Vistula. According to legend, the town was afflicted by a dragon which demanded a virgin a day to appease its appetite. Other knights tried to kill it without success. Krakus killed a sheep and filled its carcass with sulphur. The dragon consumed it in a mouthful, was tormented by thirst, plunged into the river and drank so much that it burst. Capital of the Piast kingdom (1038–1370) because Gniezno was thought to be too vulnerable to German attack and subjugation. Sacked by the Mongols (1241). Capital of Poland (1320–1609). Under Austrian rule (1795–1918). The Republic of Cracow, or Free City of Cracow, was established by the Congress of Vienna and was the only part of Poland to be independent (1815–46) after the Third Partition (1795); it was, however, under the joint protection of Austria, Prussia and Russia. Incorporated the separate town of Nowa Huta, 'New Steelworks' (1951).
Częstochowa	Tschenstochau	May be derived from *częstokoł*, palisade, which originally provided protection. Poland's national shrine. The Paulite monastery of Jasna Góra ('Bright Mountain') houses the painting of the Black Madonna, the presence of which, according to religious belief, successfully protected the monastery against a siege by the Swedes (1655).
Dzierżoniów	Reichenbach	For a time named after Baron von Reichenbach (1788–1869) who discovered paraffin (1830) and creosote (1833). Renamed (1945) after the Polish priest and apiculturalist, Jan Dzierzon (1811–1906). Having been given to Bohemia (1335), it passed to the Habsburgs, then to Prussia (1742) and finally back to Poland (1945).

Elbląg	Elbing	Named after the River Elbląg.
Gdańsk	Gyddanyzc, Kdaɪzc, Danzig	Derived from *Gatisk-anja*, end of the Goths, to signify the limit of their territory. Under the rule of the Teutonic Knights (1308–1466). Incorporated into Prussia at the Second Partition (1793). Although largely populated by Germans, a free city under a League of Nations mandate (1919–39), and thus not included in the Polish Corridor, until seized by Nazi Germany (1939). The Second World War started at the entrance to the port, at Westerplatte, when a German battleship, the Schleswig-Holstein, opened fire on a small Polish military depot (1 Sep 1939). The Solidarity movement, which led to the collapse of the Communist regime (1989), was founded here (1980).
Gdynia		Lying only 15 km north-west of Gdańsk, the city's name has the same origin as that of its sister city. Developed as an alternative port to Gdańsk in the Polish Corridor which gave Poland a narrow access to the Baltic Sea after the First World War.
Gliwice	Gleiwitz	The Germans faked a Polish raid (31 Aug 1939) on the German radio station at Gleiwitz (just inside Germany) to demonstrate Polish aggression and create an excuse for an attack on Poland the next day.
Gniezno	Gnesen	The first capital of Poland (966–1038). According to legend, Lech, a leader of the Polanie tribe, discovered the nest (in Polish, *gniazdo*) of a white eagle while out hunting. This was a good omen. The place was given its name and the eagle became the emblem of the Polish nation.
Grünwald	Tannenberg, Stebark	Means 'Green forest'. At the Battle of Grünwald (15 Jul 1410) the Poles and Lithuanians defeated the Teutonic

Jelenia Góra	Knights. The Germans defeated the Russians at the Battle of Tannenberg (26–31 Aug 1914).
Hirschberg	Means 'Deer Mountain' from *jelen*, deer, and *góra*, mountain. The name, however, is derived from another source: the king, Boleslaw the Brave, ordered a knight, Jelnek, to build a castle as a defence against the Czechs (1004). The settlement round the castle became known as Jelnek's Mountain – Jelenia Góra.
Katowice / Kattowitz, Stalinogród	Named Stalinogród, Stalin's town, two days after his death (5 Mar 1953) until the overthrow of the Stalinist regime in Poland (Oct 1956). Only became Polish (1922) as a result of a plebiscite in German Upper Silesia when many communes voted to join Poland.
Kazimierz Dolny	Named after Kazimierz (Casimir) III the Great (1310–1370), the last Piast king (1333–70). *Dolny* means 'lower' to distinguish it from Kazimierz, now a part of Cracow.
Ketrzyn / Rastenburg	Hitler's Wolf's Lair (in Polish, *Wilczy Szaniec*), the headquarters from which he directed the war against the Soviet Union, is 8 km to the east. The site of an assassination attempt against him (20 Jul 1944).
Legnica / Liegnitz, Wahlstatt	The Mongols (Tatars) annihilated the armies of Prince Henryk Pobozny (Henry II the Pious) and the Teutonic Knights at the Battle of Liegnitz (1241). Frederick II the Great (1712–86), King of Prussia (1740–86), defeated the Austrians here (1760) during the Seven Years' War.
Malbork / Marienburg	Means 'Fortress of Mary'. Capital of the Teutonic Knights (1309–1457).

Name	Alternative	Description
Malopolska		A region meaning 'Little Poland' which incorporated the tribes of southern Poland (c.1000).
Masuria		A region named after the Mazurs. They invented the *mazurka*, a Polish folk dance.
Mazovia		A region also named after the Mazurs.
Podlasie		An area in eastern Mazovia meaning 'Land close to the forest'.
Pomerania	Pommern	A region meaning 'By the sea' or 'coastland'. Sweden received Western Pomerania at the Peace of Westphalia (1648), but all of Pomerania fell into Prussian hands (by 1815). Eastern Pomerania was annexed by Prussia (1772) at the First Partition and became West Prussia. It and Central Pomerania were transferred to Poland (1945) while Western Pomerania became part of the German Democratic Republic, now eastern Germany, (1945). Gave its name to a breed of very small dog.
Poznań	Posen	May be derived from the personal name of a bygone landowner together with the title *pan*, lord. The Second Polish Republic was declared here (1918). A huge strike followed by rioting (Jun 1956) led to the overthrow of the Stalinist government and more moderate rule. Became joint capital with Gniezno (1025).
Silesia	Schlesien	A region in the south-west named after the Ślęzanie. Passed to Bohemia (1335), the Austrian Habsburgs (1526), seized by Prussia (1742) and only returned to Poland (1945).
Szczecin	Stettin	May be derived from *szczotka*, undergrowth, a reference to the local scrub. Under Swedish rule (1648–1720) and German rule (1720–1945).

Toruń	Thorn	Ceded to the Teutonic Knights (1230) who named it after Toron, the name given by the Crusaders to the town of Tibnin, previously Taphnith, now in southern Lebanon.
Warsaw	Warszawa	May be derived from a personal name such as Warsz. According to legend, a mermaid led a prince to the site and commanded him to build a city. Capital transferred from Cracow (1596–1609). Occupied by the Swedes (1655). Under Prussian domination (1795–1807). Duchy of Warsaw (1807–15). Under Russian control (1815–1917). Russian attack repulsed (Aug 1920). Captured by the Germans (27 Sep 1939). Uprising against the German occupation (Aug-Oct 1944). Liberated by the Russians (Jan 1945). Treaty of Friendship, Co-operation and Mutual Assistance, aka the Warsaw Pact, signed (May 1955; dissolved Jul 1991).
Wielkopolska		A region meaning 'Great Poland', founded as a result of the unification (10th century) of various tribes around Poznan.
Wrocław	Breslau, Boroszló	Named after Braslav, the last Slav leader of the Great Moravian Empire (as Bratislava). Burned by the Mongols (1241). Under the domination of Bohemia (1335), Hungary (1477), Austria (1526) and, by the Treaty of Breslau, Prussia (1741). Returned to Poland (1945).
Zabrze	Hindenburg	Named (1915–45) after Gen von Hindenburg (1847–1934), German President (1925–34), who defeated the Russians at the Battle of Tannenberg and at the Masurian Lakes (1915). In Prussian/German hands (1742–1945).
Zamość	Himmlerstadt	Named after Jan Zamoyski (1542–1605), chancellor and Grand Hetman of the Crown (commander-in-chief of the

armed forces). With Częstochowa and Gdansk, one of only three cities to have withstood the Swedish sieges (1655–57). Named briefly after Heinrich Himmler (1900–45), Nazi leader and Chief of the Gestapo (*Geheime Staatspolizei*, secret police).

PORTUGAL

Until it emerged as an independent kingdom in 1139, Portugal shared the history of the Iberian peninsula with Spain. There is no major physical barrier separating the two countries and the cultural differences are such that some Portuguese historians have argued that their country's independence was virtually accidental. Celts and Iberians were the first to arrive, a branch of the latter being known as the Lusitani. The peninsula was overrun by the Romans after the Second Punic War (218–01 BC), but the west was still largely occupied by the Lusitani who resisted the Romans for some decades. Only in 27 BC did the central part of Portugal today become the westernmost Roman province of Lusitania.

Although rather a provincial backwater during the Roman occupation, Portugal experienced the arrival of Christianity, the development of a road network and the emergence of a distinctive language, Portuguese, strongly derived from Latin. Different groups of barbarians crossing the Pyrenees at the beginning of the fifth century hastened the decline of Roman power. The Germanic Suevi (or Swabians) settled between the Douro and Minho rivers, forming a state in about 410; this lasted until 585 when it was forcibly incorporated into the Visigothic Empire.

Only the northern part of Portugal escaped the Muslim invasion of 711. The Moors, who included Berbers from Morocco, Syrians and Egyptians, treated their new subjects with more grace than the Visigoths. Nevertheless, the Christian Reconquest began as early as 718 (q.v.Spain). During the tenth century the rulers of Portucale – the area between the Douro and Minho roughly corresponding to the old Suevi state – began to expand their territory southwards. By the end of the 11th century Portucale had become a county and given to Count Henry of Burgundy by the King of León-Castile to defend it from a new wave of invaders, the Almoravids (Saharan Berbers), who entered the peninsula in 1086.

In 1139 Afonso Henriques, Henry's son, defeated the Muslims at Ourique and thereafter used the title of King Afonso I of Portucale. Four years later King Alfonso VII of León-Castile recognized the independence of Portugal under the Burgundian dynasty at the Treaty of Zamora. However, it was only in 1179, at the disintegration of León-Castile following Alfonso VII's death, that Portugal was recognized internationally as a kingdom.

Despite setbacks, the struggle to remove the Moors from the south continued, Lisbon being taken in 1147. By 1240 Sancho II had recaptured Alentejo and Eastern Algarve. The reconquest of Western Algarve was completed by Afonso III, Sancho's younger brother, who moved the capital from Coimbra to Lisbon in 1260. By the time of his death in 1279 the modern boundaries of mainland Portugal had been established; in 1297 the frontier between Portugal and Castile was agreed at the Treaty of Alcañices, which also confirmed Portugal's possession of Algarve and made allies of the two countries. During his 46 years' reign (1279–1325), Dinis, son of Afonso III, did much to consolidate Portuguese independence, create a distinctive Portuguese culture and encourage the use of the Portuguese language.

Having seized the throne of Castile, Henry of Trastamara, now Henry II, invaded Portugal in 1369, forcing the Portuguese king, Ferdinand, to renounce his claim to Castile through his mother, who had been the daughter of the Castilian *infante*. Henry invaded again three years later. One of the results of this antagonism was Ferdinand's agreement to the marriage of his daughter Beatriz to John I of Castile, Henry having died in 1379. Portugal's independence was now in jeopardy. When Ferdinand died in 1383, the last of the legitimate male line of Burgundy, his unpopular widow, Leonor Teles, assumed the reins of government with her Spanish lover and Castile claimed the Portuguese crown.

In answer, the Portuguese chose the illegitimate son of Peter I the Cruel, John of Aviz, as regent. Two years later he was proclaimed king, becoming the founder of the House of Aviz. Within four months John I of Castile invaded, but was heavily defeated at the Battle of Aljubarrota with the help of English archers. This victory consolidated the independence of Portugal; the help given by the English was acknowledged and an alliance, unbroken to this day, was enshrined in the Treaty of Windsor in 1386. A year later John married Philippa, daughter of John of Gaunt, the Duke of Lancaster. Peace with Castile was finally achieved in 1411.

Inspired by the half-English Prince Henry 'the Navigator', the third son of John and Philippa, the great era of Portuguese exploration began with the capture of Ceuta (now Spanish) on the North African coast in 1415. For the rest of the century Portuguese ships explored the oceans and helped to create an empire with possessions in South America, Africa and Asia. In 1494 the Treaty of Tordesillas demarcated the spheres of colonial interest between Spain and Portugal: the latter's rights extended out to 370 leagues west of the Cape Verde Islands, the Spanish beyond. Thus Brazil, discovered in 1500, became Portuguese. A similar dividing line was drawn in the Pacific in 1529. As the 16th century began Portugal was the wealthiest country in Europe and the future seemed bright.

However, despite its wealth, Portugal found it difficult to maintain its huge empire. As the great uncle of King Sebastian, who had been killed leading a crusade against Morocco in 1578, the 66-year-old Cardinal Henry became king. Celibate, Henry had no heirs and it was clear that when he died the Aviz line would also become extinct. Philip II of Spain, an uncle to Sebastian, saw his opportunity and when Henry died in 1580, he sent the Duke of Alba and an army into Portugal. Resistance was quickly quelled and Philip was proclaimed Philip I of Portugal.

The Habsburg Philip respected Portuguese autonomy; his successors did not. Spanish self-interest and neglect of Portuguese sensibilities soon generated fierce resentment. This came to a head in 1640 with a nationalist uprising. The Spanish were driven out and the Duke of Braganza, head of the most powerful family in the country, was proclaimed king as John IV. The Spanish, however, did not recognize Portuguese independence until 1668. In the meantime, Anglo-Portuguese ties were renewed with the marriage of Charles II to Catherine, daughter of John IV, in 1661.

Portugal was dragged into European affairs when Napoleon demanded that it should sever connections with Britain and help with his naval blockade. With Britain as its most important trading partner and relying on the British fleet to keep trade routes open, it is no surprise that the Portuguese refused. In 1807 the French General Junot began his march on Lisbon. The British evacuated the Royal Family to Brazil (where it remained until 1821). When Junot entered Lisbon he declared that the Braganzas had been deposed. Within a year Sir Arthur Wellesley (later the Duke of Wellington) had twice defeated the French and forced on them the Convention of Sintra whereby Junot was to leave Portugal. Twice more the French invaded and twice more they were evicted by the British, the last time in April 1811.

By 1815, with John VI still refusing to return, Portugal had become virtually a colony of Brazil. That same year Brazil had been given the status of a kingdom and Portugal, administered by General Beresford, had become effectively a protectorate of Britain. To the Portuguese this was an unsatisfactory situation. In 1820 growing unrest led to a new constitution and John VI was persuaded to return and accept its terms. Britain's authority was terminated.

Towards the end of the century republicanism came into fashion. By trying to rule as a dictator from 1906, Charles I ensured his unpopularity. In 1908 he and his eldest son were assassinated. Only 18, his younger son, Manuel II, was unequal to the task of upholding the monarchy. In general elections two years later majorities in Lisbon and Oporto voted in favour of a republic. But it was a naval and military revolt which brought the monarchy to an end on 5 October 1910.

Although it had been conducting military operations in Africa since 1914 in further-ance of its alliance with Britain, Portugal only entered the First World War on the Allied side in 1916. The years after the war were characterized by political instability, violence and high inflation. The inadequacies of the republic led to its replacement by a provi-sional military government in 1926 in a bloodless coup. Two years later Dr António de Oliveira Salazar was appointed finance minister and in 1932 he became prime minister, a post he was to hold until 1968. His regime was authoritarian and he ruled as a virtual dictator, isolating Portugal from the rest of Europe. During the Second World War Portugal remained neutral.

Portuguese attempts to slow the rush to independence in Africa alienated the army and led to a virtually bloodless military coup in 1974. While the military wrestled with the anarchic political situation, decolonization was rapid and largely completed the following year.

Area:	92 082 sq km (35 553 sq miles)
Population:	9.79 million; Portuguese
Language:	Portuguese
Religion:	Roman Catholic
Capital:	Lisbon
Administrative Districts:	2 autonomous regions (Azores, Madeira), 18 continental districts
Dates of Independence:	1640
Neighbouring Country:	Spain

English-speaking Name	Local Name	Former Names	Notes
PORTUGAL	Portugal	Lusitania	The Republic of Portugal (República Portuguesa) since Oct 1910. Previously the Kingdom of Portugal (1139–1910). The name is derived from the Latin *portus cale*, warm harbour. This referred to Oporto at the mouth of the River Douro and the fact that the port was never ice-bound. In due course the name came to represent the 'land of Portucale' between the Rivers Minho and Douro and eventually the whole country.
Alcácer do Sal		Salacia, Qasr Abi Danis	Means the 'Castle of salt', the town being named after the salt pans of the River Sado. The word *alcázar* is derived from the Arabic *al-qasr*, castle, fortress or palace.
Alentejo			An historical province, the name for which is derived from *além Tejo*, 'beyond the (River) Tagus'.
Algarve (The)		Cyneticum, al-Gharb	An historical province and at one time an independent Moorish kingdom. The name is derived from the Arab name and means 'The West', a reference to the region's location on the western edge of the Muslim lands until it fell (1249) to King Afonso III (1210–79, r.1248–79).
Aveiro		Talabriga	A district. *Briga* is a Celtic word meaning fort.
Azores	Ilhas dos Açores		Means the 'Islands of the Hawks'. The group of islands in the Atlantic Ocean form an autonomous region of Portugal. Although known to the Phoenicians (6th century BC), they were forgotten and not re-discovered until 1427 by the Portuguese. Settlement began (1439). Annexed by Spain (1580–1640).

Beja	Pax Julia	The present name is derived from the Roman name which means the 'Peace of Julius (Caesar)', that is, its pacification.
Braga	Bracara Augusta	Named after the Celtic Bracarii tribe. Capital of the Suevi (Swabian) Kingdom of Gallaecia (411). Occupied by the Moors (716–1040).
Braganza	Brigantia, Julióbriga Bragança	The present name was taken from the Celtic city which later became the seat of the ruling dynasty of Portugal (1640–1910) and the emperors of Brazil (1822–89), the House of Braganza. Catherine of Braganza married (1661) Charles II (1630–85), King of England and Scotland (1660–85). The Roman name meant the 'Fortress of Julius'.
Caldas da Rainha		Means the 'Queen's hot baths' after Queen Leonor discovered the locals bathing in the mineral springs. She sold some of her jewels and used the money to found a hospital (1484).
Castelo Branco		Means 'White castle', a Templar castle of great strategic importance near the Spanish frontier around which the town developed (from 1209).
Chaves	Aquae Flaviae	The site of a spa, it meant the 'Baths of Flavius' after the local hot springs.
Coimbra	Aeminium Conimbriga	After Roman occupation it took the name of the nearby Conimbriga when the see of a bishop there was transferred to Aeminium. Means 'Fort on the heights' from the Celtic *cun*, height, and *briga*, fort. Capital (1139–1260). Sacked by the French Marshal Masséna (1758–1817) after the Battle of Busaco (1810).

Name	Portuguese	Former names	Description
Évora		Ebora, Jabura	To the original Roman name for the city Julius Caesar added the official title 'Liberalitas Julia' (by which it was sometimes known) to denote the privileges he gave it. Renamed by the Moors (c.712–1166).
Funchal			Derived from *funcho*, fennel. Capital of the island of Madeira.
Guarda			Founded (1197) by King Sancho I (1154–1211, r.1185–1211) to guard against Moorish incursions, hence its name.
Guimarães		Vimaranes	Possibly named after Vimara Peres who founded the town (868). First capital (c.1127).
Lagos		Lacóbriga, Zawaya	Means 'lakes', a somewhat inappropriate name; however, the town lies to the west side of a bay of the River Alvor.
Lisbon	Lisboa	Olisipo, ak-Oshbuna/ Lishbuna/Ulixbone/ Olissibona	The original name may come from the Greek hero of Homer's Odyssey, Ulysses (an alternative spelling of Olisipo was Ulyssipo); more likely, it is derived from the Phoenician *alis ubbo*, attractive or good harbour, or *água boa*, good water, the city having a wonderful natural harbour. Julius Caesar added the official title Felicitas Julia to the name. The Romans were in occupation (205 BC–407 AD) until driven out by the Alani, who in their turn were replaced by the Suevi and then the Visigoths (585–715). The Moors held the city (715–1147). Capital (1256). Spain recognized Portugal's independence at the Treaty of Lisbon (1668). Devastated by an earthquake (Nov 1755). Occupied by the French (1807–09). The Portuguese king and crown prince (Luis Filipe, who had the shortest reign of any monarch, dying 20 minutes after his father) having been assassinated (1 Feb 1908) in the city, the new king, Manuel II (1889–1932, r.1908–10), abdicated and a republic was proclaimed (1910).

Madeira Islands	Arquipélago da Madeira	Insulae Purpuriae	A group of islands, an autonomous region of Portugal, in the Atlantic Ocean. Only two are inhabited: Madeira and Porto Santo. The group takes its name from the Portuguese *madeiro*, wood. since the main island, Madeira, was once covered in a variety of tropical and sub-tropical trees and shrubs, a few peculiar to the island. Known to the Phoenicians; tc the Mauritanians (c.1st century BC) they were known as the Insulae Purpuriae on account of the purple dye produced. Re-discovered by the Portuguese (1419) and given their present name. Under Spanish rule (1580–1640) and British rule (1807–14). Give their name to the rich dark brown wine and the cake.
Oporto	Porto	Portus Cale, Castrum Novum	Means 'The harbour' from *o*, the, and *porto*. Gave its name to Portugal and the rich sweet dessert wine. The Alani name meant 'New Camp'. The district is known as Porto.
Portalegre		Amoea/Ammaia	Lying on the slopes of the Serra de São Mamede near the Spanish border, the town's name means 'Happy gate'.
Queluz			Means 'What light!'
Santarém		Scalabis, Shantariya	Named after St Irene (Santa Iria), martyred (653), by a Visigoth king who had converted to Christianity. Given official title of Praesidium Julium by Julius Caesar.
Setúbal		Cetobriga	A derivation of the name of the Roman fort, but it is not in exactly the same location. The ruins at Tróia are thought to be on the site of Cetobriga which was destroyed by a tidal wave (412).
Torres Vedras		Turres Veteres	Derived from the Latin, meaning 'Old citadel'. Some 40km long, the Lines of Torres Vedras were constructed (1809–10)

		by the Duke of Wellington (1769–1852) as a defence system to protect Lisbon from the French.
Viana do Castelo	Velobriga, Diana, Viana da Foz do Lima	Its penultimate name means 'Town at the Mouth of the (River) Lima'. Viana is derived from Diana.
Vila do Conde		Means 'Town of the Counts'.
Vila Real		Means 'Royal town' after it was granted (1272) royal privileges by King Afonso III.

ROMANIA

Greek colonists settled along the Black Sea coast of modern Romania as early as the seventh century BC and by the first century BC had founded the independent state of Dacia in opposition to the encroaching Romans. Nevertheless, the Romans prevailed and in 106 AD Emperor Trajan made Dacia a province of the Roman Empire. In the face of increasing barbarian attacks Emperor Aurelian decided to abandon Dacia in 271, but not before significant Romanization had taken place: the assimilation of the Latin language and Roman civilization. Thereafter, until the 12th century, the country was submerged in consecutive waves of invading Visigoths, Huns, Avars, Slavs, Pechenegs and Bulgars. It formed part of both the First and Second Bulgarian Empires. The Magyars followed, conquering Transylvania. Surprisingly, however, they were absorbed into the local 'Roman' culture and adopted Latin as their language – unlike all their neighbours. During the 14th century Wallachia (now south-west Romania), named after the Vlachs, and Moldavia (now north-east Romania and Moldova) emerged as independent principalities.

Despite resistance to the Ottoman advance during the 15th century, both had succumbed to Turkish rule by 1512, although they were allowed to retain autonomy; Transylvania also had vassal status. Following the unsuccessful siege of Vienna in 1683, the Ottoman Empire began to lose ground and western Wallachia (Oltenia) was lost to the Habsburgs between 1718–39.

The idea of independence began to gain ground before the end of the 18th century. It received a setback, though, at the Treaty of Adrianople in 1829 when Wallachia and Moldavia became a Russian protectorate. But the Romanian *boyars* were in no mood to swap Ottoman authority for Russian. Liberal ideas from Western Europe began to spread, but they achieved little during the European revolutions of 1848, largely due to tight Russian control. Nevertheless, within a few years, the idea of an independent Romania acting as a buffer between Russia and Turkey began to appeal to the French and the British. Russian authority was brought to an end after defeat in the Crimean War in 1856. The Western Powers decreed that Wallachia and Moldavia would henceforth be two autonomous principalities with their own rulers under Turkish suzerainty. This solution was swept aside in 1859 when both principalities chose the same man, Alexandru Ion Cuza, as their prince. With Wallachia and Moldavia united, the new state took the name Romania in 1862. With Turkish approval, the authoritarian Cuza was deposed in 1866 by Charles of Hohenzollern-Sigmaringen who took the title of Prince Carol I.

Romanian independence, granted by the Treaty of San Stefano in 1878 following another Russo-Turkish war, was confirmed at the Congress of Berlin four months later. Romania also gained northern Dobrudzha, but was forced to cede southern Bessarabia (now Moldova less Transdniestria) to Russia. This rankled deeply. In 1881 Romania was recognized internationally as a kingdom. However, not all Romanians lived in Romania: about three million remained in Hungarian-ruled Transylvania, two million in Bessarabia and a substantial number in Bulgarian-controlled southern Dobrudzha. The king's aim was to unite all Romanian-inhabited lands into a Greater Romania. Having gained a little Bulgarian territory by not participating in the First Balkan War, the Romanians believed they might acquire more if they were to join in the Second Balkan War which began at the end of June 1913. They were right: at the Treaty of Bucharest in August they gained southern Dobrudzha.

Neutral for the first two years of the First World War, the Romanians joined the Entente (Britain, France and Russia) in August 1916[1], nine days after they were promised the return of predominantly Romanian-speaking Transylvania, Bukovina and the Banat. As a result of the treaties following the Bolshevik Revolution and the end of the war, Romania acquired Bessarabia from Russia, Bukovina from Austria and part of the Banat and Transylvania from Hungary. These gains more than doubled the country's size and, for the first time in their history, Romanians lived together in their own nation-state. But, by incorporating 1.7m Hungarians, at least 28 per cent of the population consisted of non-Romanian minorities. The Romanians did little to integrate them economically or culturally.

By failing to give complete support to Hitler when he began to execute his plans for territorial aggrandizement, Romania risked losing some of its territory. The Secret Protocols of the Nazi-Soviet non-aggression pact of August 1939 recognized the Soviet desire for Bessarabia and in June 1940 the Soviet Union annexed the province and northern Bukovina. Worse was to follow on 30 August with the Second Vienna Award which transferred northern Transylvania to Hungary at Hitler's insistence. Yet more territory was lost when the Romanians were forced to return southern Dobrudzha to Bulgaria in September. Such humiliations led to the abdication of King Carol II in favour of his son Michael and the appointment of General Ion Antonescu as *conducator* (leader) of Romania. Convinced that Germany would win the war, he allowed the entry of German troops into Romania which, henceforth, provided troops to fight with the Germans against the Soviet Union.

However, as Soviet armies gained the upper hand and approached Romania, the king led a coup d'état on 23 August 1944. Antonescu was overthrown and Romania changed sides. Assigned to the Soviet sphere of influence at the Yalta Conference in February 1945, Romania became a Soviet satellite after the war. A communist puppet government was installed, the king was forced to abdicate in December 1947 and Romania became a 'people's democracy'. By this time Bessarabia and northern Bukovina, recovered in the wake of German advances during the war, had been ceded to the USSR, southern Dobrudzha had been confirmed as Bulgarian and the Hungarians had returned northern Transylvania to Romania.

A member of the Warsaw Pact, Romania only had to endure the presence of Soviet troops until 1958. It was the only Pact member not to participate in the invasion of Czechoslovakia in 1968. The treatment of Hungarians in Transylvania where, in some areas, they are in the majority, remains an issue between Romania and Hungary. The Hungarians, who comprise nearly a tenth of the population of Romania, wish to preserve their nationality and keep their distance, while the Romanians fear Hungarian irredentism.

---

1. This came as a particular shock to the Kaiser because King Ferdinand of Romania was not only of German blood but also of the same royal house, the Hohenzollerns, as himself.

Area:	237 500 sq km (91 699 sq miles)
Population:	22.5 million; Romanians, Hungarians
Languages:	Romanian, Hungarian
Religions:	Romanian Orthodox, Roman Catholic, Greek Orthodox, Pentecostal
Capital:	Bucharest
Administrative Districts:	40 counties and Bucharest municipality
Date of Independence:	1878
Neighbouring Countries:	Ukraine, Moldova, Bulgaria, Yugoslavia, Hungary

English-speaking Name	Local Name	Former Names	Notes
ROMANIA		Dacia	Romania (România) since Dec 1989. Previously the Socialist Republic of Romania (Aug 1965), the People's Republic of Romania (Dec 1947) and the Kingdom of Romania (1881). The name comes from the Romans who settled (2nd century) in Dacia, now northern and central Romania, after the Emperor Trajan (53–117, r.98–117)) had finally subdued the whole country (106). Earlier the Dacians (Getae to the Greeks) had challenged Roman rule in Moesia. Their aggressive nature may have given them their name, taken from the Greek *daos*, wolf.
Alba Iulia		Apulon, Apulum, Bǎlgrad, Weissenburg, Karlsburg, Gyulafehérvár	Bǎlgrad and Weissenburg mean 'White town', perhaps because of the town's pale walls. Named Karlsburg after Emperor Charles VI (1685–1740), Holy Roman Emperor (1711–40), Archduke of Austria and King of Hungary. The present name is derived from the Latin *albus*, white, and Julius (Gyula), a mid-tenth century Hungarian prince; it is merely a translation of the Hungarian Gyulafehérvár, the 'White Town of Julius'. The Dacian Apulon gave way to the Latin Apulum, the name of a Roman camp where the 13th Roman Legion was based. Capital of Transylvania (16 & 17th centuries). Prince Michael the Brave proclaimed the city the capital of the three Romanian provinces of Moldavia, Transylvania and Wallachia, thus uniting them very briefly for the first time (1600). It was also here that the union of Transylvania and Romania was announced (Dec 1918).
Arad		Ziridava	*Ara*, of Thracian-Dacian origin, means a curve made by a river. The city is situated in a curve in the River Mures valley. Was a small Roman garrison. In Turkish hands (1552–1699). Then fell to the Austrians and later Hungarians. Ceded to Romania (1920).

Bacău		According to legend, named after Bako. He was a Hungarian outlaw from Calugara who was caught and condemned to death. At that time the condemned could escape by agreeing to become executioners themselves. The Hungarian word for executioner is *bako*. When he returned to Calugara, Bako opened a bar on a trade route and from this the 'Town of Bako' evolved.
Baia Mare	Neustadt, Nagybáriya	First, 'New town'. Now it means 'Big mine'. It was the Hungarians' chief source of gold bullion.
Banat		A name historically used to describe any frontier district under a *ban*, governor. From a Persian word meaning 'lord'. The best known is the Banat of Temesvár (now Timişoara), the historic term for what is now the western marches of Romania. The Banat in eastern Europe was divided between Hungary, Romania and Yugoslavia (1920).
Botoşani	Targul Botas	Takes its name from a Romanian nobleman, Botas or Botos, who founded and owned the town.
Braşov	Kronstadt, Brassó, Oraşul Stalin	A Saxon colony, it meant 'Crown Town' in German. Renamed 'Stalin City' (1950–60) after the Soviet leader Joseph Stalin (1879–1953). The present name derives from the Slav name Braš and the suffix *ov* to give the 'Town of Braš'.
Bucharest	Bucureşti	According to tradition, named after a shepherd named Bucur who established a forest settlement from which the city was developed (15th century). Capital of the Principality of Wallachia (1659) and of Romania (1862). The Treaty of Bucharest (Aug 1913) partitioned Macedonia at the end of the Second Balkan War and established an independent Albania.

Caransebeş	Tibiscum	The Hungarian *sebes*, referring to the river Sebeş, means fast flowing.
Cernavodă		A port on the Danube, the name literally means 'Black water'.
Cluj-Napoca	Napoca, Culus, Castrum Clus, Klausenburg, Kolozsvár	*Clus* is from the Latin *claudo*, to enclose, a reference to the surrounding hills. Historic capital of Transylvania. The Roman name Napoca was added (1974) by Nikolai Ceauşescu (1918–89) after he became president to draw attention to its Daco-Roman origin.
Constanţa	Tomis, Constantiana, Köstence/Köstendje	Named after himself by Roman Emperor Constantine the Great (274–337, r.312–37). Under Turkish rule (early 15th century-1878).
Craiova		The name is derived from the Slavonic *kraj*, edge or margin, and the suffix *ov* to give the meaning a 'border land'. This refers to the border between the Vlachs-Bulgarians and the Byzantines (11th and 12th centuries).
Hunedoara		Takes its name from the Hunaides (or Hunyadi in English), an Hungarian family, two sons from which became Kings of Hungary – Janos (1446–52) and Máryás (1458–90).
Iaşi	Jassy	The name may be derived from the Cuman *jäger*, huntsman. Capital of the Principality of Moldavia (1565–1859) and of Romania (1859–62).
Oneşti	Gheorge Gheorghiu-Dej	Originally named after Gheorge Gheorghiu-Dej (1901–65), Romanian Communist leader and President (1961–65). Built as a new town (1953) and renamed (1992).

Oradea	Grosswardein, Nagyvárad	Means 'Big town' in Hungarian. Oradea is a variation of the Hungarian *nagy*, big, and *város*, town. Ceded to Romania (1919). Destroyed by the Tatars (1241), occupied by the Turks (1660–92) and then by the Hungarians until given to Romania (1919).
Piteşti	Pirum	Derives its name from the old Roman fortification called Pirum. The suffix *esti* has the meaninf of belonging to a community with the same ancestors.
Ploeşti		According to legend, named after Father Ploaie, its founder, who had fled from Transylvania.
Reşiţa		Situated on the River Brzava, the name is derived from the Slavonic *reka*, river.
Satu Mare	Szatmárnémeti	Means 'Big village'. Ceded to Romania (1920).
Sibiu	Cibinium, Hermannsdorf, Hermannstadt, Nagyszében	Originally a Roman city and named after the River Cibin. Named Hermannsdorf, village, and then Hermannstadt, town, by the Saxons who refounded it and named it after Hermann I (1156–1217), Count Palatine of Saxony. Destroyed by the Mongols (1241). Capital of Transylvania (1703–91, 1849–65). Ceded to Romania (1918).
Sighişoara	Castrum Sex, Schässburg, Segesvár	Originally a Roman fort, it was colonized by Saxons (13th century), hence the name Castrum Sex. The present name may mean 'Sigismund's fort' after the Holy Roman Emperor Sigismund (1368–1437), King of Bohemia and Hungary, the Hungarian *vár* meaning fort. Birthplace of Vlad Ţepeş, Prince of Wallacha (c.1431–76).

Suceava		Named after the River Suceava. Derived from *soc*, common elder. Suceava can also mean a working part of a weaving loom when it is made of elder.
Timişoara	Temesiensis, Temesvár, Temeschburg	Means 'Fort by the (River) Temes'. Held by the Turks (1552–1716), then settled by Swabian Germans. Occupied by the Serbs (1919) and allocated to Romania by the Treaty of Paris (Trianon, 1920). Anti-Government demonstrations (Dec 1989) developed into a revolution that led to the fall of the Communist regime.
Tîrgu Jiu		Means 'Market town on the (River) Jiu', from *tîrg*, market or fair.
Tîrgu Mureş	Agropolis, Neumarkt, Marosvásárhely	Means 'Market town on the (River) Mureş'.
Transylvania	Transilvania, Siebenbürgen, Erdély	An historic region. Means 'Beyond the forest' from the Latin *trans*, across, and *silva*, forest; Erdély comes from the Hungarian *erdo*, forest, to mean 'The Land of Forests'. For 1,000 years a part of Hungary, although an autonomous principality of the Ottoman Empire (16th and 17th centuries), and populated by Saxons who founded seven towns (12th century), hence the German name. The seven were Bistrita (Bistritz), Braşov (Kronstadt), Cluj (Klausenburg), Medias (Mediasch), Sebeş (Muhlbach), Sibiu (Hermannstadt) and Sighişoara (Schässburg). Seized by Romania (1918) and possession confirmed by the Treaty of Trianon (1920). Roughly two-fifths were regained by Hungary (1940), but the whole region was finally ceded to Romania (1947).
Vlad Ţepeş		Named after Vlad Ţepeş (c.1431–76), otherwise known as Vlad the Impaler or Dracula. The name Dracula comes from

Dracul, devil or dragon. Vlad's father, Vlad II, Prince of Wallacia (r. 1436–47), was invested with the Order of the Dragon by King Sigismund of Hungary and Bohemia and thereafter called himself Vlad Dracul. His son inherited his father's name. His own title comes from his habit of impaling his captured enemies.

Wallachia   Tara Românesca   Muntenia, Eflak

An historic principality. Muntenia means 'Land of the mountains'. Wallachia means 'Land of the Vlachs', nomadic shepherds and fighters and were so called by the Slavs. Vlach has the meaning of 'foreigner' or 'stranger'. Tara Românesca means 'Romanian land'.

RUSSIA

Arguably, the origins of Russia may be derived from a Varangian (Viking) tribe who were known as the Rus[1], a name that came to be synonymous with the place in which they settled. They were merchant adventurers, probing eastwards into the lands of the East Slavs and down the main rivers towards the Black Sea and Constantinople in search of trade. In doing so they established permanent settlements on the way. According to the *Primary Chronicle* of the 12th century, quarrelling Slav tribes invited the Rus to come and restore order and rule over them. In 862 Rurik of Jutland, the eldest of three brothers who led the Rus, founded Novgorod. Twenty years later the Rus capital was moved to the more strategically placed Kiev on the River Dnepr and the state of Kievan Rus was established. Rather than impose a Scandinavian culture, the Rurikids absorbed that of their subjects.

During the next century efforts were made to unite the land of Rus and expand its territory. By the time Vladimir I the Great, the Rurikid leader, converted to Orthodox Christianity in 988, Kievan Rus had become a federation of principalities stretching from the Polish frontier to the River Volga and down to the Caucasus. Prone to rivalry between the princes, separatist tendencies, and raids by Turkic nomads, Kievan Rus began to fragment during the 12th century as the periphery grew stronger at the expense of the centre. By 1169 Kievan power had declined to such an extent that the court was transferred to Vladimir, 900 kilometres to the north-east.

In 1237 the Mongol Tatars[2] arrived, exploiting the divisions of the princes of Rus, destroying their towns, and within three years imposing a domination that was to last for 250 years. The princes became subject to the Khan and his Golden Horde, paying tribute and needing their approval of the right to rule. The Khan did not, however, interfere in internal or religious affairs.

Although still paying tribute to the Tatars, one of the smaller principalities, Muscovy – based on the city of Moscow – gradually began to emerge as the most assertive, largely because of its geographical position astride trade routes and comparative distance from the Tatar hordes, and also because of its loyalty to the Khan. Despite the rise of the Grand Duchy of Lithuania, which had begun to expand into the old lands of Kievan Rus in the 14th century, Muscovy grew stronger. It was at this time that the Muscovites began to call themselves Russians and their state Rossiya from the Greek name for Rus. More territory, including the Khanate of Crimea, was acquired while the Golden Horde began to split into smaller, less threatening khanates. Russians first crossed the Ural Mountains in large numbers in 1478. Eventually, in 1480, Ivan III felt confident enough to renounce Tatar supremacy and a new Russian state came into existence. During his reign (1462–1505) Muscovy grew fourfold in size and assimilated the independent principalities of Novgorod and Tver. Pskov and Ryazan were absorbed by 1521.

The first prince to assume the title of Tsar (the old Slavic word for Caesar) was Ivan IV the Terrible. Coming to the throne in 1533, he was crowned 'Tsar of Muscovy and all Russia' 14 years later. He turned on the Tatars, annexing the Khanates of Kazan and Astrakhan. The Russian arrival on the Caspian Sea heralded a struggle for supremacy with the Ottoman Empire and opened the way to Russia's expansion into Siberia. In the

West, however, he was much less successful. The Russian invasion of Livonia (q.v.Estonia and Latvia) in 1558 led to war with Sweden and Poland which had absorbed Lithuania in 1569. In 1584 Ivan IV died and both Poland and Sweden tried to take advantage of his weak successors. During the Time of Troubles (1598–1613), with the connivance of some *boyars*, the Russian nobility, the Poles occupied Moscow and the Swedes extended their hold on territory in the north-west. Following ejection of the Poles, the 16-year-old Mikhail Romanov was elected Tsar in 1613. Although a compromise choice, he was to found a new dynasty that was to rule until 1917.

Once peace had been made with both Poland and Sweden in 1618, although the western lands of Kievan Rus remained in Polish hands, Russian expansion could restart. In 1654 eastern Ukraine was annexed when the leader of a Cossack revolution against Polish rule appealed to Moscow for help denied to him by the Swedes and Turks. The Russians also struck out eastwards and in 1649 they reached the Pacific.

Peter I the Great (reigned 1682–1725) set out to modernize and westernize his country and to create a defensible state. His first objective was to secure Muscovy's southern borders against the Crimean Tatars, vassals of the Ottoman Turks. Pushing down the Dnepr and Don river valleys, he captured the Black Sea port of Azov in 1696. In 1700, allied to the Poles, he embroiled himself in the Great Northern War against Sweden which was only brought to a successful conclusion at the Treaty of Nystad in 1721: Russia acquired modern Estonia and Latvia, Ingria and Karelia. In doing so, it acquired its first large German colony. Already by 1703 Peter had gained the land on which to build a new city, St Petersburg, his 'window on the west'. On its completion in 1712 he made it his capital. Following peace in 1721, Peter assumed the title of Emperor and the Tsardom of Muscovy was retitled the Empire of All Russia. Towards the end of his reign Peter once more turned south to expand his empire along the coasts of the Caspian Sea and to pursue his ambition of establishing a trade route to India. It has been calculated by Professor Richard Pipes that Russia acquired new territory equivalent to the size of The Netherlands every year for 150 years starting in the mid-16th century.

A German princess married to a grandson of Peter the Great, Catherine II the Great succeeded her inept and unpopular husband, Peter III, six months after he came to the throne. Her reign (1762–96) was characterized by further territorial acquisitions and the arrival of more Germans to encourage agricultural development.

Five wars were fought with the Turks between 1735 and 1829 to further Russian ambitions for territory along the Black Sea coast. By 1774 most of the northern shore had been annexed and, in 1783, the Crimea; the newly acquired territory was named New Russia. In 1792 the Treaty of Jassy (now Iaşi, Romania) confirmed Russian possession of the north-west coast. Bessarabia (q.v.Romania) was acquired from the Ottoman Turks in 1812 and by 1829 Russia dominated most of the Black Sea coast from the Danube (now in Romania) to Poti (now in Georgia). In addition, large areas of Poland, Lithuania, Byelorussia (q.v.Belarus) and Ukraine came under Russian control as a result of the partitions of Poland (1772, 1793 and 1795) and gave Russia common frontiers with Austria and Prussia (q.v.Germany) for the first time. Further possessions were gained beyond the Urals and in Central Asia.

From the mid-17th century Russia had begun to acquire territory not populated by Russians or by those belonging to the Orthodox faith. The transformation of the Empire gathered new momentum during the 19th century as more peoples with different languages and cultures were absorbed. They were not all willing subjects and fierce resistance, particularly in the Caucasus, was often encountered. The Caucasus mountain barrier was crossed and in 1801 Georgia was annexed without opposition; Persia (now Iran) ceded northern Azerbaijan in 1813 and part of Armenia in 1828.

After defeat at the hands of Napoleon at Austerlitz (1805) and Friedland (1807) the Russians concluded the Treaty of Tilsit with the French in 1807. This freed the Tsar to turn on Sweden and, with his victory, Finland (then a part of Sweden) became a Grand Duchy of Russia in 1809. The Treaty of Tilsit may have bought five years of peace, but when the Russians resumed trade with England an infuriated Napoleon crossed into Russia in 1812 and advanced on Moscow. He occupied the city for six weeks before being forced to withdraw. The Russians followed him all the way to Paris.

After Napoleon's defeat at Waterloo in 1815, the Congress of Vienna established Congress Poland as a Russian dependency. The Crimean War (1853–56) originated in the power struggle between Austria, Britain, France and Russia for influence over the decaying Ottoman Empire. Diplomatic miscalculations on all sides led to the Empire declaring war on Russia after it had invaded the Ottoman territories of Moldavia and Wallachia (q.v.Romania) in 1853. Britain and France, fearful of Russian expansionism, came to Turkey's aid and sent a force to Sevastopol in the Crimea in 1854 to besiege the Russian Black Sea Fleet's home port. In a war punctuated by incompetence, the Russians suffered a bitter defeat and territorial losses in the Balkans.

After 35 years of resistance, mainly Chechen, the conquest of the North Caucasus was achieved in 1859, although spasmodic fighting continued. The Russo-Turkish war of 1877–78 yielded further gains and the Treaty of San Stefano in 1878 finally established a frontier with the Ottoman Empire and Persia that was to last until 1914. During the reigns of Nicholas I (1825–55), Alexander II (1855–81) and Alexander III (1881–94) Central Asia was annexed, the Turkmens becoming the last of the Muslims to be incorporated into the Russian Empire in 1885. In the Far East a long strip of coastline along the Sea of Japan was acquired from China between 1858 and 1860. Russian merchants had arrived in Alaska in 1784 but in 1867 it was sold to the United States for $7.2 million. The 1875 Treaty of St Petersburg established Russian control over Sakhalin Island while Japan received the Kuril Islands. During the 1880s more Germans arrived to develop Russian industry.

The main cause of the Russo-Japanese war lay in their conflicting aspirations for control of Manchuria and Korea. Russian collaboration with the Chinese, after they had been defeated by the Japanese in 1895, and dismissal of Japanese demands for recognition of each other's special interests in Korea and Manchuria provoked the Japanese into launching a surprise attack on the Russian fleet at Port Arthur in February 1904. A succession of Japanese victories on land and at sea resulted in the Treaty of Portsmouth (New Hampshire, USA) in September 1905 at which the Russians gave up the Liaotung Peninsula and Port Arthur (now in China), the southern part of Sakhalin Island and its claims on Korea.

Russia entered the First World War against Austria-Hungary and Germany in August 1914. Serious reverses humiliated the Tsar and the loss of Russian Poland was followed by two revolutions in 1917. The first, in February (Julian calendar[3], now March), led to the overthrow of the Tsar and the establishment of the Provisional Government. The second in October (now 7 November) brought a Bolshevik Government, led by Lenin, to power. Fearing serious internal resistance if the war continued, the Bolsheviks concluded a peace with Germany in 1918. At the Treaty of Brest-Litovsk in March the Russians were forced to surrender Poland, the Baltics, Finland, Ukraine, Bessarabia and the Caucasus. The treaty, however, did give an opportunity to the Bolsheviks to consolidate their revolution. They were unable to take it. The issue of peace or revolutionary war divided them and their coalition. Furthermore, they were opposed by anti-communists, the 'Whites'. In May 1918 a civil war erupted, lasting until 1920.

When the Armistice in 1919 abrogated the Treaty of Brest-Litovsk Russia moved westwards to reoccupy some of the lands lost. Conflict with Poland began in 1919 when Polish and Ukrainian nationalist forces combined to seize Ukrainian territory. The newly-formed Red Army counter-attacked and reached Warsaw by August 1920. It was driven back, but at the Treaty of Riga in 1921 most of Ukraine was confirmed as being a Soviet republic; the rest and western Byelorussia were ceded to Poland.

The formerly Russian part of Armenia and Azerbaijan were incorporated into Soviet Russia in 1920 and Georgia followed in 1921. The rest of Armenia was acknowledged as Turkish.

On 30 December 1922 the Russian, Byelorussian, Ukrainian and Transcaucasian Republics combined to form the Union of Soviet Socialist Republics. This new Soviet state was founded purely on the basis of ideology, not nationality; Russia was seen as 'first among equals', the leading nation among many within the Union[4]. In 1917 the largely Muslim populations of Central Asia had sought independence. By 1924 Soviet rule had been imposed and by the end of 1936 the establishment of five Soviet Republics was complete. The three Soviet republics of Armenia, Azerbaijan and Georgia had formed in place of the Transcaucasian Republic. A policy of divide and rule created smaller autonomous republics which often included people of different nationality, culture and language.

The Soviet-German non-aggression pact of August 1939, with its secret protocols sanctioning Soviet freedom to acquire eastern Poland, Estonia, Latvia, Finland and Bessarabia, was followed by the Soviet invasion of Poland a month later. In November 1939 the USSR invaded Finland and in June 1940 the Baltic States, including Lithuania, and Bessarabia were annexed. But in June 1941 Nazi Germany invaded the Soviet Union and reached the suburbs of Moscow and Leningrad by the end of the year. Accused of collaborating with the Germans, many nationalities, notably the Chechens, were deported to Central Asia. Their return later has been the cause of much inter-ethnic conflict. The tide turned with the Soviet victory at Stalingrad (now Volgograd) in February 1943. Soviet forces never lost another major battle as they drove the Germans back to Berlin.

By the time the war ended in May 1945 the Soviet Union had reoccupied the Baltic states, western Ukraine which was extended further westwards, western Byelorussia, and annexed the eastern part of East Prussia. In the Far East some territory lost to Japan in 1905 was recovered. In 1954 Khrushchev, the Soviet leader and a Ukrainian, transferred the Crimea to Ukraine. By the late 1960s, with the acquisition of a strategic nuclear capability, the Soviet Union had joined the United States as a super power. Russia had become more than ever the dominant core of the Union, although its separate identity as the most powerful republic was deliberately downplayed for fear of it threatening the Union's central control.

The political and social upheaval which resulted in the dissolution of the Soviet Union in December 1991 began with *perestroika* (restructuring) and *glasnost* (openness) introduced by Mikhail Gorbachev after he had come to power in 1985. The new situation encouraged Russia and other republics to press for greater autonomy. On 12 June 1991 Boris Yeltsin became Russia's first directly elected executive president. The attempted *coup* in August, when an ill-organized group of political leaders fearful of the imminent collapse of the USSR, sought to seize power and depose Gorbachev, merely hastened what it was trying to prevent. Its failure resulted in the constituent republics of the Union proclaiming their independence. As Gorbachev became a president without a country, Moscow was host to two presidents and two governments. On 8 December representatives of the Russian Federation, Belarus and Ukraine agreed to form a Commonwealth of Independent States to replace the USSR. On 21 December eight more former republics joined. Four days later Gorbachev resigned, the final act in the dissolution of the Soviet Union. It had been in existence for just 69 years and after this interregnum Russia had become once more independent. Some 25 million Russians still lived outside Russia in the former republics. Nevertheless, Yeltsin resisted the temptation of irredentism, accepting the previous inter-republican borders as the new international ones.

In 1991 Chechnya declared independence from Moscow. The break-up of Russia itself could not be countenanced – nor could the loss of land over which oil from the Caspian Sea was brought to Russia – and attempts were made to bring the recalcitrant republic to heel. When these failed Russian troops invaded in December 1994 and a brutal war ensued until August 1996. The result was defeat for Russian arms and *de facto* independence for Chechnya. Under the Khasavyurt Agreements the question of official independence for the Chechens was put in abeyance for five years. But the Russian military incursion into Chechnya in October 1999 in response to terrorist attacks in Dagestan and some Russian cities effectively annulled them. Moscow's ultimate aim is to destroy the Chechen militants' ability to launch terrorist attacks outside Chechnya and to restore its control over the republic.

1. A well-supported theory is that *Rus* is derived from the Finnish word for the Swedes, *Ruotsi*.

2. The Turkic-speaking Mongols who invaded Russia and Hungary were called Tatars by the Europeans. Their domain, the western part of the Mongol Empire, was called the Golden Horde.

3. Pope Gregory III reformed the Julian calendar in 1582. Thenceforth it was known as the Gregorian calendar. The calendar was advanced by ten days so that 4 October 1582 was followed by 15 October. The Russians, however, continued to use the Julian calendar until 31 January 1918.

4. According to official statistics, there were 140 nationalities in the USSR in 1990. Until the establishment of the Union Russian nationalism had played second fiddle to loyalty to the Tsar and the Orthodox Church. Indeed, under the law, ethnic Russians were classed, not as Russians, but as Orthodox. Even today the sense of a Russian ethnic identity is not strong.

Area:	17 075 400 sq km (6 592 800 sq miles)
Population:	146.2 million; over 100 ethnic groups; Russians, Tatars, Ukrainians, Chuvash, Bashkirs
Languages:	Russian and many others
Religions:	Russian Orthodox, Muslim, Protestant, Jewish
Capital:	Moscow
Administrative Districts:	21 republics, 49 provinces, one autonomous region, 6 territories, 10 autonomous districts, 2 federal cities
Date of Independence:	12 June 1991
Neighbouring Countries:	North Korea, China, Mongolia, Kazakhstan, Azerbaijan, Georgia, Ukraine, Belarus, Lithuania, Poland, Latvia, Estonia, Finland, Norway

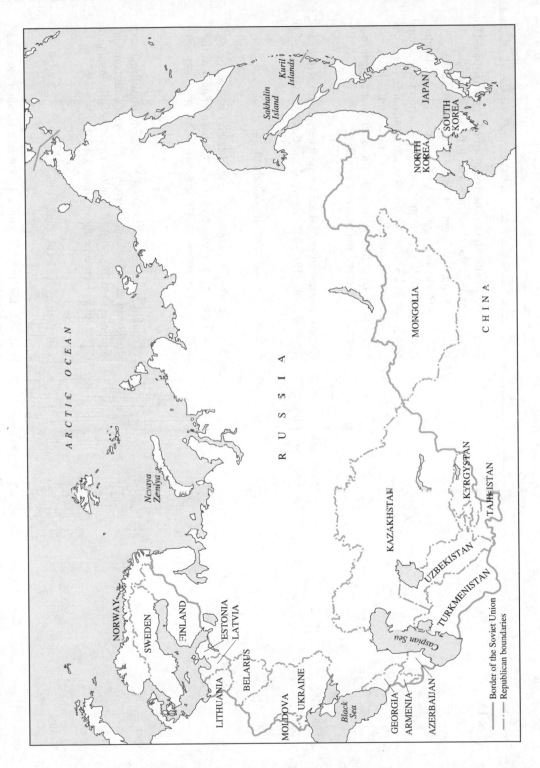

The Disintegration of the Soviet Union 1991

ARCTIC OCEAN

Kuril Islands

Sakhalin Island

JAPAN

SOUTH KOREA

NORTH KOREA

MONGOLIA

C H I N A

R U S S I A

Novaya Zemlya

KAZAKHSTAN

KYRGYZSTAN

TAJIKISTAN

UZBEKISTAN

TURKMENISTAN

Caspian Sea

NORWAY

SWEDEN

FINLAND

ESTONIA

LATVIA

LITHUANIA

BELARUS

MOLDOVA

UKRAINE

Black Sea

GEORGIA

ARMENIA

AZERBAIJAN

Border of the Soviet Union
Republican boundaries

English-speaking Name	Local Name	Former Names	Notes
RUSSIA	Rossiya	Rus	The Russian Federation (Rossiyskaya Federatsiya) since Dec 1991. Previously the Russian Soviet Federative Socialist Republic within the Union of Soviet Socialist Republics (1922). Rus was the name of a Viking tribe which migrated from Scandinavia (9th century) and established itself around Kiev (now in Ukraine) before spreading outwards to encompass what is now western Russia, Ukraine and Belarus. *Soviet* means 'council' to indicate the basic unit of government at all levels from national to village.
Abakan		Khakassk, Ust-Abakanskoye	Named after the River Abakan, *ust* meaning 'Mouth of the river'. Original name (1925–31). Changed and then shortened (1931).
Adygeya			A republic named after the Adyghian.
Akademgorodok			Means 'Academic campus'. Founded (1957) as a purpose-built town, the headquarters of the Siberian branch of the Academy of Sciences, to which some 40,000 scientists were sent.
Altai			A republic named after the Altai Mountains from the Turkic-Mongolian *altan*, golden. Formerly the Oyrot Autonomous Region named after the Mongoloid Oyrot people. Renamed (1948) and given republican status (1991).
Apatity			Named after the local apatite ore.
Archangel	Arkhangelsk	Novokholmogory, Arkhangelsky	Founded (1584) and re-named (1613) as a fortified monastery after the Archangel Michael. Allied troops landed here (Aug 1918) to intervene, unsuccessfully, in the Civil War (1918–1920).

Arsenyev	Semenovka	Named (1952) after Vladimir Arsenyev (1872–1930), explorer in the Far East, ethnographer and author.
Asbest	Kudelka	Means 'asbestos' which was first discovered here (1720).
Astrakhan	Khadzhi-Tarkhan,	The first name is said to come from the Turkish *hadji*, a Muslim who had made the pilgrimage to Mecca, and *tarhan*, untaxed, a reference to the city's exemption from taxes. Nearby Sarai was the capital of the Mongol Golden Horde (1242). Destroyed (1395) by Tamerlane (1336–1405) and captured (1556) by Tsar Ivan IV the Terrible (1530–84, r.1547–84). Was an important trading post between Europe and Central Asia and gave its name to the fur first brought to Russia by Astrakhan traders.
Azov	Tana, Azak	Derived from the Turkish *azak*, low. The Sea of Azov is the world's shallowest and gives its name to the town.
Bagrationovsk	Eylau	Named after Pyotr Bagration (1765–1812), a general who distinguished himself during the Napoleonic Wars and fought at the Battle of Eylau (Feb 1807) between the Russians and Prussians, and the French. It ended in stalemate, but the Russians and Prussians withdrew from the battlefield.
Baltiysk	Pillau	Pillau (1686–1945) was originally in East Prussia.
Barnaul		Possibly derived from *boruan*, wolves, and *ul*, river. The city lies on the River Barnaul.
Bashkortostan	Bashkiriya	A republic with a name meaning the 'Land of the Bashkirs', a Turkic people who settled here (13–15th centuries).

Bering Island		Named after the Danish-born ship's captain Vitus Bering (1681–1741) who died on the island. Tsar Peter I the Great (1672–1725, r.1682–1725) appointed him (1724) as leader of an expedition to see if Asia and North America were joined by land. The Bering Sea and the Bering Strait are also named after him.
Bilibino		Named (1958) after Yuri Bilibin (1901–52), a geologist who first discovered gold in this northerly region of Siberia.
Birobidzhan	Tikhonkaya	Renamed after the two rivers Bira and Bidzhan. In a remote part of the Russian Far East, the Birobidzhan autonomous province, also called Yevreyskaya (in Russian, *Yevrey*, a Jew), was planned (1934) as a homeland for the Jews. The Jewish population now is very low.
Blagoveshchensk	Ust-Zeysk	First settled (1644) at the confluence of the Amur and Zeya rivers, the original name meaning the 'Mouth of the River Zeya'. Ceded to China (1689–1856). Then built up as a military post. Two years later the Church of the Annunciation was erected and the town was renamed after the Church: *Blagoveshcheniye* means 'The Annunciation', literally 'Good news'.
Borisoglebsk		Named after two half-brothers, Boris and Gleb, who were murdered (1015) by their elder brother without trying to defend themselves. They were later revered as saints.
Budennovsk	Svyatoi Krest, Prikumsk	Originally meant 'Holy Cross'. Renamed Prikumsk (1924–35, 1957–73), meaning 'By the River Kuma', Budennovsk (1935–57 and since 1973) after Marshal Semyon Budenny (1883–1973), commander of the Red Army's 1st Cavalry Army during the Civil War and first deputy commissar of defence (1940).

Buryatia	Buryatiya	A republic named after the Buryats, a people of Mongol descent.
Chapayevsk	Ivashchenko*vo*	Renamed (1929) after Vasiliy Chapayev, a hero of the Civil War.
Chaplygin	Ranenburg	Renamed (1948) after Sergey Chaplygin (1869–1942), aerodynamicist.
Chechnya	Ichkeria	A republic, the Chechen name for which is the Chechen Republic of Ichkeria. *Chechen* means 'people' in Chechen and the Chechen for Ichkeria is Noxçiyçö. Derived from the name of a village on the River Argun where the Russians and Chechens fought their first battle (1732). United with Ingushetia (1934–91) as a single autonomous republic which ceased to exist (1944–53) when Stalin deported the Chechens to Central Asia having accused them of collaborating with the German army.
Chelyabinsk	Chelyabi Varagay	Founded as a fortress (1736) with a Turkic name meaning 'Ancient forest' from *chelyabi*, ancestral, and *varagay*, pinewood. Renamed (1786). Known informally during the Second World War as 'Tankograd' where the T-34 was built.
Chernyakhovsk	Insterburg	Renamed after Army General Ivan Chernyakhovsky (1904–45), commander of the 3rd Byelorussian Front who was killed during the fighting for Königsberg (now Kaliningrad).
Chita		Lies at the confluence of the Ingoda and Chita rivers, whence its name. In Evenki *Chita* itself means 'clay'.
Chuvashia	Chuvashiya	A republic named after the Chuvash, a Turkic people.

Dagestan		A republic whose name means 'Mountain country' from the Turkish *dag*, mountain, and *stan*, country.
Dimitrovgrad	Melekess	Renamed (1972) after Georgy Dimitrov (1882–1949), Bulgarian secretary-general of the Comintern (1935–43) and Bulgarian leader (1945–49).
Dzerzhinsk	Chernorechye, Rastyapino	Originally named 'Black river. Renamed Rastyapino (1919–29) and then after Felix Dzerzhinskiy (1877–1926), a fanatical Polish communist who founded (1917) the Soviet secret police, the Cheka (The All-Russian Extraordinary Commission for Combating Counter-revolution and Sabotage).
Elista	Stepnoy	The former name (1944–57) meant 'steppe'. Founded (1865) and took its name from the Kalmyk *elstia*, sandy.
Franz Josef Land	Zemlya Frantsa-Iosifa	An archipelago of nearly 200 islands in the Barents Sea discovered (1871) by an Austro-Hungarian expedition and named after the Emperor (1830–1916). Annexed by the Soviet Union (1926).
Furmanov	Sereda	The original name is taken from *sreda*, Wednesday, that is, market day. Renamed (1941) after Dmitry Furmanov (1891–1926), Civil War hero and novelist.
Gagarin	Gzhatsk	Renamed after, and birthplace of, Yuriy Gagarin (1934–68), cosmonaut and the first man to orbit the earth (Apr 1961). Originally named after the River Gzhat.
Gatchina	Khotchino, Trotsk Krasnogvardeysk	Original name (1499–1923), then Trotsk (1923–29), then Krasnogvardeysk, meaning 'Red Guards' (1929–44), to commemorate their role in the events in Petrograd (Nov 1917). In Swedish possession until returned to Russia

Name	Former/Other names	Description
Gelendzhik		(1721). *Gat* means a 'brushwood road', laid over marshes. Catherine II the Great (1729–96, Empress 1762–96) built a summer palace here (1772) for her favourite, Count Orlov.
	Toricos, Eptala	Means 'White bride' in Turkish, possibly because it was once a centre of the slave trade. Eptala was the main port for the export of local girls to the Turkish harems.
Golbshtadt	Nekrasovo	Renamed Nekrasovo (1949–91), meaning 'ugly or unsightly', when Stalin abolished the German National District in the Altai Republic in an act of revenge. The original name was only restored on the eve of the 50th anniverary of the Nazi invasion (June 1991).
Gorno-Altaysk	Ulala, Oyrot-Tura	Original Turkish name, meaning 'Great One', until 1932, then Oyrot-Tura after the Mongoloid Oyrot people (1932–48). Named after the Altai Mountains: *altan*, golden, and *gora*, mountain.
Grozny	Groznaya, Sölz-Gala	Means 'awesome' or 'terrible' or 'menacing' and was so named (1818) as a Russian fortress to inculcate fear into the Chechens and indicate Russian power in the Caucasus. Changed to Grozny (1869). Unofficially renamed (1997) by the Chechens as Dzhokar-Gala, the City of Dzhokar, after Dzhokar Dudayev (1944–96), the first self-proclaimed president of the 'independent' Republic. Badly damaged during the Russo-Chechen war (1994–96) and destroyed during the second war (1999–2000). Chechen capital but this may be moved to Gudermes.
Gusev	Gumbinnen	Renamed (1946) after Guards captain S.I.Gusev who was killed near here (Jan 1945).
Gus-Khrustalny		On the River Gus, this city is famous for its glass industry which gives it its name: *Khrustal*, cut-glass or crystal.

Gvardeysk	Tapiau	Renamed (1946) 'Guards', probably after 11 Guards Army which took part in the fighting in East Prussia (1945) in the assault group of 3rd Byelorussian Front. This army comprised various guards corps and divisions. 11 Guards Army remained in the Baltic republics until withdrawing after the disintegration of the Soviet Union. Any Red Army unit could be designated 'guards' for some act of valour or particularly meritorious action. The unit would then incorporate *gvardeiskii* in its title.
Ingushetia		A republic named after the Ingush, a Muslim mountain people in the North Caucasus. Split from Chechnya (1994). On their return from exile in Central Asia (1958) the Ingush found that some of their lands had been given to the North Ossetians. Conflict ensued (1992).
Ivanovo	Ivanovo-Voznesensk	Possibly named after Ivan IV the Terrible. Dropped Voznesensk (1932), probably because of Stalin's crusade against religion, *Vozneseniye* meaning 'The Ascension' to which the original village church was dedicated.
Izhevsk	Ustinov	Founded (1760) and named after the River Izh. Renamed (1984–90) after Marshal Dmitry Ustinov (1908–84), minister of defence (1976–84).
Kabardino-Balkaria	Kabardin	A republic named after the Kabardins and the Balkars. The Kabardin and Balkaria provinces amalgamated (1922) and became a republic (1936). After the deportation of the Balkars (1944) the republic was renamed Kabardin and the Balkar region was attached to the Georgian Republic. The republic's status was restored when the Balkars returned (1957).
Kaliningrad	Králover, Königsberg	Originally founded and named by Přemysl Otakar II (1230–78), the King of Bohemia, using the Czech word for king, *král*.

Changed to 'King's mountain' after he had led a crusade (1255) against the pagan Prussians. The town was founded to consolidate the power of the Teutonic Knights in newly acquired territories. Renamed (1946) after Mikhail Kalinin (1875–1946), Soviet President (1919–46). Capital of East Prussia (15th century-1945). The northern half of East Prussia was ceded to the USSR by the Potsdam Agreement (Aug 1945) after being almost completely destroyed by the Soviet Army (1945); the southern half was annexed by Poland. The province, *oblast*, is geographically separated from Russia by Belarusian, Latvian and Lithuanian territory.

Kalmykia		A republic named after the Kalmuck, a Mongol tribe originally from western China.
Karachai-Cherkessia		A republic named after the Karachai and Circassian peoples.
Karelia	Kareliya	A republic named after the Karelians. Eastern Karelia was acquired by Russia (1323) and Western Karelia (1721–1920). Western Karelia was re-annexed (1940) after the Russo-Finnish war ended.
Kazan		Possibly taken from the Tatar *kazan*, cauldron, referring to the strong currents in the River Kazanka. Founded (late 13th century) by the Tatars. Capital of the Tatar khanate (1445).
Kemerovo	Shcheglovsk	Deriving its name from the Turkish *kemer*, fort; it lies on the River Tom. Former name (1918–32).
Khabarovsk	Khabarovka	Founded and named (1858) after Yerofey Khabarov (c.1610–?), an explorer who completed the conquest of Siberia. Renamed (1893).

Khakasiya		A republic named after the Khakass, a Siberian Turkic people.
Kingisepp	Yam, Yamburg	Yam (1384–1707) and then Yamburg until renamed (1922) after Viktor Kingisepp (1888–1922), Estonian Bolshevik leader.
Kirov	Khlynov, Vyatka	Original name (1181–1780) after the River Khlynovitse, then Vyatka (1780–1934) after the River Vyatka until renamed after the murder (1934), probably on Stalin's orders, of Sergey Kirov (1886–1934), communist leader of Leningrad. Kirov's murder started something of a cult and many towns throughout the Soviet Union were named after him in a seeming act of penance.
Kirovgrad		Renamed (1936) after Sergey Kirov.
Kirovsk		Near the Khibiny Mountains. Renamed (1934) after Sergey Kirov.
Kislovodsk		Means 'Sour waters' from *kisliy*, sour, and *vodi*, spa or watering place.
Komi		A republic named after the Komi, a Finno-Ugric people.
Komsomolsk-na-Amure		Built (1932) by, and named after, members of the Komsomol, the Young Communist League. Lies on the River Amur, whose inclusion in the city's name distinguishes it from other Komsomolsks.
Korolev		In Moscow region. Renamed (1996) after Sergey Korolev (1906–66), missile, rocket and spacecraft designer.
Krasnodar	Yekaterinodar	Meaning 'Catherine's gift' from the Russian *dar*, gift, it was originally named (1792–1920) after Catherine II the Great.

Place name	Former/alternative names	Notes
Krasnoyarsk	Krasnyy Yar	When captured by the Red Army (1920) it was changed to 'Gift of the Reds' from *krasny*, red. Founded (1628) as a fort on the left bank of the River Yenisey, the name means 'Red bank' because of the reddish soil on the river banks.
Kronshtadt	Kronslot	Captured (1703) by Peter I the Great from the Swedes who called it 'Crown castle'. Peter built a fortress on the island to protect St Petersburg and renamed it 'Crown city'. The Kronshtadt Rebellion occurred (Mar 1921) when sailors, having played a major part in the October Revolution (1917) but now dissatisfied with the Bolsheviks, mutinied against the Soviet government.
Kropotkin	Romanovsky Khutor	Originally meant 'Romanov's village', then renamed (1921) after Peter Kropotkin (1842–1921), leading anarchist and revolutionary.
Kurgan	Tsaryovo Gorodishche, Tsaryov Kurgan	Founded (1553) on a large ancient tumulus, its original name meant 'Royal fort'. When it became a town (1782) its new name meant 'Royal burial mound'.
Kuril Islands	Kurilskiye Ostrova / Chupka, Rakkoshima, Chishima	Possibly derived from the Russian *kurit'*, to smoke, a reference to the active volcanoes. But more likely the name comes from the Ainu word *kur*, man or people, the Ainu being the first known inhabitants of the Kurils. The Ainu *Chupka* means 'When the sun rises'. The Japanese *Chishima* means 1000 islands although there are in fact only 56 and a few uninhabited rocks. The Treaty of Simoda (1855) placed the border between Russia and Japan between the islands of Iturup (called Etorofu by the Japanese) and Urup ('Salmon' in Ainu). The Japanese islands were first called Rakkoshima and then Chishima. The Treaty of St Petersburg (1875)

resulted in the Japanese giving up their claim to the southern half of Sakhalin Island (not part of the Kurils) and the Russians ceding the rest of the Kurils chain to Japan. It was from Iturup Island that Japan launched the attack on Pearl Harbour, Hawaii (Dec 1941). In accordance with the Yalta Agreement (Feb 1945), the entire chain was ceded to the USSR (1945). A joint Soviet/Japanese declaration was signed (1956) in which the USSR agreed to hand over Shikotan ('Best place' in Ainu) and the Habomai islands to Japan. The Soviet government then refused to implement the declaration. The Japanese are keen to have the south-ernmost islands of Iturup, Kunashir ('Black island' in Ainu), Shikotan and the Habomai group returned – that is, a return to the 1855 position. This dispute has so far delayed the signing of a peace treaty between Russia and Japan which would formally conclude the hostilities between them which began when the Soviet Union declared war on Japan six days before the Second World War ended (14 Aug 1945).

Kursk Named after the River Kur. Destroyed by the Tatars (1240). Site of the Battle of Kursk (Jul-Aug 1943), the largest set-piece battle in history with over 2 million men, 6,100 tanks, 29,000 guns and 4,900 aircraft involved.

Kyzyl Belotsarsk, Khem-Beldyr Originally (1914–18) meant 'White Tsar town'; second name (1918–26); present Turkic name means 'Red'.

Leninsk-Kuznetskiy Kolchugino Named after Lenin (1870–1924), founder of the Russian Communist Party and first head of the Soviet Union, and for the blacksmiths (in Russian, *kuznets*) who worked in support of the coal-mining here. Renamed (1925). Lenin's original name was Vladimir Ulyanov which he changed (1901). He took his new name from the River Lena in Siberia where he was exiled (1897–1900).

Lipetsk	Name derived from the Russian *lipa*, lime tree.
Liski	Original name (until 1943) means 'Freedom' while the second name (1965–90) was in honour of Gheorghe-Dej (1901–65), Romanian Prime Minister (1952–55) and President (1961–65).
Svoboda, Georgiu-Dezh	
Magas	Means 'City of the Sun' in Old Nakh. A city of this name was destroyed by the Tatars (1239). This city is still being built, but it replaced Nazran as the capital of Ingushetia (11 June 1999).
Magnitogorsk	The city grew up (1929–31) round the Metallurgical Combine which was constructed at the same time. It was named after the magnetic iron ore which is mined from Magnitnaya Gora, Magnetic Mountain.
Makhachkala	Built (1844) by Peter I the Great, it was at first called 'Peter's port'. Renamed (1922) after Makhach (real name Muhammed-Ali Dakhadayev) (1882–1918), a Dagestani revolutionary, with the Armenian *kala*, fortress, added.
Petrovsk Port	
Mari El	A republic named after the Mari, a Finno-Ugric people.
Marks	Originally named after Catherine II the Great. Renamed (1920) Markstadt after Karl Marx (1818–1883), German political theorist, economist and author of *Das Kapital*. Shortened to Marks (1941).
Yekaterinstadt	
Maykop	Name derived from the Adygey *myequape*, valley of apple trees.

Michurinsk	Kozlov	Original name (1636–1932). Renamed after Ivan Michurin (1855–1935), horticulturist.
Mineralnyye Vody		A spa, its name means 'Mineral waters'.
Mordovia		A republic also known as Mordvinia and named after the Mordvin, a Finno-Ugric people.
Moscow	Moskva	Possibly named after the Moscow River, itself possibly taken from the Slav word *moskva*, marshy. Founded (1147) and became the chief city of the Principality of Muscovy (13th century). Sacked by the Mongols of the Golden Horde (1237–38, 1382). Burnt by the Crimean Tatars (1571). Occupied by the Poles (1605–6, 1608–12) and the French (1812). Besieged by the Germans (1941). Capital of Russia (c.1478–1712, 1991) and of the Soviet Union (1918–91). Now has federal status. The Treaty of Moscow (1921) fixed Turkey's north-eastern border. A proposal (1937) to rename the city Stalinodar as a 'gesture of gratitude from the people' was rejected by Stalin.
Murmansk	Romanov-na-Murmane	The name may be derived from the Saami word *murman*, the edge of the earth. Founded (1915) as a port in the Arctic and originally named (1915–1917) after the Romanov dynasty.
Naberezhnyye Chelny	Chelny, Brezhnev	Chelny (until 1930) from *chelno*, boat, then (1982–88) after Brezhnev (1906–82), Soviet leader and president (1977–82). Situated on the left bank of the River Kama, *naberezhnaya* means 'embankment' or 'quay'.
Nakhodka	Amerikanka	Means 'Discovery' or 'Godsend', a sheltered bay having been found by chance (1850s) in the Sea of Japan.

Nizhniy Novgorod	Gorkiy	Renamed (1932–90) after Maxim Gorkiy (1868–1936), novelist; his real name was Alexei Peshkov whose birthplace this was. *Nizhniy* means 'lower', that is, south, to distinguish it from the more important Novgorod the Great, north-west of Moscow. *Novgorod* means 'New town'. *Gorod* was originally a word for fortification.
North Ossetia	Severnaya Osetiya	A republic named after the Ossetians whose ancestors were the Alani. Under a new constitution (1994) the name North Ossetian Alania was adopted but it is not in common usage. South Ossetia is in Georgia. Hopes were raised (1990s) that the two parts of Ossetia might unite under the name Alania.
Novaya Zemlya		A group of islands in the Arctic Ocean. The name means 'New Land'.
Novgorod	Holmgard	The Varangian name Holmgard means 'Island town'. One of Russia's oldest cities, it became known as Novgorod Velikiy (Novgorod the Great) and retained this title (until the 18th century). For long a rival to Moscow, it was finally forced to recognize Moscow's supremacy (1478) after defeat (1456) by Vasiliy II (1415–62), Grand Prince of Moscow. Renamed Novgorod the Great (1 January 2000).
Novocherkassk	Starocherkassk	Means 'New Place of the Cherkess (Circassians)'. Because of flooding, the original town, Cherkassk, was moved (1805) to its present location. When the building of New Cherkassk was complete Cherkassk was renamed Starocherkassk, 'Old Place of the Cherkess'.
Novokuznetsk	Kuznetsk, Stalinsk	Originally named 'Ironworkers' village', then renamed (1932–61) after Joseph Stalin (1879–1953), Soviet leader (1924–53); he was a Georgian and his real name was Dzhugashvili. As a professional revolutionary he chose the

		name Stalin, 'Man of Steel' (1912). Development of a new town on the opposite bank of the River Tom occasioned the new name of New Kuznetsk.
Novomoskovsk	Bobriki, Stalinogorsk	Original name (1930–34), renamed (1934–61) after Stalin as 'Stalin's city', *gorsk* being a word associated with *gorod*, city.
Novorossiysk		Founded as a fortress (1838). It and the surrounding territory were named 'New Russia' by Catherine II the Great. Passed from Turkey to Russia (end 18th century).
Novosibirsk	Novaya Derevna, Gusevka, Aleksandrovsk, Novonikolayevsk	Developed from the village of Krivoshchekovo when this location was selected (1893) as the crossing point over the River Ob for the Trans-Siberian railway. Named Aleksandrovsky (1894) after Tsar Alexander III (1845–94, r.1881–94). Renamed (1895–1925) Novonikolayevsky, 'New Nicholas', in honour of the accession of Tsar Nicholas II (1868–1918, r.1895–1918). Renamed 'New Siberia' (1925).
Omsk		Lying at the confluence of the Rivers Om and Irtysh, the name is taken from the Tatar word *om*, quiet or slow.
Oranienbaum	Lomonosov	Founded (1710) as a summer retreat for the Russian royal family with its present name which may have been taken from the Dutch *Oranje boom*, orange tree. For four months Peter I the Great worked as a ship's carpenter with the Dutch East India Company. Renamed (1948–96) after Mikhail Lomonosov (1711–65), scientist and grammarian.
Orenburg	Chkalov	Means 'Fort on the River Or'. Founded (1735) as a fortress but moved 240 kilometres west of the river (1743) although retaining the same name. The original site is now called

	Orsk. Renamed (1938–57) after Valeriy Chkalov (1904–1938), commander of the non-stop 63 hour flight in an ANT-25 from Moscow to Portland, Oregon, USA via the North Pole (1937).
Oryol	Founded (1564) as a fortress where the Orlik joins the River Oka, hence its name. The word *oryol* actually means 'eagle'.
Pechenga Petsamo	May be derived from a Finnish word meaning 'pinetree'. Named Petsamo when ceded to Finland (1920–40).
Pechory Petseri	Possibly derived from the Russian *peshchera*, cave. According to legend, a hunter suddenly heard singing from somewhere underground. On searching, he found a group of holy men chanting in a cave. Petseri (1920–40) when in Estonia.
Pereslavl- Zalesskiy Kleshchin	Named (15th century) after Pereslavl near Kiev. Pereslavl is derived from two old Russian words *pereyat*, to achieve, and *slava*, glory. The suffix Zalesskiy was added later to describe its remote position, *za lessami*, beyond the virtually impenetrable forests to which the local Slavs had been driven by hostile nomads.
Perm Yegoshikha, Peryamaa Molotov	First name (1723–80), then Peryamaa (1780–1940) from the Finnish *perya*, rear, and *maa*, land. Named (1940–57) after Vyacheslav Molotov (1890–1986), People's Commissar for Foreign Affairs (1939–49, 1953–56); his real name was Skryabin. Professional revolutionaries commonly adopted aliases; Molotov meant 'the Hammer', a name he chose to further his reputation as a tough negotiator. His alias inspired the 'Molotov cocktail', a crude home-made device consisting of a bottle filled with inflammable liquid with a rag pushed into the neck that was lit with a match. It was

invented (1940) by the Finns for use against Soviet tanks during the Winter War (Nov 1939-Mar 1940) and put into production in the Soviet Union by Molotov.

Petropavlovsk-Kamchatskiy	Named after the two ships, the St Peter and the St Paul, of Vitus Bering during his second Kamchatka expedition. Kamchatskiy was added (1924) to distinguish the town from the other Petropavlovsks in the Soviet Union. These are now outside Russia and so the Kamchatskiy has officially been dropped, although it is commonly still included in the name.
Petrozavodsk	Means 'Peter's factory', a reference to the iron foundry and armaments plant founded (1703) by Peter I the Great from *zavod*, factory or works.
Pokrovsk, Engels	Originally named after its church, the Protection of the Virgin, the ecclesiastical meaning of *Pokrov* being 'The Protection', literally 'The Covering', or the Festival of the Protection of the Virgin. Renamed (1931) after Friedrich Engels (1820–95), German socialist philosopher and close collaborator with Karl Marx. Reverted to original name (1992).
Pravdinsk	Site of the Battle of Friedland (Jun 1807) which ended in French victory over the Russians and Prussians. Renamed (1946) after the Soviet Communist Party newspaper, Pravda. *Pravda* means 'truth'.
Pugachev	Named after Yemelyan Pugachev (c.1744–1775), a Don Cossack who, claiming to be Peter III, the murdered husband of Catherine II the Great, led a rebellion to end serfdom in the south (1773–74) before being captured and executed. First name (1760–1835), then (1835–1918) after Tsar Nicholas I (1796–1855, r.1825–55).

Pushkin	Tsarskoye Selo, Detskoye Selo	Residence of the Imperial family. Meant 'Tsar's village' (1728–1918). Renamed Detskoye Selo (1918–37), 'Children's village', after the Imperial residence had been converted into a holiday camp for workers' families after the Tsar's abdication. It then became Pushkin to mark the centenary of Alexander Pushkin's death (1799–1837), poet, who studied here (1811–17).
Pyatigorsk		Means 'Five peaks' and is a Russification of the Turkish Beshtau, the name of the highest mountain in the area.
Rostov	Rostov Velikiy, Rostov-Yaroslavski	Used to mean 'Rostov near Yaroslavl' and is named after Prince Rosislav. Its first name meant Rostov the Great but it lost this title (17th century) when its power declined with the transfer of its religious power to Yaroslavl.
Rostov on Don	Rostov-na-Donu	Named after the fortress/church dedicated to St Dmitry of Rostov Velikiy near Yaroslavl. Renamed (1806) Rostov-na-Donu (Rostov-on-the River Don) to distinguish it from the older Rostov.
Rybinsk	Shcherbakov, Andropov	Renamed (1946–57) after Alexander Shcherbakov (1901–45), Chief of the Main Political Administration of the Red Army, then Rybinsk again (1957–84) and (1984–88) after Yuriy Andropov (1914–84), Soviet leader (1982–84).
St Petersburg	Sankt Peterburg, Pieterburkh, Petrograd, Leningrad	Named (1703) by Peter I the Great after his patron saint, St Peter. Renamed to the Russian-style Petrograd (1914–24), then (1924–91) after Vladimir Lenin. Cradle of the revolutions (1905, 1917). Besieged by the Germans (8 Sep 1941-27 Jan 1944). Capital of the Russian Empire (1712–1917) and of the Soviet state (Nov 1917-Mar 1918).

Sakhalin	Karafuto	An island. The name is derived from the Manchu *Sahalin Ula*, black river. The southern half was ceded to Japan under the Treaty of Portsmouth (1905). It returned to Soviet possession at the end of the Second World War (1945).
Samara	Kuybyshev	Lies at the confluence of the Volga and Samara rivers. Renamed (1935–91) after Valerian Kuybyshev (1888–1935), a Red Army commander in the Civil War who became a Bolshevik agitator while working in the city. Seat of the Soviet government (1941–43).
Saratov	Sarytau	Built (1590) as one of a line of fortresses on high ground to protect the trade route along the River Volga, the name comes from the Tatar *sary*, yellow, and *tau*, mountain.
Sarov	Arzamas-16, Kremlev	Built (1946–54) as a 'closed city' whose existence was a state secret until openly mentioned (1980s) by Andrey Sakharov, nuclear physicist. With a nuclear research centre, it manufactured weapons of mass destruction. The suffix 16 is its postcode and distinguishes it from the nearby town of Arzamas. At various times it was also known as the Volga Office of the Chief Municipal Construction Board, Design Office 11, Facility 550, Centre 300 and Kremlev. Renamed (1996).
Sergiyev Posad	Sergiyev, Zagorsk	Means 'Sergius' quarter' from the Russian *posad*, trading quarter or suburb. Originally named after St Sergius of Radonezh (1314–92). Renamed just Sergiyev (1918–30) and then (1930–91) after Vladimir Zagorsky, secretary of the Moscow Communist Party, killed by an anarchist (1919). Located some 65 km north-east of Moscow, Zagorsk could also mean 'Beyond the city', that is, beyond Moscow, from *za*, beyond, and *gorsk*, city.

Serov	Nadezhdirsk	Renamed (1939) after Anatoliy Serov (1910–39), a military test pilot who also participated in the Spanish Civil War (1936–39).
Severodvinsk	Molotovsk	Situated at the mouth of the Northern Dvina river, it means 'Northern Dvina'. Renamed (1938–57) after Molotov.
Shakhty	Aleksandrovsk-Grushevskiy	Named (until 1920) after Tsar Alexander and lying on the River Grushevka, the present name means 'pits' in recognition of the coal mines in the area.
Shamilkala	Svetogorsk	Means the 'City of Shamil' and named after Shamil (c.1797–1871), a Muslim Imam who led the Dagestani and Chechen resistance to the Russian conquest of the Caucasus (1835–59).
Sharypovo	Chernenko	Briefly renamed (1985–88) after Konstantin Chernenko (1911–85), Soviet leader (1984–85).
Shlisselburg	Oreshek. Nöteborg Schlüsselburg, Petrokrepost	Both the Russian and Swedish names mean 'nut' (in Russian, *orech*), the original fortress being so named (1323) after the hazelnut-shape of the island, also called Oreshek, on which it was built. Captured by the Swedes who built the fortress of Nöteborg (1611–1702). Recaptured by Peter I the Great and renamed Schlüsselburg (1702–1944, 1991), 'Key Fort' (in German, *schlüssel*, key) since it protected St Petersburg. Following the German siege of Leningrad (Sep 1941-Jan 1944), the German name was changed to Petrokrepost meaning 'Peter's Fortress' (1944–91). Also formerly known by the Finns as Pähkinäsaari: the Treaty of Pähkinäsaari (1323) demarcated the Russian and Swedish spheres of influence through the middle of Karelia and thus established the Finnish border with Russia. Situated on the River Neva where it flows out of Lake Ladoga, it was the key to the river and thus to the Baltic Sea 70 km away.

Siberia	Sibir	A region meaning 'Sleeping land' in Tatar, first applied to a Tatar khanate (13th century). This was overthrown by the Cossacks (1582) who then began the conquest of the region between the Ural Mountains and the Pacific Ocean and gave it the name Siberia.
Simbirsk	Sinbirsk, Ulyanovsk	Founded (1648) and renamed Simbirsk (1780–1924), meaning 'Mountain of winds'. Renamed again (1924–91) after Vladimir Ulyanov (Lenin), whose birthplace this was.
Smolensk		The name may be derived from the Russian *smoleniye*, tarring, used for the local boats. Smolensk lies on the River Dnepr, the direct trade route between the Baltic Sea and the Byzantine Empire. After being sacked by the Tatars (1237), it passed to Lithuania. Occupied by the Poles (1611) and finally taken by the Russians (1654). On the strategic route to Moscow, it was burned by the French (1812) and fought over by the Germans (1941, 1943).
Solnechnogorsk		Means 'Sunny city' from *solnechno*, sunny, and *gorsk*, city.
Sovetsk	Tilsit	Originally named after the River Tilza. Renamed (1946) after it was transferred to the Soviet Union from Prussia at the end of the Second World War. Site of the Peace Treaties between France, Prussia and Russia (Jul 1807). Napoleon and Tsar Alexander I (1777–1825, r.1801–25) agreed to become allies and divide Europe between them. The Tsar agreed to the creation of the Grand Duchy of Warsaw.
Stavropol	Voroshilovsk	Means 'Town of the Cross' from the Greek *stavros* and *polis*. Renamed (1935–43) after Voroshilov to mark his promotion to the rank of Marshal of the Soviet Union. Returned to its original name on the liberation of the Caucasus from the Germans.

Name	Former name(s)	Description
Svetlogorsk	Rauschen	There is a number of towns with this name, meaning 'City of Light'. This one, in Kaliningrad region, was renamed (1946) after its transfer from Prussia to the USSR.
Svobodnyy	Alekseyevsk	Means 'Free' or 'Unrestrained'. Renamed (1924).
Syktyvkar	Ust-Sysolsk	Lies at the mouth, *ust*, of the River Sysola. Syktyv is the Komi name for the Sysola to which has been added the Komi word *kar*, town. Renamed (1930).
Taganrog	Troitsky	Literally means 'Trivet horn', from *tagan*, trivet, and *rog*, cape or horn; the latter word in the sense of promontory. It may mean that some warning beacon was set up here using an iron tripod or bracket. Founded (1698) by Peter I the Great as a naval base.
Tatarstan	Tatariya	A republic meaning the 'Place of the Tatars'. Originally inhabited by Bulgars, the region was settled by the Mongols of the Golden Horde (13th century) and became a Tatar khanate until absorbed into the Russian Empire (1552). The adjective 'tartar', a violent and irascible person, comes from the Tatars.
Tobolsk	Kashlyk	Lies at the confluence of the Tobol and Irtysh rivers.
Tolyattigrad	Stavropol-na-Volge, Togliatti	Renamed (1964) after Palmiro Togliatti (1893–1964), Italian communist leader (1926–64). Changed from Togliatti to Tolyattigrad (1991).
Tovarkovskiy	Kaganovich	Named (1934–64) after Lazar Kaganovich (1893–1991), Communist Party leader and one of Stalin's most devoted henchmen.

Tuapse		Means 'Two waters' in the Adygey language and is so named because the Rivers Tuapse and Pauk flow into the sea here.
Tula	Taydula	The name may originate from a Baltic word meaning 'settlement'.
Tutayev	Romanov-Borisoglebsk	Originally named after Prince Roman, the great-grandson of a grand prince who was granted the area round the village of Borisoglebsk on the Volga. Renamed (1921) after a Red Army hero, Tutayev.
Tuva	Tyva	A republic named after the Tuvans. Part of the Chinese Empire for nearly 200 years until annexed by the Soviet Union (1944).
Tver	Kalinin	Lies at the confluence of the Tvertsa and Volga rivers. The name may be taken from the Russian *tvërdyy*, solid or firm, and meaning stronghold. (In Serbo-Croat *tvrdjava* means fortress.) Renamed (1931–90) after Kalinin whose birthplace this was. Developed into a principality (13th century) to rival Moscow but was annexed by Moscow (1485).
Tyumen	Chinga-Tura	Originally the Tatar Town of Genghis Khan'. Now means 'Ten thousand' from the Tatar *tyu*, ten, and *men*, thousand, perhaps a reference to the strength of Genghis' army.
Udmurtia	Votskaya, Udmurtiya	A republic named after the Udmurt, a Finno-Ugric people. The Udmurt were previously known (until 1932) as the Votyaks.
Ulan Ude	Udinskoye, Verkhne-Udinsk	Named after the River Uda, the Buryat word *ulan* meaning red. The previous name (1783–1934) means 'Upper Uda' to distinguish it from Nizhne-Udinsk, Lower Uda, these being two different rivers.

Ussuriysk	Nikolskoye, Nikolsk-Ussuriysk, Voroshilov	Named after the Ussuri River. First named (1866) after Tsar Nicholas I (1796–1855, r.1825–55). Then Nikolsk-Ussuriysk (1926), Voroshilov (1935), Ussuriysk (1957).
Uyar	Olginsk	Previously named after the Grand Duchess Olga Nikolayevra (1895–1918), eldest daughter of Nicholas II.
Vladikavkaz	Ordzhonikidze, Dzaudzhikau	Renamed (1931–44, 1954–90) after Ordzhonikidze and Dzaudzhikau (1944–54). Means 'Ruler of the Caucasus' from the Russian *vladet*, to possess or control and so named (1783) when a fort was built here. The Ossetian name Dzaudzhikau is still legally valid.
Vladimir		Named (1108) by Vladimir II Monomakh (1053–1125), Grand Duke of Kiev, possibly after St Vladimir the Great (956–1015). Capital of Kievan Rus (1157–c.1328). The name could be translated as 'Ruler of the World' or alternatively 'Possessor of Peace', *mir* meaning both world and peace.
Vladivostok		Means 'Ruler (or Possessor) of the East' from *vostok*, east.
Volgograd	Tsaritsyn, Stalingrad	Means 'Town on the (River) Volga'. Originally a fortress, possibly named Tsaritsyn (1589) in honour of Ivan IV the Terrible (1530–84), to protect the newly acquired territory along the River Volga; more probably, Tsaritsyn is a Tatar name meaning 'Town on the River Tsaritsa' (Yellow River) which was so called because of the golden sands at this point on the bank of this tributary of the Volga. Renamed (1925–61) after Stalin, chairman of the local Military Committee, who had organized the defence of the city (1919) against White Russian armies during the Civil War. Renamed again after Stalin's fall from grace. Site of the Battle of Stalingrad (Aug 1942–Feb 1943) in which the German 6th Army and elements of the 4th Panzer Army were forced to surrender.

Name	Alternate name	Description
Voronezh		Named after the River Voronezh, the name for which is taken from the Russian *voron*, raven, here to mean probably black.
Vostochnyy	Aleksey Orlovskiy	Means 'Eastern'. Previously named after Prince Aleksey Orlov (1786–1861), statesman and authoritative adviser to Tsars Nicholas I and Alexander II (r.1855–81).
Vyborg	Viipuri	Means 'Holy fort' in Swedish. Originally built and held by the Swedes (1293–1710), then the Russians (1710–1918). While in Finnish hands (1918–40) renamed. By the Treaty of Moscow (Mar 1941) ceded back to the USSR at the end of the Soviet-Finnish Winter War. Occupied by German and Finnish forces (1941–44).
Yakutia	Sakha	A republic named after the Yakut. This name is a Russian corruption of the Yakut name Sakha.
Yakutsk	Yakutsky Ostrog	May be derived from the Yakut *yekot*, strangers. The original name means the 'Fortress of the Yakut', founded (1632).
Yantarnyy	Palmnicken	Noted for amber products, it means 'Town of amber'.
Yaroslavl	Medvezhy Ugol	The original name means 'Bear's corner'. Founded (1010) by Yaroslav (980–1054), then Grand Duke of Novgorod, after he had killed a bear which had attacked him. He ordered a fortress to be built on the spot. Named after Yaroslav when he became Grand Prince of Kiev (1019); later he was known as Yaroslav I the Wise. Capital of an independent principality (1218–1471). Sacked by the Tatars (1238), by Ivan I Kalita (1332), and captured by Novgorod (1371). Recovered quickly each time until submitting to Moscow (1471).

Yekaterinburg	The *burg* indicates a fortified settlement which was originally named (1723) after Catherine I (1684–1727), wife of Peter I the Great and Empress (1725–27). Catherine was originally a Lithuanian servant girl whose name was Martha Skavronskaya. Captured by the Russians (1702), she entered the service of Prince Menshikov (1660–1729), Peter's principal minister, through whom she met Peter. Renamed (1924–91) after Yakov Sverdlov (1885–1919), a Bolshevik leader and the first editor of *Pravda*, the official daily newspaper of the Communist Party. Site of the assassination by Red Guards of Tsar Nicholas II (1868–1918) and his family (Jul 1918).
Sverdlovsk	
Yoshkar-Ola	Located on the River Kokshaga, it was originally named for tsardom, the 'Tsar's town on the River Kokshaga' (1584–1919), then for communism as the 'Red town on the Kokshaga' (1919–27).
Tsaryovokokshaysk, Krasnokokshaysk	
Yuzhno-Sakhalinsk	Means 'South Sakhalin'. Originally named (1881) after a military officer. Renamed Toiohara by the Japanese during their occupation (1905–45).
Vladimirowka Toiohara	
Zelenogradsk	Means 'Green town'. (There are quite a few cities in Russia whose names begin with *zelen*).
Kranz	
Zheleznogorsk	Previously a 'closed city', now meaning 'City of iron'. Previous names meant 'Atom city' and 'Group of nine'.
Krasnoyarsk-26 Atomgrad, Devyatka	
Zheleznovodsk	Means 'Iron waters'. (As with *zelen*, there are a number of Russian cities beginning with *zhelezno*).

Zlatoust

Named (1754) after the church dedicated to St John Chrysostom (c.347–407), archbishop of Constantinople (398–403), Zlatoust being the Russian for Chrysostom.

1. The names of the nationality-based Republics come largely from the peoples populating them. Quite often the name of the people is derived from the local word for 'man' or 'people', e.g. *murt* or *sakha*.

SLOVAKIA

Slovakia was first settled by a Celtic tribe called the Cotini. It was not until the sixth century that the Slav Slovaks arrived from the east to contest the land with the Avars. Only in 830, by which time the Avars had been driven out, was the Great Moravian Empire founded (q.v.The Czech Republic). In time it came to include western Slovakia. At the Battle of Bratislava in 907 the Hungarians defeated the Moravians and their empire collapsed. During the course of the next two centuries Slovakia became a land of the Hungarian crown. It was to remain under Magyar rule for the next thousand years, although some areas were claimed by Bohemia.

The Hungarian Árpád dynasty encouraged German settlement. Known as Upper Hungary, the area was a virtual backwater though subject to a policy of Magyarization. When the Turks defeated the Hungarians at the Battle of Mohács in 1526 and Louis II died, the Austrian Habsburgs inherited the Hungarian throne. Their kingdom, however, was divided, but it included much of modern Slovakia which formed part of what came to be known as Royal Hungary.

Under complete Hungarian control for so long, it is no surprise that the Slovak advance to nationhood was so slow. Not until the late 18th century did a Slovak national renaissance begin; and not until the 1840s did a literary Slovak language emerge. Nevertheless, the rebellion against Hungarian rule in 1848 failed. But in gratitude for their assistance in helping him to suppress an Hungarian revolt aimed at achieving complete independence for the Kingdom of Hungary, the Emperor gave the Slovaks some limited autonomy. Shortly afterwards, in 1867, this was lost when the Dual Monarchy of Austria Hungary was created. Full Hungarian control of Slovakia was re-established and Magyarization returned with a vengeance.

When the First World War broke out the Slovaks, the most deprived minority in the Hungarian part of the empire, found themselves in the unenviable position of having to fight for Austria-Hungary against fellow Slavs. Many deserted and joined the Czechoslovak Legion which fought against the Austrians. Keen to throw off the Austro-Hungarian yoke, Slovaks and Czechs combined during the war to campaign for an independent state. In May 1918, Czech and Slovak exiles, in Pittsburgh, USA, issued a manifesto on the creation of an independent Czech-Slovak state in which the Slovaks would have their own parliament. On 28 October 1918 the Republic of Czechoslovakia was proclaimed. Slovakia had never possessed precise historical borders and ethnic boundaries were indistinct. Only by the Treaty of Trianon in 1920 was the Slovak-Hungarian border fixed. The result was a majority Hungarian population in southern Slovakia. Although they had no link at all with the Czechs, the Ruthenians[1] in the north-eastern Carpathians also joined the new republic.

In October 1938, following the Munich Settlement when the Sudetenland was annexed by Germany, Slovakia and Ruthenia gained autonomy; a month later Hungary seized parts of Hungarian-inhabited southern Slovakia with German approval – the 1938 First Vienna Award. The day before German troops began their occupation of the Czech Lands, on 14 March 1939, the Slovaks felt secure enough to declare independence. Hitler agreed to the creation of a separate Slovak state and Slovakia became independent for the first time in its history, although in reality it was no more than a German puppet state. Ruthenia, however, was lost to Hungary.

During the Second World War the Slovaks were initially sympathetic to Germany, but as Nazification began to gather pace a resistance movement formed.

The Slovak National Uprising in August 1944 tied down a few German divisions, but it also alerted Stalin to Slovak nationalism. As the Red Army entered Slovakia a couple of months later Stalin reneged on his promise of recognizing Czechoslovakia's pre-Munich borders. He took Ruthenia as the spoils of war as well as a great swathe of Polish territory, thereby gaining a common Soviet-Czechoslovak border for the first time. Slovakia and the Czech Lands were reunited in 1945 and, after only six years, Slovakia's so-called independence came to an end. The territory seized by Hungary in 1938 was returned. Most of the German-speaking population was expelled, but attempts to get rid of the Hungarian-speaking minority were obstructed by Budapest.

After the communist *coup d'état* in 1948, every effort was made to extinguish any sense of Slovak nationhood. However, following the Warsaw Pact invasion of Czechoslovakia in August 1968 (q.v. The Czech Republic), a new federal system was introduced on 1 January 1969. Two equal republics, the Czech and Slovak Socialist Republics, were created with their own governments – but under centralized communist rule and underpinned by the presence of Soviet troops.

The Velvet Revolution (q.v. The Czech Republic) in November 1989 brought an end to communist rule and ushered in a new democratic era. Soviet troops which had been stationed in Slovakia since 1968 were withdrawn by June 1991. Cultural, political and economic differences between the Czechs and Slovaks soon reached a point where the Slovaks began to demand changes to their status within the federation. For centuries the Slovaks had felt themselves to be second-class citizens and, when part of Czechoslovakia, had resented Czech domination. As the Slovaks moved away the Czechs did little to stop the drift towards independence. Measures to divide the two formally began and proceeded amicably, partly because there was little ethnic intermingling. The so-called Velvet Divorce took place on 1 January 1993 when the Slovak Republic came into existence as an independent state. When the borders of Czechoslovakia were drawn in 1918 some 750 000 Hungarians had found themselves in the new state. Today some 600 000 remain in Slovakia, largely unassimilated.

---

1. The Ruthenians are Ukrainians whose ancestors were subjects of the Poles, Austrians and Austro-Hungarians. They have never had a state of their own, but they claim to be a nation. Ruthenia was a name given to that part of Czechoslovakia in which they lived in 1938. At the end of the Second World War Ruthenia was ceded to the Soviet Union, becoming a part of Ukraine. Other Ruthenians live in other countries of eastern Europe and the Balkans.

Area:	49 035 sq km (18 933 sq miles)
Population:	5.37 million; Slovaks, Hungarians, Gypsies, Ukrainians
Languages:	Slovak, Hungarian
Religions:	Roman Catholic, Protestant, Greek Orthodox
Capital:	Bratislava
Administrative Districts:	8 regions
Date of Independence:	1 January 1993
Neighbouring Countries:	Poland, Ukraine, Hungary, Austria, Czech Republic

English-speaking Name	Local Name	Former Names	Notes
SLOVAKIA	Slovenská	Upper Hungary	The Slovak Republic (Slovenská Republika) since 1 Jan 93 when Czechoslovakia split in two. Part of the Czech and Slovak Federative Republic (1990) and before that (1969) the Slovak Socialist Republic within Czechoslovakia. Named after the Slovaks, a Slav tribe which came from the East. Their language is called *Slovenský*, which should not be confused with *Slovenski*, the language of the Slovenes.
Bratislava		Posonium, Istropolis, Pressburg, Pozsony	Named after Bratislav (or Braslav), the last Slav leader of the Great Moravian Empire. The Hungarian defeat of the Moravians at the Battle of Brezlauspurg (907) terminated the Empire. The Peace of Pressburg (1491) acknowledged Habsburg claims to the Hungarian throne. Capital of Hungary (1536–1784). At the Treaty of Pressburg (Dec 1805), after the French victories at Ulm and Austerlitz, Austria was forced to surrender large swathes of territory and agree to a number of conditions demanded by Napoleon. Capital of Slovakia within Czechoslovakia (1918) and as capital of independent Slovakia (1993).
Kežmarok		Caesareoforum, Kesmarkinum, Käsmark	Derived from the German 'Cheese market'.
Košice		Cassovia, Kassa, Kaschau	Became part of Czechoslovakia (1920) having been founded by Saxons. The Polish nobles agreed to accept the king's choice of successor provided he guaranteed their rights and privileges at the Pact of Koszyce (1374). First seat of the pro-Soviet provisional Czechoslovak government after Soviet troops liberated it (Apr 1945). Succeeded by a new government in Prague (May 1946).

Prešov	Fragopolis, Preschau, Eperjes	Site of the declaration of Bela Kun's short-lived Slovak Socialist Republic (1919). His Hungarian Red Army occupied much of the country until it was ejected by the Czechoslovak Legion (1919).
Šturovo	Parkany	Named after the Slovak nationalist leader Ludovit Štur (1815–56), the seminal influence in the creation of a written Slovak language as distinct from Czech. As a deputy in the Hungarian Diet, he kept the Slovak opposition on the side of the Habsburgs against the Magyar revolutionaries (1848).
Trnava	Tyrnavia, Tyrnau, Nagyszombat	Called the 'Slovak Rome', the town became the religious centre of Hungary when the bishop of Esztergom moved his see here (1541–1820) after the Turks had overrun most of Hungary. Nagyszombat means 'Holy Saturday', the Saturday before Easter.

SLOVENIA

In 1991 the Slovenes freed themselves from external control for the first time in over 1200 years and established their first unified and independent state. With the Serbs and the Croats, the Slovenes (sometimes referred to as Alpine Slavs) began to occupy their present homeland and another area as big to the north and west in the sixth century. By 750 this area had been incorporated into the Frankish empire of the Carolingians. After that empire was partitioned the land of the Slovenes, which had already been reduced by Bavarian and Magyar invasion to the southern part of the area, was incorporated in the tenth century into the German kingdom. The Slovenes themselves were dispersed in the border regions of Carinthia, Carniola and Styria. A process of Germanization was begun. However, the Slovenes were fortunate in that they were able, with great difficulty, to preserve their identity and culture.

By the start of the 13th century the first tentacles of Austrian Habsburg authority had spread south. Styria was acquired in 1278; Carniola, a substantial part of present-day Slovenia, and Carinthia became crown lands in 1335; and Istria, on the Adriatic, fell to the Habsburgs in 1374. While German influence waxed during the next four centuries Slovene culture was cherished by an evolving Slovene literary language. Furthermore, the region was spared serious depredations by the Ottoman Turks and the Slovenes never had to live under their rule.

Germanization was halted temporarily between 1797 and 1814 when French troops advanced into Slovene lands; between 1809 and 1814 the French ruled in the 'Illyrian Provinces' (Carniola, western Carinthia, Istria, Dalmatia and parts of Croatia), so created by Napoleon. Material conditions improved significantly and the French encouraged the growth of Slovene culture. The Russian defeat of the French in 1812 and the subsequent Congress of Vienna in 1814–15 brought an end to French rule and the restoration of Austrian authority. Nevertheless, the new-found sense of a Slovene national identity did not dissipate.

The crisis within the Habsburg Empire in 1848–49 provided an opportunity for the Slovenes to demand an autonomous and united Slovene province under Habsburg rule. The only response was to allow Carniola and other Austrian provinces to function as crown lands with their own parliaments; autonomy was severely limited. Of the 15 crown lands Slovenes lived in four. When the Dual Monarchy of Austria-Hungary was created in 1867 they found themselves in the Austrian part; this did nothing to dampen Slovene nationalist aspirations and activities.

Slovene discontent grew and was further strengthened by increasing contact with the Croats. When Bosnia-Hercegovina was annexed to Austria-Hungary in 1908 there was widespread resentment. Nevertheless, Slovene troops fought in the Austro-Hungarian Army during the First World War. A hundred years earlier the idea that the South Slavs shared common interests had been put forward. Thus, with the fall of the Habsburg Empire in 1918, the Slovenes joined with their fellow Slavs to form the Kingdom of the Serbs, Croats and Slovenes. The former Austrian provinces of Carniola, southern Styria and a small part of southern Carinthia were incorporated into the kingdom as was Hungarian Prekomurje. The Italians received practically the whole of Istria and Primorska on the Adriatic coast; the Slovenes in the rest of Carinthia joined Austria. Thus the Slovenes were still not united, about a third of their population living beyond the borders of the new kingdom. On the other hand, Slovenia itself was almost completely ethnically homogeneous.

When King Alexander suspended the constitution, declared a royal dictatorship, reorganized the country into nine *banovine* (governorships) and changed the name of the country to Yugoslavia in January 1929, Slovenia (a name only coined in 1918) became the Dravska *banovina*. So it remained until 1941 when it was partitioned between Germany, which took the northern half, and Italy which gained Ljubljana and the southwest; Hungary recovered Prekomurje.

As victory beckoned in 1945 Tito's Partisans did not stop at Yugoslavia's pre-war borders. They pushed on into Austria and Italy. Having captured the greater part of Istria, Tito's aim was to acquire Trieste. However, under strong Allied pressure, he was forced to withdraw from the city and from the territory his troops had occupied in Austria. Apart from the so-called Free Territory of Trieste, Yugoslavia was allowed to retain what had been captured from Italy; on average, the border was moved 40 kilometres westwards and various Adriatic islands which had fallen to Italy after the collapse of the Austro-Hungarian Empire in 1918 also became Yugoslav. Slovenia's possession of most of Istria was legally recognized in 1954, thus giving it 47 kilometres of coastline and including those Slovenes who had been under Italian rule since the 1920 Treaty of Rapallo; at the same time Tito agreed that Trieste should return to Italy. Only in 1975 at the Treaty of Osimo was the Italo-Yugoslav border finally formalized.

The 1946 federal constitution of the new Yugoslavia recognized Slovenia as one of the six republics. The death of President Tito in 1980 ultimately led to the creation of a multi-party system in the republic which declared its right to secede from Yugoslavia in September 1989. In April 1990 Slovenia held the first fair and free multi-party elections in Yugoslavia since before the start of the Second World War. They were won by a party calling for an independent Slovenia. On 25 June 1991 Slovenia declared its independence from Yugoslavia. This was immediately contested by the Yugoslav Army which was ordered to secure the country's international borders and other critical points. The fighting that broke out lasted only ten days; the Slovenes emerged victorious. On 8 July a three-month moratorium on implementing the declaration of independence was agreed. On 18 July 1991 the decision was taken in Belgrade to withdraw the Yugoslav Army from Slovenia by 25 October – a move which signalled the end of the Socialist Federal Republic of Yugoslavia.

Area:	20 251 sq km (7818 sq miles)
Population:	1.99 million; Slovenes, Croats, Serbs, Bosniaks
Languages:	Slovenian, Serbo-Croat
Religions:	Roman Catholic, Orthodox, Protestant, Muslim
Capital:	Ljubljana
Administrative Districts:	12 statistical regions
Date of Independence:	8 October 1991
Neighbouring Countries:	Austria, Hungary, Croatia, Italy

English-speaking Name	Local Name	Former Names	Notes
SLOVENIA	Slovenija		Republic of Slovenia (Republika Slovenija) since Jun 1991. A republic within Yugoslavia (1946). Named after the Slovenes.
Carniola	Kranjska	Krain	A region named after the Carni. Became an Austrian crownland (1849), with most of it being absorbed into the Kingdom of the Serbs, Croats and Slovenes (1918) and all of it into Slovenia (1947).
Celje		Claudia Celeia, Cilia	Named by the Celts. *Kel* means refuge or place of residence.
Izola		Haliaetum	Although now linked with the mainland, it takes its name from *isola*, island.
Kobarid		Caporetto	The site of the famous Battle of Caporetto (Oct 1917) during which the Austro-German Army routed the Italians and broke through to the River Tagliamento.
Koper		Aegida, Insula Caprea, Justinianopolis, Caput Histriae Capodistria	Formerly an island; Insula Caprea means Goat Island. Then named after Justinian I (483–565), Byzantine Emperor (527–565). Captured (1279) by the Venetians after which it was made the capital of Istria, hence Caput Histriae and then Capodistria. Part of the Free Territory of Trieste until passing to Slovenia (1954).
Ljubljana		Emona, Lubigana Laibach, Lubiana	May be derived from *ljub*, love, to mean a well-loved place. Capital of Napoleon's Illyrian Provinces (1809–13) and of Slovenia (1918).
Maribor		Marburg/Marchburg	The name refers to the construction of a defensive wall round the town from the German *burg*, fortified town.

Murska Sobota		Means 'Mura Saturday', i.e. the day of the market.
Novo Mesto	Gradac, Rudolfswert, Neustadt	Means 'New place' though it has been continuously inhabited since 1000 BC. Renamed (1365) after the Habsburg Rudolf IV. Renamed again (early 15th century).
Prekmurje		A region meaning 'Beyond the Mura River'.
Primorska		A region meaning 'Littoral' although Slovenia's coastline is only 47 km long.
Velenje	Titovo Velenje	The Slovenian town named (1946–92) after President Tito (1892–1980). Partisan leader during the Second World War, Prime Minister (1945) and President (1953).

SPAIN

Narrowly separated from the continent of Africa, the Iberian peninsula has been subject to invasion from north and south for centuries. The original Iberians, Phoenicians and Greeks (along the coastal regions to the east and south), Celts (in the centre and north), Carthaginians (in the south-east), Romans, Visigoths and Moors all imposed their cultures on the peninsula before Christianity gained the upper hand in the 15th century.

Pursuing their interest in trading, the Phoenicians are thought to have established a colony at Gadir (now Cadiz) around 1100 BC to exploit Spain's mineral wealth. They were followed three or four centuries later by the Greeks who set up their trading posts along the east coast. The Carthaginians arrived in the third century BC, having been expelled from Sicily by the Romans as a result of their defeat in the First Punic War (264–41 BC). Roman troops were despatched to the peninsula in 218 BC, initially to prevent reinforcements reaching the Carthaginian general Hannibal, who, by then, had launched his invasion of Italy (Second Punic War, 218–01 BC). By 210 BC the Carthaginians had been evicted from all but Cadiz and from then on the Romans gradually began to extend their hold on the entire peninsula. This took another 200 years and only in 19 BC was Emperor Augustus able to claim the subjugation of the peninsula (called Hispania) and divide it into three provinces: Baetica, Lusitania and Tarraconensis. By the fourth century AD nine provinces had been created, all with their own culture and administered separately. Apart from Bilbao and Madrid, there is hardly a town in Spain which was not either founded or developed by the Romans.

This stable and prosperous situation was shattered by the arrival of the Alani, Suevi and Vandals, Germanic tribes, who crossed the Pyrenees early in the fifth century. The Alani went south to Lusitania, the Suevi north-west to Galicia and the Vandals also south to Andalusia (and then onto North Africa). In 415 the Visigoths, Arian Christians and allies of Rome, followed, eventually emerging as the dominant power in Spain and establishing a kingdom. However, Roman influence remained strong, and it was not until the second half of the sixth century that the Visigoths were able to impose themselves on the Suevi and the Basques.

In 589 King Recared renounced Arianism for Catholicism. This cleared the way for an alliance between the minority Visigoths and the Hispano-Romans and confirmed the superiority of Roman civilization. Nevertheless, without a well-established system of hereditary succession to the Visigothic throne, succession was always liable to present a problem. When King Witiza's son was prevented from succeeding him in 710 by Roderick, Duke of Baetica, Witiza's family invited the Moors – Arabs who had spread the Islamic faith across North Africa – and Moroccan Berbers to come to its aid. The Muslim governor of Tangier, Tariq ibn Ziyad, took full advantage of this opportunity to enter Spain, seizing Gibraltar the next year and routing Roderick's army. The subsequent conquest was conducted at amazing speed. By 716 the whole peninsula, except for the Pyrenees and a strip of Asturian coastline, had been conquered and Spain had been declared an Emirate of Damascus.

The Visigothic hold on these remnants of territory enabled the Christians to begin the reconquest (*Reconquista*) before the conquest itself had been completed. As early as 718 the Moors were defeated at Covadonga and this inspired the Christians to continue their opposition to Moorish rule. Complete victory, however, was not to be achieved for nearly 800 years – in 1492. The victory at Covadonga led to the formation of the Kingdom of Asturias.

In 756 Abd al Rahman I assumed leadership of the Moors and made Córdova his capital. He founded the Emirate of Córdova, subject to the caliphate (empire) which had moved to Baghdad. In 929 Abd al Rahman III raised the Emirate to the status of a caliphate. Moorish unity broke down following the death in 1002 of al-Mansur, who had exercised dictatorial control in the young caliph's name. As hostility between the Arabs and the Berbers grew the caliphate disintegrated into a number of small independent kingdoms, or *taifas*.

Meanwhile, Asturias had expanded until it had become too large to be controlled from Oviedo, its capital. This was moved to León in 910 and eventually the name was changed to the Kingdom of León. At much the same time a Basque Kingdom of Navarre was formed round Pamplona. Founded as a county within León was Castile. By 1034 Sancho III the Great, King of Navarre, had imposed his rule over all the Christian states save Catalonia. This unity was broken on his death the next year when his three sons were awarded Aragón, Castile and Navarre. All bearing a royal title, their acquisitions became kingdoms. To Castile, Ferdinand I, the eldest brother, added León in 1037 and claimed the title of Emperor of Spain.

The capture of the Moorish capital of Toledo in 1085 by Alfonso VI of León-Castile caused panic among the Muslim kings and the ruler of Seville asked for help from the Islamic Berber Almoravids. Arriving in 1086 from Morocco, they overwhelmed Alfonso and restored unity among the petty Muslim kingdoms. But they soon lost their missionary fervour and their rule lasted only until 1147. The Almohads, another group of militant Berbers and less tolerant of Christianity, invaded in 1146. Defeating the Almoravids, they remained in power until the rulers of Aragón, Castile and Navarre combined to defeat them at the great battle of Las Navas de Tolosa in 1212.

Meanwhile, during the 12th century, as Christian armies took advantage of internecine fighting among the Muslims to regain more territory, León was divided into the three kingdoms of León, Castile and Portugal, and Navarre had split from Aragón. When Alfonso IX of León died in 1230, his son Ferdinand III, already King of Castile, succeeded him as King of León; thereafter they remained permanently united. Ferdinand continued the onslaught against the Moors. When he took Seville in 1248 only the Kingdom of Granada remained in Muslim hands; even so, the Muslims had to pay an annual tribute. Five kingdoms now shared the peninsula: Muslim Granada, Aragón (including Catalonia and Valencia), Castile-León (including Asturias and Galicia), Navarre and Portugal. Aragón became involved in Italian affairs when the king was elected to the throne of Sicily in 1282 (q.v.Italy). Majorca was annexed permanently to Aragón in 1343.

In 1469 second cousins Isabella of Castile and Ferdinand of Aragón married, succeeding to their respective thrones as Isabella I in 1474 and Ferdinand II in 1479; ruling together in both kingdoms, the two most powerful crowns in Spain were thus joined in a personal union. That year Spanish sovereignty was established over the Canary Islands. In 1482 the campaign to drive the Muslims out of their Kingdom of Granada was launched. It lasted ten years and finally brought Muslim power in Spain to an end. In that same year Christopher Columbus discovered America, heralding the start of the Spanish penetration into Central and South America.

To Sicily, Aragón added the Kingdom of Naples in 1503. Isabella died in 1504, leaving Castile to her surviving daughter, Joan the Mad, who was married to Philip, ruler of the Netherlands. Joan's mental illness worsened and in August 1506 Philip assumed control of Castile as Philip I. However, Philip died within a month, leaving Ferdinand, Joan's father, as King of Aragón and Castile. When he died in 1516, his grandson, Charles I, ruler of the Netherlands and heir to Habsburg lands in Austria and southern Germany, came to the Spanish throne. Three years later he was elected Holy Roman Emperor as Charles V. By this time, with Castile having annexed the Spanish part of Navarre (south of the Pyrenees) in 1512, the boundaries of modern Spain had been set and it could boast the largest European empire since Rome. When Charles abdicated in 1556, his son, Philip II, succeeded to all his dominions save Germany[1]. Keen to unify the entire Iberian peninsula, Philip despatched an army under the Duke of Alba to attack Portugal in 1580 in support of his familial claims. For the next 60 years the Portuguese were subject to Spanish rule.

The maintenance of empire proved costly and ultimately caused Spain's decline. Spanish power began to erode during the foreign wars of the late 16th and 17th centuries. In 1700 Charles II died childless, thus ending the Spanish Habsburg dynasty. In his will he had nominated Philip, the Bourbon Duke of Anjou and grandson of Louis XIV of France, as his successor. It was conceivable that Philip might also inherit the French crown. This possibility alarmed the British and the Dutch; having a Bourbon King of Spain would tilt the balance of power in Europe towards France, particularly if it acquired the Spanish Netherlands. Austria retained the hope of placing its candidate, Archduke Charles of Habsburg, on the throne. The War of the Spanish Succession began in 1701 when Austrian and British armies invaded Spain in an attempt to replace Philip, now Philip V, with Charles. Defeat for the Spaniards led to the Treaty of Utrecht in 1713. Spain lost the Spanish Netherlands and its possessions in Italy to the Austrian Habsburgs and was forced to cede Gibraltar and Minorca (won back in 1783) to Britain which had been seized during the war; Philip, however, retained the throne of Spain and Spanish colonies overseas.

In 1796 Spain elected to join revolutionary France against England. It proved to be a disastrous decision: the Franco-Spanish fleet was decisively defeated by Nelson at the Battle of Trafalgar in 1805. Nevertheless, Napoleonic victories on land convinced Godoy, the Spanish prime minister, that he should not desert the French. On the understanding that he would be given half of Portugal and be crowned its king when Napoleon realized his ambition of annexing the country, Godoy sent the French funds and 15,000 Spanish troops. In November 1807 French troops crossed the Pyrenees on their way to Portugal, still a close ally of Britain which responded with an expeditionary force. Many Spaniards, however, were alienated when French troops occupied fortresses in northern and central Spain and Napoleon demanded some Spanish territory. Godoy was dismissed and Charles IV and then his successor, Ferdinand VII, forced to abdicate. In deposing both kings in 1808 and installing his brother Joseph in their place, Napoleon miscalculated.

The Spanish rose in revolt, but within a year the French were in control of most of the peninsula. While the Spanish began the first-ever guerrilla operations, it was the British under Sir John Moore and then the Duke of Wellington who defeated the French in what the Spanish called the War of Independence and the British the Peninsular War. In 1814 Ferdinand VII was restored to the throne. The American colonies now began to assert their drive for independence and by the end of the century Spain had lost most of its possessions overseas (Cuba, Puerto Rico and the Philippines alone in 1898).

Civil strife and war marked the years between 1820 and 1873 when King Amadeo, second son of the King of Italy, abdicated (having been elected in 1870) and Spain's First Republic was inaugurated. It lasted only to 1874 when the British-educated Alfonso XII came to the throne as 'a good Spaniard, a good Catholic, and a Liberal'.

Spain remained neutral during the First World War. But gathering political and social chaos encouraged General Primo de Rivera to head a military *coup* in 1923. Republicanism began to gain ground. The 1931 election results demonstrated a groundswell of anti-monarchist feeling and Alfonso XIII decided to abdicate. The Second Republic was declared. The next year Catalonia achieved a degree of autonomy.

Between 1936 and 1939 Spain was racked by civil war. While the Republicans received aid from the Soviet Union and International Brigades of communists and socialists, the Nationalists were helped by Nazi Germany and Fascist Italy. In 1939 the Nationalists claimed victory and General Francisco Franco became head of state.

Spain was too weak to be anything but neutral during the Second World War. In 1947 the Law of Succession was passed by referendum. Included among its clauses were that Spain was a *kingdom*, that Franco was Head of State, and that he could choose the next king. In 1969 he declared Don Juan Carlos de Borbón, the grandson of Alfonso XIII, to be his successor. When Franco died in 1975 Juan Carlos became king.

1. From 1554 until her death in 1558, Philip was also joint sovereign of England as the husband of Mary I.

Area:	504 782 sq km (194 896 sq miles)
Population:	39.8 million; Spaniards
Languages:	Spanish, Catalan, Galician, Basque
Religion:	Roman Catholic
Capital:	Madrid
Administrative Districts:	19 autonomous communities
Date of Independence:	1492 (final defeat of the Moors)
Neighbouring Countries:	Andorra, France, Gibraltar, Portugal

English-speaking Name	Local Name	Former Names	Notes
SPAIN	España	Iberia, Spania, Hispania al-Andalus	The Kingdom of Spain (Reino de España) (1492–1931) and since Nov 1947 (although without a king until 1975). Previously under the personal rule of Gen Franco (1892–1975, leader 1939–75). One theory is that the name Spain comes from the Punic (the language of Carthage) *span*, rabbit, which were numerous in the peninsula, or from a Phoenician word for a type of hare peculiar to the peninsula. Iberia was first used by the Greeks for the country of the Iberians who dwelt along the Iberus (River Ebro). The peninsula was called Hispania by the Romans; in due course the *H* was dropped and the short *i* became *e* to give España. Al-Andalus, which eventually became Andalusia, means 'The Isle of the Vandals'
Albacete			Derived from the Arabic *al-Basit*, the Plain.
Alcalá de Henares		Complutum, al-Qala an-Nahr	Means 'Fortress on the River Henares'. Destroyed by the Moors (1000) and rebuilt (1038). Regained (1088) by King Alfonso VI (c 1040–1109), King of León (1065–70) and Castile-León (1072–1109).
Alcántara			Means 'The bridge' from the Arabic *al-Qantarah*, a reference to the old Roman bridge built (105) over the River Tagus. Gave its name to the Order of Alcántara, the military order founded (1167) to defend Christian Spain against the Moors; it was given the city (1218).
Alcázar de San Juan		Alces, al-Qasr	Means the 'Castle of St John', the 'St John' being added after the castle was captured by the Knights of St John of Jerusalem (1186). The Arabic *al-Qasr* means 'castle', 'fortress' or 'palace'.

Name	Alternative names	Description
Algeçiras	Portus Albus, al-Jazira al-Khadra	The Arab name means 'Green island' taken from the offshore Isla Verde. Destroyed (1368) by Muhammed V of Granada (r.1354–59; 1362–91).
Alicante	Akra Leuke, Lucentum, al-Lucant/al-Akant	Founded (325 BC) by the Greeks, the original name means 'White peak' while the Romans who captured it (201 BC) named it Lucentum, the 'Place of light'.
Almadén	Sisapo	The Arabic *al-Madin* means 'The mine', a reference to the very rich mercury deposits in the area.
Almeria	Campus Spartarius, Urci, Portus Magnus, al-Mariyah	The original Roman name means 'Land of esparto grass'. The Arab name means 'Mirror of the sea'. Held by the Moors (1157–1489).
Andalusia / Andalucia	al-Andalus	An autonomous community meaning the 'Land of the Vandals' and which is largely equivalent to the Roman province of Baetica. The Arabic name referred to the whole Iberian peninsula. As the Christian re-conquest of the peninsula progressed the meaning changed to cover only the territory under Muslim control. Occupied by the Moors (711–1492) until it was incorporated into the Christian Kingdom of Castile.
Aragón		An autonomous community named after the river of the same name. It uses the Indo-European root word *ar*, water. A former kingdom in north-east Spain which was conquered by the Visigoths (5th century) and the Moors (8th century). An independent kingdom (1035–1479). United with Catalonia (1140) and with Castile (1479).
Asturias		An autonomous community (and also a principality) whose name derives from the Basque *asta*, rock, and *ur*, water. An independent kingdom (718–910) established by Visigothic nobles who had been driven out of their lands by the

		Moors. Absorbed by León (12th century). The title 'Prince of Asturias' is held by the eldest son of the king.
Ávila	Abula, Avela	The full name is Ávila de los Caballeros, 'Ávila of the Knights'. The town was named after them in recognition of their exploits in the recapture of Saragossa (12th century) and Córdoba, Jaén and Seville (13th century).
Badajoz	Pax Augusta, Batalyaws, Bax Augos Badaljoz	Means the 'Peace of Augustus', the first Roman emperor (63 BC–14 AD) and corrupted by the Moors. Recaptured from them (1229) and then captured by the Portuguese, (1385, 1396, 1542, 1660). The British successfully besieged and captured the town (Mar/Apr 1812) but suffered severe losses in doing so.
Balearic Islands	Islas Balearas	An autonomous community with a name meaning the 'Pine Islands'. There are two groups, the smaller, western group being called Islas Pitiusas. See Ibiza/Majorca/Minorca. An independent kingdom (1298–1349) at which time it joined Aragón. Established as a Spanish province (1833).
Barcelona	Barcino, Faventia Jula, Barcinona, Barshaluna	Originally believed to owe its name to the family of the Carthaginian general Hamilcar Barca (c.270–228 BC). However, there is now doubt about this and the name may come (15 BC) from a Roman, Faventia Julia Augusta Paterna Barcino. Founded by the Phoenicians and thereafter held by the Carthaginians, Romans, Visigoths (capital 415) and Moors (717); captured by the Carolingian Franks (801). Sacked by the Moors (985). Occupied by the French (1808–13). Seat of the Republican Government (1938–39).
Basque Country	Euskardi (Basque), Pais Vasco (Spanish)	An autonomous community named after the Euskaldunak whose language is Euskara. The Basque terrorist group has the name ETA (*Euzkadi Ta Azkatasuna*, Basque Homeland and Liberty). Part of the Basque lands are in France.

Place	Alternative names	Description
Bilbao	Bilbo (Basque), Bellum Vadum	A corruption of the Roman name which means 'Beautiful ford', a reference to its position on the Nervión estuary. Sacked by the French (1808).
Burgos		Founded (822) as a fortress (German *burg*, fortress) to protect the eastern reaches of the Kingdom of Asturia. Capital of the County of Castile and then Kingdom of Castile (951–1492).
Cáceres	Norba Caesarina, Alkazares/Quazri	Possibly founded (74 BC) and named after the Roman consul Quintus Caecilius Metellus.
Cádiz	Gadir, Gades, Julia Augusta Gaditana, Jaziret Qadis	Founded (c.1100 BC) by the Phoenicians as Gadir meaning an 'Enclosure' which came to mean a 'Walled place' or 'fortress'. Destroyed by the Visigoths (5th century), held by the Moors (711–1262). Ships at anchor in the harbour were burnt by Sir Francis Drake (1587). Capital of all that part of Spain not under French control (1810–12).
Calatayud	Qalat Ayub	Means 'Ayub's castle'.
Canary Islands	Islas Canarias	An autonomous community possibly named after the many large dogs (Latin *canis*, dog) on the island of Grand Canary (Gran Canaria). Called the Blessed or Fortunate Islands by Pliny the Elder (23–79). Spanish sovereignty recognized (1479).
Cantabria	Santander	A mountainous autonomous community whose people, the Cantabri, derived their name from the root word *kanto*, rock.
Cartagena	Mastia, Carthago Nova	Means 'New Carthage'. Re-founded (227 BC) by the Carthaginian leader Hasdrubal Barca (d.221). Sacked by Sir Francis Drake (1585).

Castile-La Mancha	Castilla-La Mancha	al-Qila	An autonomous community famous for its castles built by Alfonso III (838–910?, r.866–910), King of Asturias, to defend his frontiers against Muslim assaults. It was thus named 'Land of castles'. The Arab name means 'The castles'. La Mancha (in Arabic, *al-Manshah*, wilderness,) was added to Castile (1982) to form the autonomous community. A former independent kingdom (1029), although Ferdinand III (1200–52, r.1217–52) was also King of León (1230–52), until united with Aragón (1479).
Castile-León	Castilla-León		An autonomous community formed (1983). See Castile above and León below.
Catalonia	Cataluña		An autonomous community whose name means the 'Land of *castlans*' (castellans – the governor of a castle). Occupied by the Visigoths (5th century), the name might perhaps mean 'Land of the Goths'. Known as the Spanish March by Charlemagne (742–814), King of the Franks (768–814), when part of his empire. United with Aragón (1137).
Ceuta		Septem Fratres, Sebta	Originally meant 'Seven brothers' from the seven hills of the Jebel Musa range on the slopes of which the town stands. Ceuta, a limited autonomous community, is an exclave on the Moroccan coast. Held by the Moors (711–1415), captured (1415) by John I, King of Portugal (1357–1433, r. 1385–1433) and became Spanish as a result of the union of Spain and Portugal (1580).
Ciudad Real		Villa Real	Means 'Royal city'. Founded (1262) by Alfonso X the Wise, King of Castile and León (1221–84, r.1252–84) and elevated to the status of Royal City (Spanish, *ciudad*) (1420).
Ciudad Rodrigo			Named after Count Rodrigo González who founded it (1150).

Corunna	La Coruña	Brigantium, Ardobirum Coronium	Derived from Coronium by which it was known in the Middle Ages. The Spanish Armada sailed from here (1588) and the port was sacked by Sir Francis Drake (1589). The British army, led by Sir John Moore, was forced by the French to withdraw to Corunna (1809) and embark for England.
Extremadura			An autonomous community meaning the 'Land beyond the River Duero', a reference to the territory beyond Moorish control during the period of the Christian Reconquest.
Galicia		Gallaecia	An autonomous community named after a Celtic tribe, the Gallaeci. A former kingdom of the Suevi in the north-west until conquered by the Visigoths (585). Largely ignored by the Moors.
Granada		Elibyrge, Illiberis, Gharnatah	Pomegranates, and other fruit, are plentiful in this province and it and the city may be named after the Latin *granatum*, pomegranate. Originally an Iberian settlement, then Roman. The Moorish name may mean 'Hill of strangers'. Capital of the Moorish Kingdom of Granada (1238–1492). Besieged by the armies of Ferdinand and Isabella for seven months (1491) before the Moors surrendered (Jan 1492).
Guadalajara		Arriaca, Wadi al-Hajarah	Means the 'River of stones', a reference to the Henares River. The Republicans defeated Franco's troops and an Italian expeditionary corps (Mar 1937).
Huesca		Ileosca, Osca	Derived from its original name given to it by its ancient Iberian founders. The Roman name probably comes from the Oscan tribe.
Ibiza	Eivissa	Ebusus	The third largest of the Balearic Islands. Inhabited at one time by the Phoenicians, the name may be of Punic origin to mean 'Island of perfumes'.

Place Names of the World – Europe

Place	Former names	Description
Jerez de la Frontera	Asido Caesariana, Scheris, Xeres	Derived from the Roman Caesar with the added 'of the frontier', a reference to the Christians' advancing border with the Moors. Has given its name to sherry (17th century) from the Moorish version of Jerez, Scheris. Xeres is the same – the Spanish used to pronounce 'x' as 'sh' (16th century).
La Rioja		Named after a tributary of the River Ebro, the Rio Oja. It gives its name to the famous wine.
Lanzarote		One of the Canary Islands. Named after Lancelloti Malocello.
Las Palmas		The largest city on Gran Canaria island, it is so-named because of the large number of palm trees.
León	Legio	Named after the Roman 7th Gemina Felix Legion, *Legio septima*, which was deployed in the area (1st century AD).
Lugo	Lucus Augusti	Derived from the original Roman name, possibly meaning the 'Sacred place of Augustus'.
Madrid	Majerit	May mean 'Big ford' from *mago*, big, and *rito*, ford, although it is located on the unimpressive Manzanares River. Built as a fortress (1083) from the Moors by King Alfonso VI. Republic proclaimed here (1931). Civil War siege (1936–39). Officially capital (1607) although the court moved here (1561).
Majorca	Majorica	The largest of the Balearic Islands, the name is derived from the Latin *major*, greater. Known in Italy as Maiolica. The tin-glazed earthenware introduced into Italy from Majorca was given the name majolica.
Mallorca		

Málaga	Malaca	Derived from the Phoenician *malac*, to salt, on account of the trade in salt fish. Established as a kingdom (early 11th century–1487).
Melilla	Rusaddir	An exclave on the Moroccan coast captured by Spain (1497). Its border with Morocco was fixed at the maximum range of a cannonball fired from the fortress. Attacked and besieged many times by local tribesmen (which led to the War of Melilla, 1893–94) and Moroccan forces. The first town to rebel against the Popular Front government (17 Jul 1936).
Mérida	Augusta Emerita	This was a town that was founded (25 BC) for Roman veterans, *emereti*, and dedicated to Augustus, hence its original name which has been corrupted to Mérida.
Minorca	Menorca	The smaller of the two main islands of the Balearics, the name comes from the Latin *minor*. Its capital Mahón gave its name to mayonnaise, formerly mahonnaise.
Murcia	Mursiyah	Rebuilt and fortified, *mursah*, by the Moors. An independent Moorish kingdom (1063–1243).
Navarre	Navarra	An autonomous community and former Franco-Spanish kingdom. The Spanish part joined Spain (1515) and the French part France (1589). Taken from the Basque 'Nafarroa', the name for two groups of the Vascones living around Pamplona.
Orense	Aurium	Probably derived from the legendary gold, *oro*, of the River Miño. However, it may be derived from its hot springs which the Romans called Urentae or Aquae Originis.

Place	Other names	Notes
Palencia	Pallantia, Campi Gotici	The second name acknowledges the presence of Visigoths in the area.
Palma		Means 'Palm tree'. Capital of the Balearic Islands.
Pamplona	Pompaelo, Pampilona	Built (75 BC) by the sons of the Roman statesman and general Pompey (106–48 BC) who named the city after their father. Capital of the Kingdom of Navarre (11th century-1841). Pompaelo was also known as Pompeiopolis, Pompey's city, by the Romans and by the Moors as Pampilona or Banbalunah.
Plasencia	Ambroz	Renamed (12th century) by Alfonso VIII (1155–1214), King of Castile (1158–1214), to mean 'That it may please God', *eut Deo placeat*.
Pontevedra	Pons Vetus	Means the 'Old bridge' after the Roman bridge over the Lérez River.
Sagunto	Saguntum, Murbiter, Murviedro	Possibly founded by Greeks from the island of Zakinthos from which the name is derived. Besieged and destroyed by Hannibal (219 BC) after the refusal of Rome to come to its rescue. Hannibal's action was considered a declaration of war against the Roman Empire and triggered the Second Punic War (218–201 BC). The Moorish name, Murbiter, was derived from *muri veteres*, old walls, and later corrupted to Murviedro. Renamed Sagunto (1868).
Santander	Portus Victoriae	A corruption of the Spanish *Santa Irena*, St Irene.
San Sebastián	Donostia (Basque)	Named (1603) after St Sebastian, a fourth century martyr about whom virtually nothing is known. He may have been an officer of the Imperial Guard in Rome.

Santiago de Compostela		Named after the Apostle St James the Great (known in Spanish as Sant Iago), the patron saint of Spain, whose body, according to legend, was brought to Spain after his martyrdom (44); it was then lost and rediscovered (813) on the site of the present cathedral. Compostela is believed by some to be a corruption of the Latin *campus stellae*, a reference to the star which, according to legend, shone above the tomb to reveal its position. It may, however, come from *compos stellae*, possession of the star, to indicate that the city has the tomb.
Saragossa	Zaragoza	Salduba, Caesar Augusta, Saraqustah	Corrupted from the Roman name which it was given by the Emperor Caesar Augustus when it became a Roman colony (27 BC). Corrupted further to Sarakosta or Saraqustah. Capital of the Kingdom of Aragón (1118–1479). The Treaty of Saragossa (1529) solved a dispute between Spain and Portugal concerning possession of the Moluccas, a group of islands now a part of Indonesia, and provided for a line demarcating Spanish and Portuguese interests in the Far East.
Segovia	Segóbriga	Derived from the Celtic *sego*, strong, and *briga*, fort, built to thwart Muslim raiders; nevertheless, it was occupied by the Moors (8th century–1079).
Seville	Sevilla	Hispalis, Colonia Julia Romua, Ihbiliya/Ixvillia	Situated in the middle of a plain, its name is derived from the Punic *sefela*, plain. Capital of the Almohads (12th century–1248) until recaptured by Ferdinand III (1201?–1252, canonized 1671), King of Castile (1217–52) and of León (1230–52). Gives its name to the bitter orange used to make marmalade.
Tarragona	Tarraco, Colonia Julia Victrix Triumphalis	Capital of the Roman province of Hispania Tarraconensis. Renamed by Julius Caesar in honour of his victories.

Tenerife	Nivaria	The largest of the Canary Islands. Means 'Snow-capped mountain' after its highest peak which is over 3,700m (12,000 ft) high. The Roman name means 'Snowy'.
Toledo	Toletum, Tolaitola/Tulaytulah	The name may be derived from the Celtic *tol*, elevation or tor. It is situated on a promontory overlooking the River Tagus on three sides. Capital of the Visigoths (554–711), and of Spain (1087–1560).
Torremolinos		*Torres* means towers and *molinos* windmills around which the town was developed.
Valencia	Valentia, Valentia Edetanorum, Medina bu-Tarab	The Roman name means the 'Fortress of the Edetani'. The Arab name means 'City of joy'. Seat of the Republican Government (1936–37). Capital of the medieval Kingdom of Valencia (1010–1238).
Valladolid	Valad-Olid/ Balad-Walid	Means the 'Town of Olid', who was possibly the Moorish governor.
Vigo		Taken from the Latin *vicus*, settlement.
Vitoria	Gasteiz	Victoriacum (Basque)	Founded and named by the Visigothic King Leovigild to commemorate his victory over the Basques (581). The Duke of Wellington defeated the French under Napoleon's brother, Joseph Bonaparte, at the Battle of Vitoria (1813), thus forcing the French to leave Spain.

SWEDEN

Ironically, famous for its neutrality during the World Wars, Sweden was in earlier centuries a byword for violence: Europe quivered before the Vikings, King Gustav II Adolf and King Charles XII. Although the Roman historian Tacitus mentioned the Sviones (Swedes) or Svear in his famous *Germania* of 98 AD, little is known of Swedish history before the tenth century or how the Swedes in Svealand around modern Uppsala and the Goths further south in Götaland came to be united in an independent state. More is known of the Swedish warrior-trader Vikings who cultivated the trade routes through the Baltic and Russia to the Black Sea and the Byzantine Empire. Some settled and one tribe, the Rus, gave their name to Russia.

Although Olof Skötkonung was the first chieftain to declare himself to be the King of all Sweden at the beginning of the 11th century, it was a spurious claim. Not until the mid-12th century did Svealand and Götaland unite to form the Kingdom of Sweden. Skåne, Halland and Blekinge in the south were Danish and other provinces in the west and north belonged to Norway. By the end of the first half of the 14th century the Swedish king, Magnus II (meaning 'Great', a name which became popular from the Latin name of Charlemagne – Carolus Magnus) Eriksson, who had already inherited the Norwegian crown in 1319, ruled over Sweden, Norway and Finland. By the Treaty of Varberg in 1343 the Danish king recognized Magnus' possession of Skåne, Halland and Blekinge which the Danes had earlier mortgaged to Sweden; until Sweden renounced its claims to these provinces in 1570 sovereignty alternated between Denmark and Sweden. It was during Magnus' reign, in about 1350, that the principle of electing the king was codified.

Peace and stability were hard to come by. For the next 50 years territory was lost and won as the family of Magnus vied with each other and the nobles for power. In 1363 Margaret, daughter of King Valdemar IV of Denmark, married Magnus' younger son, Haakon (who had become Haakon VI of Norway in 1355 and joint King of Sweden with his father in 1362). Already the ruler of Denmark in her son's name since 1375, when her husband died in 1380 Margaret also became ruler of Norway; her son, Olav, was only ten. Olav died in 1387 and Margaret became regent of Denmark and Norway. In 1388 the Swedish nobles proclaimed her the 'rightful ruler' of Sweden and the following year her troops defeated Albert of Mecklenburg who had been offered the Swedish throne after driving off Magnus and Haakon in 1364. Margaret was given the right to choose the next king. With no heir she chose her nephew, Erik of Pomerania, and in 1397, at the age of 17, he was crowned King of Denmark, Norway and Sweden (which included Finland) in Kalmar, thus heralding the start of the Kalmar Union.

Erik was deposed in 1439 and within a few years open confrontation flared up between Sweden and Denmark. Christopher of Bavaria was chosen as king, but when he died in 1448 Sweden and Norway chose a Swedish nobleman, Karl Knutsson, as Charles VIII in the hope that he would become the Union king; the Danes, however, chose Christian of Oldenburg. War between Denmark and Sweden followed; twice Charles was forced to flee, once being replaced by Christian as the Swedish king. Not until 1483, when

Christian's son, Hans, was accepted as King of Sweden (John II) did the three kingdoms have a single king once more. In 1501 John was deposed and various regents ruled Sweden until 1520. That year a Danish mercenary army, on behalf of their king, attacked and overran Sweden, killing the regent. The Swedes were forced to acknowledge the authoritarian Dane, Christian II, as hereditary king. By now it was clear that Danish interests within the Union had become paramount and this only served to antagonise the Swedes further.

To stamp out dissent Christian proceeded to execute some 80 people, including nobles, in what became known as the Stockholm Bloodbath. This proved to be completely counter-productive, merely sparking more unrest and stronger demands for Swedish independence. Gustav Vasa, the nephew of a former regent, became the leader of the national cause and in 1521 was elected regent. Two years later Christian was deposed as the Danish king and the same year Gustav Vasa was elected as the Swedish king. Making the monarchy hereditary, he founded the Vasa dynasty. The Kalmar Union was dissolved and Sweden became independent.

From 1611 to 1718 Sweden was fortunate in having three particularly strong kings and the country enjoyed its 'age of greatness'. When Gustav II Adolf ascended the throne in 1611, Sweden was already at war with Denmark-Norway, Poland and Russia. Peace was concluded with all of them by 1629 (Truce of Altmark) by which time Sweden had acquired Ingria and Livonia (q.v.Latvia) and the control of some German and Polish ports in the Baltic. He was free now to enter the Thirty Years' War (1618–48) on the side of the German Protestants against their emperor who posed a threat to Sweden's security and religious freedom. In 1630 Gustav landed in northern Germany and swept as far south as Bavaria. He was killed in 1632. His heir, his daughter Christina, was only five. The governing council of nobles chose to continue fighting. At the Peace of Westphalia Sweden was awarded Western Pomerania and other territory along the Baltic and North Sea coasts, but the Polish ports had to be surrendered.

Christina came to the throne in 1644 but, determined not to marry and keen to convert to Roman Catholicism, she abdicated in 1654. She was succeeded by her first cousin who became Charles X Gustav. In 1655 he invaded Poland and succeeded in conquering most of the country. However, as resistance grew the Danes sensed an opportunity to humble the Swedes and they declared war. At once, Charles turned on them, occupied Jutland and threatened Copenhagen. At the Peace of Roskilde in 1658 Sweden won the provinces of Blekinge, Halland, Skåne, and Bohuslän (from Norway) and established the borders which exist today. In addition, the Swedish Empire included Estonia, Finland, Ingria, Karelia, Livonia and Western Pomerania.

When Charles XII, only fifteen, came to the throne in 1697 Sweden was not short of enemies. The Great Northern War began in February 1700 when Sweden's Baltic provinces were attacked by Russia, Poland and Denmark. Charles, a great warrior king who thrilled to danger, quickly forced the Danes to sue for peace and then decisively defeated the Russians at the Battle of Narva in Estonia in November. After four years of confrontation with Poland (1702–06), including the occupation of Warsaw and the dethronement of the king, Charles began his offensive against Russia in 1707 with the aim of destroying tsardom and the threat posed by Russia. Diverted from attacking Moscow by Peter the Great's scorched-earth policy, he was lured into Ukraine where he suffered a heavy defeat at Poltava in 1709. Charles escaped to Moldavia, then part of the Ottoman Empire, from where he still managed to rule over his distant realm. He did not return to Sweden until 1715. He was killed in 1718 fighting in Norway. The Peace of Nystad (now Uusikaupunki, Finland) in 1721 formally ended the Great Northern War. Sweden relinquished Estonia, Ingria, most of Karelia and Livonia to Russia which returned Finland except for Vyborg. Other territorial concessions were made in the west to Denmark, Hanover and Prussia. The Peace brought to an end Sweden's standing as a great European power and its interference in the affairs of Denmark, Germany, Poland and Russia.

In 1805 Sweden entered the Napoleonic Wars against the French, but found itself politically isolated when France and Russia signed the Treaty of Tilsit in 1807. The loss of a third of its territory (Finland, the Åland Islands and land in the north) to Russia followed a Russian attack in 1808. Succession to the throne became another issue when the incompetent Gustav IV Adolf was deposed in March 1809 and the senile and childless Charles XIII was elected. By now there was a growing feeling, particularly among the junior ranks of the officer corps and the government, that Sweden should join France. It was proposed that Marshal Bernadotte, one of Napoleon's commanders, should be offered the crown in the hope that he might be able to get Napoleon's help against Russia. As crown prince from October 1810, Bernadotte took the name Charles John. In 1818 he became king as Charles XIV John, founding the existing Swedish royal dynasty.

Charles John's relations with Napoleon were uncomfortable. They were further strained in January 1812 when French troops marched into Swedish Pomerania to secure their rear areas before marching on Moscow. This was an act of war. In April Charles John concluded an agreement with Russia whereby Sweden would support Russia against Napoleon and allow Finland to remain a Russian possession, but only after the Russians had helped Sweden to seize Norway from Denmark. With Napoleon's defeat at the Battle of Leipzig in October 1813, Charles John was free to lead an army against Denmark. Resistance was light and at the Treaty of Kiel in January 1814, Denmark had to surrender Norway. Having just declared their independence, the Norwegians were resentful and the new union had to be backed up by a Swedish invasion. It was not until November that final agreement on the absorption of Norway was made. The forced union did not prove a success and in 1905 it was peacefully dissolved.

During the two world wars Sweden adopted a policy of neutrality and strong defence of its own territory; despite this, it allowed German troops to transit Swedish territory *en route* to and from Norway.

Area:	449 964 sq km (173 731 sq miles)
Population:	8.9 million; Swedes, Finns, Yugoslavs
Language:	Swedish
Religions:	Evangelical Lutheran, Roman Catholic, Pentecostal
Capital:	Stockholm
Administrative Districts:	23 counties
Date of Independence:	1523
Neighbouring Countries:	Norway, Finland

English-speaking Name	Local Name	Former Names	Notes
SWEDEN	Sverige	Svithiod	The Kingdom of Sweden (Konungariket Sverige) from about 1000 with a constitutional monarchy since Jun 1809. Named after a powerful Germanic people, the Svear or Sviones. *Svea rike* means the Kingdom of the Svear. Gave its name to suede, originally gloves made in Sweden, *gants de Suède*.
Bofors			May mean 'To live', *bo*, 'by the rapids', *fors*. Alfred Nobel (1833–896), founder of the Nobel prizes, owned an armaments factory here which gave its name to a rapid firing anti-aircraft gun, the Bofors.
Eskilstuna		Tuna	Renamed to commemorate St Eskil, an English priest who became a bishop and who was stoned to death (c.1080) after protesting against an idolatrous festival. He was buried here.
Filipstad			Named after Karl (Charles) Filip, son of King Charles IX (1550–1611, r.1604–11).
Gothenburg	Göteborg		Means 'Fort of the Goths'. The city was founded (1603) on the site of a Gothic settlement.
Gotland			A county and an island. Means 'Land of the Goths'. Originating in southern Scandinavia, they migrated to the Black Sea (end 2nd century) and later split into two: the Visigoths, Western Goths, and the Ostrogoths, Eastern Goths.
Helsingborg	Hälsingborg		Takes its name from the Danish port Helsingør lying opposite with *borg*, fort, added. Ceded to Sweden (1658) by Denmark.

Jokkmokk		Means 'Bend in the river' from the Lapp *jokk*, bend.
Jönköping		*Köping* means 'Market town' but the meaning of Jön is unknown.
Karlskrona		Means 'Charles' crown'. Founded by and named (1680) after King Charles XI (1655–1697, r.1660–97).
Karlstad	Tingvalla	Originally named after the *ting* (or *thing*), the local or provincial assembly of freemen, the highest decision-making authority. It was usually held in the open air on raised ground; *valla* means mound. Then (1584) after King Charles IX. The site of the Treaty of Karlstad (1905) which confirmed the dissolution of the union with Sweden.
Kristianstad		Founded by and named (1614) after Christian IV (1577–1648), King of Denmark and Norway (1588–1648) when Skåne belonged to Denmark. Ceded to Sweden (1658), recovered (1676) by Christian V (1646–99, r.1670–99) and finally taken by Sweden (1678).
Kristinehamn	Bro	Renamed (1642) after Queen Christina (1626–84) who reigned for only ten years (1644–54) before abdicating.
Linköping		Means 'Flax market' from *lin*, *flax*, and *köping*.
Luleå		Named after the River, *å*, Lule.
Malmö	Malmhaug, Elbogen	Means 'Mineral island' from *malm*, mineral, and the Danish *ö*, island. Originally meant 'Sandpile' and renamed by German merchants 'Elbow' because of the curve in the coastline. Ceded to Sweden (1658).

Norrköping		Means 'Northern market town'.
Nyköping		Means 'New market town'.
Örebro		Means 'Gravel bridge', a reference to the medieval ford across the royal route from Oslo to Stockholm.
Stockholm	Agnefit	Built on numerous islands, it means 'Pole island' from *stak*, pole or log, and *holm*, island. According to legend, its first name comes from a Viking warrior king, Agne, who went to Finland. There he snatched a chieftain's daughter, Skjalf, and killed her father. Returning home, he stopped on one of Stockholm's islands, drank heavily and fell asleep. Skjalf freed her fellow prisoners and together they killed Agne before setting off for Finland. Capital since 1523. Gave its name to the 'Stockholm syndrome', a condition experienced by some hostages who come to identify themselves with their captors even to the extent of refusing to cooperate with the police when released.
Umeå		Means roar, *uma*, from the sound of the rapids on the River Umeå.
Uppsala	Östra Aros	Means 'Upper Sala', that is, above Sala, previously a village. Originally founded as a trading post at the point just east of where the River Fyris becomes navigable; Aros means 'river mouth'.
Västerås	Aros, Västra Aros	The name has evolved from *Väst*, west, *å*, river, and *os*, mouth, the original site to the west side of the river mouth.
Växjö		Named for its location where trading routes, *väg*, met on Lake, *Sjö*, Växjö.

Visby

Means 'Settlement or hamlet', *by*, at the sacred (pagan) site, *vi*. S was added later.

SWITZERLAND

After Julius Caesar had defeated one of the strongest Celtic tribes, the Helvetii, in 58 BC, Roman settlement and peace prevailed for the next 300 years. In 259 Germanic tribes began their penetration of the region, the Burgundians to the west and the Alemannians (from the 11th century called Swabians) to the south. In 843 at the Treaty of Verdun Switzerland was divided between the sons of Charlemagne (q.v.France). The Burgundian area went to the Middle Kingdom of Lothair I while the Alemannian became part of the East Frankish kingdom of Louis II the German. During the 11th century modern Switzerland passed to the Holy Roman Empire but, as it declined, a group of semi-independent states began to emerge which, by the end of the 13th century, had come to be dominated by the Habsburgs.

In 1291 the Emperor Rudolf died. Concerned about the possibility of greater Habsburg control and wishing to establish some sort of separate political identity to include a common defence against all enemies, the German-speaking forest communities of Uri, Schwyz and Unterwalden signed an agreement of mutual assistance – the Everlasting League – whereby an attack on one was to be considered an attack on all; any disputes between them would be resolved without external mediation. From this 1291 agreement stemmed the Swiss Confederation.

When the Habsburgs tried to take control of the Reuss valley in Uri and the St Gotthard Pass in 1315, they were roundly defeated at Morgarten. The League was confirmed and a new clause introduced to deny to a member the right to negotiate peace or enter a new alliance without the consent of the others. The aim now changed to abolish all Habsburg feudal rights within the area. By the end of the 14th century five more communities, or cantons, wary of the Habsburgs, had joined; for the first time, three were urban cantons based on Bern, Lucerne and Zürich. Although the association was looser than at the time when the original three had come together, common hostility towards the Habsburgs ensured uniform aims and action.

When the Emperor Maximilian I tried to reassert Habsburg control over Switzerland from the east, the eastern-most cantons joined with Romansch-speaking Graubünden to counter him and the Swabian League, an alliance of south German principalities. In their war against Burgundy in 1474–76, the Swiss proved their military prowess and this, combined with the mountainous terrain of their homeland, erased any doubts about their military security. The short-lived Swabian War, fought to prevent further Swiss expansion, followed in 1499 when Graubünden was attacked and the cantons came to its aid. Maximilian yielded within months and in so doing tacitly accepted Swiss independence. With another five cantons joining between 1481 and 1513, 13 independent statelets now formed the Confederation. In addition, it had 'subject' territories such as Thurgau and Italian-speaking Ticino. Nevertheless, no formal declaration of independence had yet been made.

Despite conflict between Catholic rural cantons and Protestant urban cantons during the Reformation, a full-blown civil war was avoided. The ties binding the Confederation proved strong enough to enable it to remain aloof from the Thirty Years' War (1618–48). At its conclusion the Peace of Westphalia recognized Swiss independence of the Habsburg

Empire and within 30 years the Swiss had embarked on a formal policy of neutrality, the only way, in their view, to maintain that independence. Nevertheless, the 13 cantons still had no central government or joint army.

Switzerland was thus poorly placed to face the consequences of the French Revolution. Basle was annexed by France in 1793 and in 1798 Napoleon invaded, bringing an end to the Confederation and establishing the Helvetic Republic. It lasted only to 1803 by which time anarchy was widespread. By means of the Mediation Act Napoleon imposed a new constitution. The 13 cantons were joined by six more to form the Helvetic Confederation. In 1815 the Congress of Vienna guaranteed the neutrality of the Confederation as another six cantons, including the predominantly French-speaking Geneva, Neuchâtel and Valais, were added to it. The 26th canton, Jura, was created out of the Bern canton in 1974, becoming a canton in its own right in 1979.

Despite differences of opinion between French and German speakers, Switzerland remained neutral during the Franco-Prussian War and the First World War. In 1920 Switzerland joined the League of Nations but, when it became enfeebled in the 1930s, left it in favour of total neutrality. This was proclaimed when the Second World War broke out. Switzerland has not joined the United Nations in the belief that this would compromise its neutrality.

Area:	41 293 sq km (15 943 sq miles)
Population:	7.41 million; Swiss (German, French, Italian)
Languages:	German, French, Italian, Romansch
Religions:	Roman Catholic, Protestant, Muslim
Capital:	Berne
Administrative Districts:	26 cantons
Date of Independence:	1648
Neighbouring Countries:	France, Germany, Austria, Liechtenstein, Italy

English-speaking Name	Local Name	Former Names	Notes
SWITZERLAND	Die Schweiz (Ger), La Suisse (Fr), Svizzera (It)	Helvetia	The Swiss Confederation (Schweizerische Eidgenossenschaft (German), Confédération Suisse (French), Confederazione Svizzera (Italian), since 1291. Takes its name from Schwyz (1315), one of the three original cantons, whose inhabitants became known as Schwyzers. Schweiz became the official name (1803).
Aargau	Argovie (Fr)		A canton which joined the Confederation (1803). Takes its name from the River Aare.
Appenzell		Abbatis Cella	Means the 'Abbot's cell'. Also the name for two demi-cantons (1513) whose name comes from their abbot-rulers, the prince bishops of St Gallen.
Basel/Basle	Bâle (Fr)	Robur, Basilia	Originally a Celtic settlement and given the Roman name Robur from the Latin *roburetum*, oak grove. The present name is derived from the Greek *basileia*, royal, when Emperor Valentinian I (321–75, r.364–75) developed the settlement into a fort (374). It is also the name for two demi-cantons, one urban (1501) and one rural, which joined the Confederation (1833) when the original Basel canton was split in two.
Berne	Bern		According to legend, Berthold V, Duke of Zähringen, killed a bear, *bär*, while hunting in the local area (1191) and the name is taken from the German word. Capital (1848). Also a canton (1353).
Fribourg	Freibourg (Ger)		Means 'Free fort' and comes from the Latin *Friburgum*. Also a canton (1481).

Geneva	Genève (Fr), Genf (Ger)	Genava	Its name may be derived from the Indo-European root *gen*, bend, a reference to the shape of Lake Geneva at its southern end where the city lies. The Geneva Convention is a series of treaties signed between 1864 and 1949 which seek to establish the way soldiers (prisoners and the wounded) and civilians are treated in war. By means of the Geneva Accords (1954) a ceasefire was arranged between the French and the Viet Minh along the 17th parallel which effectively cut Vietnam into two states, North and South. Also a canton (1815).
Graubünden	Grisons (Fr), Grigioni (It)		A canton (1803). Means 'Grey Leagues', formed (1395) to resist the growing power of the Habsburgs in the Upper Rhine valley. So called because the men wore grey cloth. Grey is also to be found in the French and Italian names and the Romansch Grishun.
Interlaken			Means 'Between the lakes' (Brienz and Thun).
Jura			A canton (1979) whose name means 'Forested mountain' from the Gaulish *jor* or *juria*. The discovery of fossils in the Jura Mountains gave rise to the geologic time from about 210 to 145 million years ago known as the Jurassic Period. Also the name of a French *département*.
Lausanne		Lausonium, Lausonna, Lausodunum	The Roman name means 'Fort on the (River) Laus'. The Treaty of Lausanne (1923) was the final First World War treaty. It was signed between Turkey, six European nations and Japan, and established the boundaries of modern Turkey; it also recognized British possession of Cyprus.
Lucerne	Luzern (Ger)		The name is derived from the Benedictine monastery of St Leodegar (Luciaria). Capital of the Helvetic Republic (1798–1803). Also a canton (1332).

Lugano		Derived from the Gaulish *lacvanno*, lake dweller, to describe the people, living along the shores of Lake Lugano.
Martigny	Forum Claudii Vallensium, Octodurum	Originally named after the Roman Emperor Claudius (10 BC-54 AD, r.41-54).
Montreux		Derives its name from the Latin *monasterium*, monastery, around which the town developed.
Neuchâtel	Neuenburg (Ger), Novum Castellum	Means 'New Castle'. A personal principality of the King of Prussia (1707-1857). Also a canton (1815).
Nidwalden		A demi-canton (1291). Part of the former Unterwalden canton, one of the original three, whose name means 'Below the forest' (of Kerns).
Obwalden		A demi-canton (1291). Like Nidwalden, part of the former canton of Unterwalden. Means 'Above the forest'.
Sankt Gallen	St Gall (Fr)	Named after St Gall (?- c.640) an Irish monk, who built a hermitage here (612). Having split from St Columban (612), another Irish monk, legend has it that he stumbled into a briar patch which he took to be a sign from God that he should settle here. Also a canton (1803).
Schaffhausen	Schaffhouse (Fr), Villa Scafhusun	Means 'Sheep house' but it is not known why. Also a canton (1501).
Schwyz	Suittes	The valley, in which the original village of Suittes is situated, is highly forested and it may take its name from the Old High German *suedan*, to burn, indicating forest clearance. Also a canton (1291) whose leading role in the Everlasting

			League caused the Confederation to be named after it.
Uri			A canton (1291). Probably derived from the Latin *urus*, aurochs, a wild ox now extinct.
Valais	Wallis (Ger), Vallese (It)	Vallis Poenina	A canton (1815) whose name comes from its earlier Latin name meaning Upper Valley of the Rhône.
Vaud	Waadt (Ger)		A canton (1803) whose name may be derived from what the inhabitants were called by their neighbours, *walho*, strangers.
Winterthur		Vitodurum	Corrupted from the Roman name which means, probably, 'Vito's fort'.
Zürich		Turicum	The name may be derived from the Celtic *dur*, water, a reference to the Helvetii's settlement on the edge of Lake Zürich. The personal name Turus also appears in documents. Also a canton (1351).

UKRAINE

During the first millennium BC Cimmerians, Scythians and Sarmatians all occupied what is modern Ukraine while Greeks colonized the northern shores of the Black Sea. In the first centuries of the next millennium the Greeks were absorbed into the Roman Empire while Ostrogoths and then Huns supplanted the early settlers to the north. They were followed by the Slavs who appeared in the sixth and seventh centuries. It was not until about 880 that a Varangian (Viking) tribe called the Rus arrived to join with the Slavs to found the state of Kievan Rus with Kiev as its capital. It was strategically located on the River Dnepr, a waterway which facilitated trade between the Baltic Sea and Byzantium. Within 100 years Kievan Rus had expanded to stretch from the Baltic in the north to the Black Sea steppes in the south, and from the River Volga in the east to the Danube, becoming the dominant power in Eastern Europe.

During the twelfth century, weakened by internecine struggles, Rus declined, unable to defend itself from its enemies. In 1237, out of the depths of Asia, came the Tatars. Under Batu Khan, grandson of Genghis Khan, they laid waste to the territory for the next three years.

From Rus there emerged three ethnic branches: Russians in Muscovy to the northeast, White Russians (Byelorussians) to the north and Ukrainians – those people that lived on the edge, *okraina*, of Rus (and sometimes known as Little Russians by the Russians who considered them to be no more than a separate tribal grouping). The most important principality on what is now Ukrainian territory was Galicia-Volhynia which could claim to be the first Ukrainian state. In 1349 it ceased to exist, Galicia passing to Poland and Volhynia to Lithuania; southern Ukraine remained in the hands of the Tatar Golden Horde (q.v.Russia). A century later the Golden Horde was in disarray; from it emerged the Crimean khanate which submitted to the Ottoman Empire in 1475, opening the way to Ottoman domination of the Ukrainian coast.

In 1569 at the Union of Lublin Lithuanian control of Ukrainian territories was transferred to Poland. By tying the peasantry to the land, the Poles succeeded in alienating them. Some fled southwards to become known as Cossacks ('outlaws' or 'adventurers'), hostile to the Polish authorities who regarded them as rebels. Cossack resistance gradually became more focused and open rebellion, led by the Hetman (chieftain) of the Zaporozhian Cossacks, Bohdan Khmelnytskyy, broke out in 1648. In various encounters both the Poles and the Cossacks suffered defeat. To overcome the Poles Khmelnytskyy realized that he would need external assistance and so, in 1654, he concluded the Treaty of Pereyaslav with Muscovy whereby the Cossack state of Eastern Ukraine, Kiev and the lands east of the River Dnepr, came under Russian suzerainty in return for protection from the Poles.

In 1660 Russia and Poland came to blows over control of Ukraine. At the Treaty of Andrusovo in 1667 Ukraine was divided along the River Dnepr: the Poles took the western region while Russia kept Eastern Ukraine. The presence of large numbers of Russians in Eastern Ukraine today is a hangover from that agreement. Trying to unite both halves of Ukraine, Hetman Ivan Mazeppa concluded an alliance with Charles XII of Sweden following his invasion of Poland in 1708: Ukraine was to become independent under Mazeppa's control. But after the Russian defeat of the Swedes at the Battle of Poltava in 1709, Ukrainian dreams of independence withered. At the first partition of Poland in 1772 Galicia in the far west of Ukraine came under Austrian rule to be followed by Bukovina two years later, while Russia took the rest of Western Ukraine, including Volhynia, in the second (1793) and third partitions (1795). With the Russian annexation of the Crimean khanate in 1783 Ukrainians were encouraged to move south and settle, thus increasing the area of ethnic Ukrainian territory. Nevertheless, these lands were called Novorossiya, New Russia.

In an attempt to suppress any ideas of Ukrainian nationalism and strengthen the integration process the Tsarist authorities banned the public use of the Ukrainian language in schools, in the theatre and in books as a result of an abortive Polish uprising against Russian rule in 1863. This was reinforced in 1876, but an unexpected outcome was the appearance of an embryo Ukrainian nationalist movement outside Russia – in Austrian-ruled Galicia.

Following the Bolshevik revolution in 1917 and the end of the First World War, Ukraine had a chance to win independence: a Ukrainian National Republic was proclaimed in November 1917 and two months later independence was declared. The Bolsheviks, however, announced that Ukraine was a Soviet republic and occupied Kiev. But at the Treaty of Brest-Litovsk in March 1918 they were forced to recognize Ukrainian independence and the Red Army withdrew from Kiev in April. After the defeat of Germany the same year the Republic merged with the Ukrainian territory, by this time known as the West Ukrainian National Republic, given up by Austria-Hungary in October 1918. The occupation of Lviv by Ukrainian troops in November sparked off a war with a Poland keen to include Galicia in the new Polish state. Thereafter Ukraine became a battleground over which Bolsheviks, anti-Bolsheviks, Ukrainians and Poles fought. By the end of 1919 almost the whole of Ukraine was occupied by the Red Army.

In May 1920 Polish and anti-communist Ukrainian troops turned the tables and entered Kiev. The Bolsheviks counter-attacked and reached the gates of Warsaw. The Poles threw them back but the Russians had by now secured Eastern Ukraine and in December 1920 a Ukrainian Soviet Socialist Republic (SSR) was formed. Western Ukraine, however, was a different matter. In March 1921 the Treaty of Riga brought an end to the Russo-Polish war. Galicia and part of Volhynia were confirmed as belonging to Poland, Transcarpathia was awarded to the new state of Czechoslovakia and Bukovina to Romania. The new Polish-Soviet Russian border now followed a line from the River Dvina in the north, west of Minsk in Byelorussia to the Romanian border on the River Dnestr; about five million Ukrainians and Byelorussians were left in Poland. Ukraine became a founder member of the USSR in December 1922.

Following the Nazi-Soviet non-aggression pact in August 1939, the Red Army moved into Polish Ukraine, and eastern Galicia and western Volhynia were seized. In June 1941 the Germans invaded the Soviet Union and remained in occupation of Ukraine until October 1944. They did not meet wholesale resistance, a Ukrainian Insurgent Army being formed in 1943 which opposed Soviet rule until the 1950s.

Victory in 1945 brought its own rewards. Eastern Galicia and Volhynia, Czechoslovak Transcarpathia (under Hungarian control during the Second World War), northern Bukovina and southern Bessarabia were all ceded to the Soviet Union and most ethnic Ukrainians were thereby scooped up into an expanded Ukrainian SSR. By extending

Ukraine westwards, the Soviet Union gained a common border with both Czechoslovakia and Hungary which facilitated the move of Soviet troops into both these countries. Ukraine was granted its own seat in the United Nations.

In 1954 Crimea was transferred from Russia to Ukraine in commemoration of the 1654 Treaty of Pereyaslav. With its largely Russian population Crimea became a bone of contention in the 1990s, the Crimean Parliament even voting to declare independence from Ukraine in May 1992; the vote was quickly annulled in Kiev. In May 1997, however, a treaty was signed between Russia and Ukraine recognizing current borders and thus Ukrainian sovereignty over Crimea; simultaneously, the Russians were allowed to lease certain Crimean facilities, not territory, for its Black Sea Fleet for 25 years.

Mikhail Gorbachev's reforms in the late 1980s provoked a new national awareness in Ukraine and sovereignty was proclaimed in July 1990. Following the attempted *coup* in Moscow in August 1991, the Ukrainian Supreme Soviet declared independence, subject to approval by the population in a referendum in December. This was forthcoming. A week later the Russian Federation, Belarus and Ukraine agreed to form the Commonwealth of Independent States in anticipation of the dissolution of the USSR which occurred two weeks later.

Area:	603 700 sq km (231 990 sq miles)
Population:	50.8 million; Ukrainians, Russians, Jews
Languages:	Ukrainian, Russian, Romanian, Polish
Religions:	Russian Orthodox, Uniate Catholic
Capital:	Kiev
Administrative Districts:	24 provinces and the Autonomous Republic of Crimea
Date of Independence:	5 December 1991
Neighbouring Countries:	Belarus, Russia, Moldova, Romania, Hungary, Slovakia, Poland

English-speaking Name	Local Name	Former Names	Notes
UKRAINE	Ukraina	Malorossiya	Previously the Ukrainian Soviet Socialist Republic (1922–91) with its own seat in the United Nations. Derived from the Russian *okraina*, land on the edge or borderland, from *u*, beside, and *kray*, region, to denote the territory between the open steppes of Russia to the east and the populated lands of the Polish-Lithuanian Commonwealth to the west. The present name was not in wide use until the 19th century; until then Ukrainians were often called Ruthenians and the country was known as Little Russia.
Alchevsk		Voroshilovsk, Kommunarsk	Founded (1895) and named after Alchevsky who founded the ironworks. Renamed (1931–61) after Marshal Kliment Voroshilov (1881–1969), Red Army commander and Soviet president (1953–60), and Kommunarsk (1961–92).
Artemovsk	Artemivsk	Bakhmut	On the River Bakhmut. Renamed (1924) after Fedor Sergeyev (1883–1921), an early Bolshevik leader, nicknamed Artem.
Bakhchysaray			Means 'Garden palace' in Turkish. Capital of the Crimean khans (15th century-1783).
Balaklava		Symbalon, Cembalo	A Turkish name meaning 'Fish nest'. Conquered by the Genoese (1357), Turks (1475) and Russians (1783). Site of the indecisive Battle of Balaklava (Oct 1854) during which the charge of the British Light Brigade took place. Gave its name to the balaclava, a knitted woollen covering for the head and neck worn by the British against the cold.
Belgorod-Dnestrovsky	Bilhorod-Dnistrovskyy	Tyras, Akkerman, Cetatea Alba	Means 'White City on the (River) Dnestr'. Akkerman (Turkish) and Cetatea Alba (Romanian) both mean 'white

fort'. Tyras was the Greek name for the Dnestr. In Turkish hands (1484–1812), Russian (1812–1918), and Romanian (1918–40).

Berdyansk	Osipenko	Previously named (1939–58) after Polina Osipenko (1907–39), a fighter pilot in the Soviet Air Force and a Hero of the Soviet Union who set five aviation world records for women.
Bukovina		Part Ukrainian, part Romanian, it means the 'Land of the beech trees' from the Slavonic *buk*, beech tree. After acquiring its own name and identity (1775), it was ceded by the Turks to Austria (1777). Became a duchy and a crown land. Part of Romania (1918–40, 41–44).
Chernivtsi	Czernowitz, Cernauti Chernovitsy	Under Turkish rule (18th century), then within the Austro-Hungarian Empire (1775–1918), Romania (1918–1940), and then taken over by the USSR as Chernovitsy, and becoming Chernovtsi (1944), a result of the Nazi-Soviet non-aggression pact (1939).
Chervonograd	Krystynopol	Means 'Red town' from *chervonnyy*, red – meaning communist, and *hrad*, town. Renamed (1953).
Crimea	Tauric Peninsula (or Taunda) Gotland	An autonomous republic. Derived (15th century) from the Turkish *kerim*, fort. When the Tatars of the Golden Horde entered the peninsula they made their first encampment beneath the remains of a stone tower which they called a *kerim*. One of the three most important Tatar khanates (founded 1443). Subdued and annexed by Russia (1783). The Crimean War (Oct 1853-Feb 1856) was fought principally on the peninsula between the Russians and the British, French and Ottoman Turks. Unofficially called Gotland during the Nazi occupation (1942–44). Formerly part of Russia until transferred (1954) to Ukraine to commemorate

			the Treaty of Pereyaslav (1654) which incorporated much of Ukraine into Russia. There may be a link between *kerim*, *Krim* and *kreml*, Kremlin.
Dneprodzerzhinsk	Dnipro-dzerzhynsk	Kamyansk/Kamenskoye	Renamed (1936). The first part of the name refers to the fact that the city lies on the River Dnepr (or Dnipro in Ukrainian). The second part is named after Felix Dzerzhinsky (1877–1926), a Pole who founded the Soviet secret police, the Cheka.
Dnepropetrovsk	Dnipropetrovsk	Yekaterinoslav, Novorossiysk	Lying on the River Dnepr, the second part is named after Grigory Petrovsky (1878–1958), leading figure in the Ukrainian Communist Party until Khrushchev arrived (1938). Originally named (1783–96) after Empress Catherine II the Great (1729–96, r.1762–96) by its founder, Grigory Potemkin (1739–91), Catherine's right-hand man and possibly her secret husband, to mean 'To the Glory, *slava*, of Catherine'. It then became Novorossiysk (New Russia – 1796–1802); Yekaterinoslav was then restored (1802–1926) until the present name was adopted.
Donetsk		Yuzovka, Trotsky, Stalino	Originally named (1862) after John Hughes, a Welsh engineer who established what later became the largest steelworks in Imperial Russia. Renamed (1920) after Leon Trotsky (1879–1940), Bolshevik commissar for foreign affairs and for war during the Civil War. When Trotsky fell from favour named (1924–61) after Stalin. Reverted to Yuzovka during the German occupation (1941–43) and, near the River Donets, became Donetsk (1961). It is the main city in the Donbas(s), a shortening of *Donetsky Ugolny Basseyn*, the Donetsk Coal Basin.
Ivano-Frankovsk	Ivano-Frankivsk	Stanisławów, Stanislav	Founded (1661) as a Polish frontier town and named after a Polish prince, Stanislav. Held by Austria (1772–1918), then

		Poland until annexed by the Soviet Union as a result of the Soviet-Nazi non-aggression pact (1939). Ceded to Russia (1945) and renamed Stanislav. Renamed (1962) after Ivan Franko (1856–1910), poet and novelist.
Kadiyevka	Sergo, Stakhanov	Former name (1978–992) after Alexei Stakhanov (1906–77), a coalminer who (30 Aug 1935) voluntarily dug 102 tons of coal during a single night compared to the norm of 6.5 tons during a 5¾ hour shift. 102 tons represented twice the amount expected from a squad of eight men. Later (19 Sep) he produced 227 tons in a single shift. He was emulated by other 'shock workers' whose new working methods – largely improved technique and huge exertion with new equipment – to increase personal productivity led to the creation of the Stakhanovite movement. Previously named after Grigoriy Ordzhonikidze (1886–1937), a Bolshevik leader who was known as Sergo.
Kerch	Panticapaion Panticapaeum, Korchev	Founded by the Greeks (6th century BC), it was the chief city of the Kingdom of the Bosphorus and was later absorbed into the Roman Empire. Ceded by the Tatars to the Genoese (1318) and renamed Korchev from which the present name is derived. This may have been chosen because of the iron mines in the area, *kerch* being a Slavonic root word meaning metal worker. Passed to the Turks (1475) and the Russians (1771).
Kharkov	Kharkiv	Thought to be derived from the name of its Cossack founder, Kharko, who built a military fortress here. Capital of Soviet Ukraine (1917–34).
Kherson	Chersonesos, Korsun, Korsun–Shevchenkovsky	Settled by the Greeks (5th century BC), Chersonesos means peninsula from *khersos*, dry land, and *nësos*, island, and at that time referred to the whole of the Crimea. It subsequently gave its name to the port which lies some 25 km

		from the mouth of the Lower Dnepr River and is not in the Crimea. The original ruins of Chersonesos lie some 5 km west of Sevastopol in the Crimea. Shevchenkovsky added (1944) after Taras Shevchenko (1814–61), poet.
Khmelnitsky	Khmelnytskyy Ploskurov, Prokuriv/Proskurov	Founded as a Polish fort (15th century) and named after the River Ploskaya, *ploskii* meaning flat. Renamed as Proskurov when ceded to Russia at the second partition of Poland (1793). Renamed again (1954) after the great Cossack leader Bohdan Khmelnytskyy (c.1595–1657), Hetman of Ukraine (1648–57), who organized an uprising against the Poles which led to the transfer of Ukraine east of the Dnepr River from Polish to Russian control under the Pereyaslav Agreement (1654). The province of Khmelnytsky was known as Kamenets-Podolsky (also the name of a city, now Kamyanets Podilskyy) until 1954 at which time Proskurov also gave way to Khmelnytskyy.
Kiev	Kyyiv	According to legend, named after the eldest of three eastern Slav brothers, Kiy, who together with their sister founded a city on the heights above the Dnepr River. Capital of Kievan Rus (882), of Soviet Ukraine (1934) and of Ukraine (1991). Devastated by the Tatars (1240), captured by the Lithuanians (1362) and given to Poland by the Union of Lublin (1569). Within an autonomous Cossack state, came under Russian protection (1667) until given to Russia (1686). Briefly occupied by the Poles (1920) and by the Germans (1941–43). Gives its name to Chicken Kiev: when the Hotel Kiev was opened (1957) in Moscow a new chicken dish was served.
Kirovograd	Yelizavetgrad, Zinovyevsk, Kirovo Kirovohrad	Originally named (1775) 'Elizabeth's town' after Empress Elizabeth Petrovna (1709–62, r.1741–62). Renamed (1924–36) after Grigoriy Zinoviev (1883–1936), Bolshevik revolutionary and one of Lenin's principal colleagues, whose birthplace this was. Renamed Kirovo (1936) and Kirovograd (1939), 'Kirov's town', after Sergey Kirov (1886–1934), communist

leader of Leningrad, for complicity in whose murder Zinoviev was executed.

Kremenchug	Kremenchuk		May be derived from the Russian *kremen'*, flint or the Turkic *kermen*, fort. Kremenchug was founded (1571) as a fortress.
Krivoy Rog	Kryvyy Rih		Literally means 'Crooked horn', and is taken to mean 'curved bend'. The city lies at the confluence of two rivers.
Lugansk		Yekaterinoslavsk, Voroshilovgrad	Originally named (1795–97) after Catherine II the Great. Then (1797–1935) Lugansk after the River Lugan and Voroshilovgrad (1935–58) after Marshal Voroshilov. Reverted to Lugansk (1958–70) after Voroshilov took part in an unsuccessful attempt to overthrow the Soviet leader, Nikita Khrushchev (1894–1971). Became Voroshilovgrad again (1970–89) after Voroshilov's death and subsequent rehabilitation.
Lvov	Lviv	Lwów, Lemberg	Built by Danylo (Daniel), Prince of Galicia-Volyn (mid-13th century), as a castle-town and named after his son Lev, meaning lion (Lemberg, lion's fortress). Came under Polish rule (1349) and, although seized by the Cossacks (1648) and the Swedes (1704), remained so until given to Austria at the first partition of Poland (1772). Seized by the Poles (1918), captured by the Russians (1939) and, after occupation by the Germans (1941–44), annexed by the USSR (1945).
Mariupol		Pavlovsk, Zhdanov	Renamed Mariupol (1779–1948), 'Mary's city', after Maria Fyodorovna (she was originally Sophia Dorothea of Württemberg), wife of the future Emperor Paul I (1754–1801, r.1796–1801). Renamed (1948–89) on the death of Andrey Zhdanov (1896–1948), leader of the Leningrad Party organization and of that city's defences against the German siege (1941–44), whose birthplace this was.

Mukachevo	Mukacheve	Munkács	Founded as a fortress to guard a pass across the Carpathian mountains and named after a certain Mukach with the possessive suffix *evo* added. In Hungarian hands (1018–1920), ceded to Czechoslovakia (1920) and to the Soviet Union (1945).
Nikolayev	Mykolayiv	Olbia	Founded near the ancient Greek city of Olbia and then renamed after St Nicholas because the modern town was founded (6 Dec 1788) on his feastday.
Nikopol		Slavyansk, Nikitin Rog	The original name means 'Slav town'. Founded on a bend on the River Dnepr, it was then developed and renamed 'Nikita's horn' (1630s). Finally renamed (1782) as 'Victory town' from the Greek *nikē*, victory, and *polis*, town.
Odessa	Odesa	Odessos	A Turkish fort called Khadzhibei was captured (1789) and a new fortress and port were built around it. On the orders of Catherine II the Great, after a proposal by the Academy of Sciences, the name was changed to Odessos after a nearby ancient Greek fishing village. Having founded the port and being a woman, Catherine decreed (1795) that the city should have a feminine name and so it was changed to Odessa.
Pereyaslav-Khmelnytskyy		Pereyaslav-Russky, Pereyaslav	Named after Bohdan Khmelnytskyy who signed the Pereyaslav Agreement (1654) whereby Ukraine accepted the Tsar's overlordship, effectively uniting the two countries. Khmelnytskyy added (1943).
Pervomaysk		Olviopol, Petromaryevka	Means 'First of May' after the international, originally communist, holiday. Petromaryevka means Peter and Mary.
Polesskoye		Kabany, Kaganovich	Named (1934–57) after Lazar Kaganovich (1893–1991), Bolshevik leader who supervised the policy of collectivization (1930s), whose birthplace this was.

Poltava	Oltava/Ltava	Lying along the River Vorskla, the name may be taken from the Slavonic *pal*, marsh. Site of the Battle of Poltava (1709) when Peter I the Great of Russia (1672–1725, r.1682–1725) annihilated the Swedes under Charles XII (1682–1718, r.1697–1718). This defeat ended Charles's attempt to destroy tsardom and brought to a close Sweden's status as a great power.
Radivilov	Chervonoarmeysk	Previously in Poland, named after the princely Polish-Lithuanian Radziwill family. Renamed after the 'Red Army' (1940–?).
Sevastopol	Chersonesos, Theodorichafen	Means 'City of glory' from the Greek *sebastos*, noble, and *polis*, and built just to the east of the ancient Greek colony of Chersonesos. After a steady decline, redeveloped by the Russians after their annexation of the Crimea (1783); a fortress and naval base were built and the new city renamed Sevastopol. Site of the Siege of Sevastopol (17 Oct 1854–11 Sep 1855) during the Crimean War. The unofficial German name only lasted while the Germans occupied the city (1942–44). The Russians claim now that when the Crimea was transferred to Ukraine (1954) Sevastopol was not included, having republic status within Russia. The Ukrainians dispute this, arguing that the Soviet Constitution did not mention Sevastopol's status as a federal city.
Simferopol	Neapolis, Ak-Mechet	On the site of the ancient Scythian capital of Neapolis, 'New City'. Renamed (1784) from the Tatar settlement named 'White Mosque' to 'Useful City' from the Greek *sumpheron* and *polis*.
Slavgorod	Slavhorod	Means 'Town of the Slavs'.
Soledar	Karlo-Libnekhtovsk	Previously named after Karl Liebknecht (1871–1919), one of the German founders of the Spartacus League, an underground group in Berlin that was the forerunner of the German Communist Party.

Place name	Alternative names	Description
Sverdlovsk		Founded (1938) and named after Yakov Sverdlov (1885–1919), a Bolshevik leader and the first editor of Pravda the Communist Party newspaper; *Pravda* means truth.
Svitlovodsk	Kremges, Khrushchev	Previously named after Nikita Khrushchev (1894–1971), Soviet Communist leader (1953–64) and first secretary of the Ukrainian Communist Party (1938–49).
Torez	Chistyakovo	Renamed (1964) after Maurice Thorez (1900–64), French communist leader.
Uzhgorod	Uzhhorod, Ungvár	Means 'Town on the River Uzh', although the Hungarian name means 'Fortress on the river'. In Austria-Hungary until passed to Czechoslovakia (1919), to Hungary (1938), to Czechoslovakia again (1945) and to the USSR (1945).
Volodymyr-Volynskyy	Vladimir-Volinskiy	Founded by and named after Volodymyr I the Great (c.956–1015), saint and first Christian ruler of Kievan Rus. Volynskyy (Volyn province) added to differentiate between it and the Vladimir in Russia. Came under Polish control (1347), but returned to Russia at the third partition of Poland (1795). In Polish hands again (1919–39) until seized by the USSR (1939) and annexed (1945).
Yalta	Dzhalita, Healita	Now a popular holiday resort on the southern Crimean coast, the present name is derived from the Polovtsian original itself derived from the ancient Greek *gialos*, shore. The Polovtsians were a Turkic-speaking tribe of the Kipchak confederation living north of the Black Sea (11th century). Yalta came under Russian control (late 18th century). Site of the Conference between Churchill, Roosevelt and Stalin (Feb 1945) (because Stalin refused to leave Soviet territory) to plan the final defeat of Germany and decide their respective spheres of influence; in particular, with regard to the

		countries of Eastern Europe (especially Poland) occupied by the Germans where Soviet military strength was already overwhelming. A secret protocol allowed the USSR to regain the territory lost to Japan in the 1904–05 war provided that it entered the war against Japan within two or three months of the defeat of Germany.
Yevpatoriya	Kirkinitida, Gezlev, Kozlov	Renamed after Mithridates VI Eupator the Great (?–63 BC), King of Pontus (120–63 BC). Captured by the Turks (14th century) who named it Gezlev. This became Kozlov when Russia annexed the Crimea (1783).
Zaporozhye	Zaporizhzhya Aleksandrovsk	Originally named after Alexander Golitsyn (1718–83), the commander of an army in the area at the time the town was founded (1770). Renamed (1921). Means 'Below the rapids' from *za porohi* and is so called because it was a place to which peasants, tied to the land in the service of a particular landlord, fled to escape the Polish authorities. Lies on the River Dnepr just below its former rapids which have been submerged. On Khortytsa Island the Cossacks (from the Turkic *kazak*, free man or outlaw) first formed their militaristic society in a *sich*, fortified camp, and hence some are known as Zaporozhian Cossacks.
Zholkva	Nesterov	Previously named after Peter Nesterov (1887–1914), a pioneer military pilot.
Zmiyëv	Gotvald	Renamed Gotvald (1976–91) after Klement Gottwald (1896–1953). Czechoslovak communist leader and president (1948–53).

Note: The spelling of place names has changed slightly since the Ukrainian form superseded the Russian in 1991, e.g. Lvov has become Lviv.

UNITED KINGDOM

Cut off from continental Europe by the Channel, Britain remained for long more or less inaccessible to migrants in large numbers and thus largely immune to external influences. Thus the British Isles have only experienced five major invasions by foreigners, the last being in 1066: the Celts, one of whose tribes was the Britons, in the first millennium BC, the Romans, the Anglo-Saxons, the Vikings and the Normans.

Apart from the short-lived invasions of Julius Caesar in 55 and 54 BC, the Romans did not conquer Britain until 43 AD. As Roman civilization was introduced it became the Roman province of Britannia, incorporating modern England and Wales; the conquest of Wales was complete by 78, but Scotland was never tamed, largely due to a shortage of manpower. A wall built by the Emperor Hadrian between 122 and 130 along the Tyne-Solway line became the northern frontier of Roman Britain (although the Wall of Antoninus briefly advanced it further north).

Roman rule ended in 410 and four decades later the Germanic tribes of Angles, Saxons and Jutes began to sweep over the eastern half of England. The Britons succumbed and by the end of the seventh century had been subsumed into the 'nation of the English' which comprised seven principal kingdoms, Northumbria, Mercia and Wessex being the strongest. There were times when these kingdoms acknowledged a single overlord.

Bands of marauding Vikings, principally Danes, began raiding England and Scotland in 787. During the next century raiding gave way to conquest, and the raiders began to settle. These incursions did not stop Kenneth I MacAlpin, King of the Scots, also becoming King of the Picts in 843 and embarking on the unification of Scotland. The Kingdom of Wessex offered the greatest opposition to the Vikings and in 878 they suffered a heavy defeat at the hands of Alfred the Great, King of Wessex. The ensuing Treaty of Wedmore divided England along a rough line that joined London and Chester. Alfred ruled south of this line and was recognized as overlord of the north, an area which came to be known as the Danelaw. Alfred's successors busied themselves with the reconquest of the Danelaw and were so successful that by 927 King Athelstan, having received the allegiance of the Kings of Wales and Scotland, was calling himself 'King of all Britain'.

Viking raids became more intense again in 980. The English response was disorganized and in 1013 they were forced to accept the Danish king, Sweyn, as king; in 1019 England became a part of the Danish Empire. This Danish dynasty, however, was short-lived. In 1042 Hardecanute was succeeded by his half-brother, the chaste half-Norman Edward the Confessor, son of the former English King Ethelred. In 1051, according to Norman claims, Edward promised the succession to Duke William of Normandy, but on his deathbed in 1066 Edward reneged and instead offered the throne to his brother-in-law who became king as Harold II.

Edward's duplicity and papal support encouraged William to launch an invasion of England in 1066. At the ensuing Battle of Hastings Harold was killed and William became king. The Anglo-Saxons thenceforth became subservient to the Normans, themselves originally Vikings, who proceeded to impose their form of feudal society on the country. The English language evolved as a mixture of the Norman French spoken by the nobility and the Anglo-Saxon spoken by the peasants. Offa's Dyke, the border with Wales, did not prevent Norman penetration, although complete control was not achieved.

Meanwhile, the Scots consolidated their kingdom and brought the British Kingdom of Strathclyde into the fold in 1034. Attempts to expand further into northern England failed, but Scotland remained an independent monarchy. The English King William II's conquest of Cumbria in 1092 did much to determine the Anglo-Scottish border from which followed two centuries of friendly relations.

Norman and Plantagenet kings proceeded to rule England for the next 300 years. In 1171 Henry II invaded Ireland, received the allegiance of the Irish kings and became Lord of Ireland. Welsh independence was ended with Edward I's conquest and Wales became a principality of England in 1284. When succession to the Scottish throne became uncertain in 1290 (13 claimants came forward), the English King Edward I saw his opportunity to intervene and take control of Scotland as a fief. In 1296 he marched north and quickly subdued the country. But the Scots fought back and in 1306 Robert I the Bruce was crowned King of Scotland. The following year Edward died. He was succeeded by his son, Edward II, who was decisively defeated at the Battle of Bannockburn in 1314. This did not end the fighting. Only in 1328 at the Treaty of Northampton did the English recognize Robert as king of an independent nation. Even before the turn of the century the Scots had begun to look to the continent for allies.

Shortly after Richard III became king in 1483, his popularity began to decline as a result of the rumours that he had killed his two young nephews. At this time Henry Tudor, Earl of Richmond and sole male claimant to the throne, was in exile in Brittany. Seeing his opportunity, he began to raise a small army to which Louis XI contributed some troops. Henry landed at Milford Haven in August 1485 and three weeks later he had defeated and killed Richard at the Battle of Bosworth Field. Seizing the throne as Henry VII, he founded the Tudor dynasty.

Despite continuing hostility, the decisive defeat of the Scots at the Battle of Flodden in 1513, and various plots to replace the Protestant Elizabeth I with Mary, Queen of Scots, which culminated in her execution in 1587, a personal union of the two kingdoms was effected in 1603. Elizabeth died that year without a direct heir and the throne passed to her Stuart cousin, James VI of Scotland (whose great-grandmother, a daughter of Henry VII, married James IV of Scotland) who became James I of England. The following year a union between the two countries was proposed; it was not realized, however, until 1654. Finally, in 1707, the Act of Union between the two kingdoms created, together with Wales, the Kingdom of Great Britain; the Scottish Parliament was dissolved.

English control of Ireland had weakened during the 15th and 16th centuries and was reduced to an area round Dublin. In an attempt to strengthen it Henry VIII declared himself 'King of Ireland' in 1534. A little over a century later, in 1649–51, Ireland was conquered by Oliver Cromwell. Three years later it united with England and Scotland.

Following the execution of Charles I in January 1649, the monarchy was abolished and Britain became a republic called the Commonwealth. It lasted only until 1660 when the monarchy was restored under Charles II, the son of Charles I.

Intending to restore Catholicism when he came to the throne in 1685, James II soon came into conflict with the Church and Parliament. This was exacerbated when a Catholic heir was born in 1688. Persecution of the Protestants, intrigues with the French and other ill-considered policies encouraged seven leading statesmen to invite the Protestant William of Orange (the Dutch royal house), whose grandfather was Charles I, to come to England "to rescue the nation and the religion". William arrived with an army and James fled to France. William's wife, Mary, was James's daughter. Once Parliament had declared that James was no longer king, William and Mary were crowned joint sovereigns the following year as William III and Mary II.

Matters now came to a head in Ireland as the Catholics gave their support to James while the Protestants supported William. In 1689 James invaded, setting up his base in Catholic Dublin as William concentrated his forces in Protestant Belfast. At the Battle of the Boyne the following year Catholic resistance was crushed and James returned to exile in France. From this and other victories emerged today's Orange Order to emphasize and perpetuate Protestant domination over the Catholics.

On 1 January 1801 the United Kingdom of Great Britain and Ireland officially came into existence. By this time many Lowland Scots Presbyterians had been introduced into the northern and eastern parts of Ireland. When their descendants refused to join the rest of Ireland in Catholic-dominated home rule, the state of Northern Ireland was created in 1920 from six of the nine Ulster counties. In 1922 the Irish Free State came into official existence and in 1937 it was renamed Eire.

William and Mary were childless and when the son of Mary's sister, the staunchly Protestant Anne, died in 1700, the succession became a serious problem. It was settled the next year by the Act of Settlement which decreed that Britain would never have a Catholic monarch after Anne's death and that the grandchildren of James I would succeed. Anne died in 1714 without a direct heir, despite enduring eighteen pregnancies. The next Protestant heir was the great-grandson of James I and son of Sophie, the Electress of Hanover. Georg Ludwig, unable to speak English and largely ignorant of British customs and traditions, became George I in 1714. In 1840 Queen Victoria, whose mother was a Saxe-Coburg, married Albert, Prince of Saxe-Coburg-Gotha, and so, when she died in 1901, the House of Hanover gave way to the House of Saxe-Coburg-Gotha. In view of anti-German feeling during the First World War, in 1917 George V changed the family name to Windsor.

The most recent attempt to impose foreign rule on Britain was during the Battle of Britain when Hitler unleashed the German Air Force to attack targets all over the British Isles between June 1940 and April 1941 as a prelude to a seaborne invasion.

In 1969 violent clashes between Catholics and Protestants broke out in Northern Ireland. Five years later home rule gave way to direct rule from London. In May 1998 a majority of voters in Northern Ireland backed the 'Good Friday' Peace Agreement on the sharing of power in the province in a referendum. Britain undertook to repeal the Government of Ireland Act 1920, while the Irish government agreed to amend articles of its constitution to remove its territorial claim on Northern Ireland; the province was to remain part of the UK as long as the population so desired. Fundamental disagreements between the Unionists and Sinn Féin delayed the transfer of political power from London to a power-sharing executive in Belfast until 2 December 1999. The failure to decommision any weapons, however, resulted in the suspension of devolved government and a return to direct rule from London after only ten weeks (11 February 2000).

In September 1997 referenda in Scotland and Wales resulted in positive votes for devolution and the creation of a Scottish Parliament and a Welsh Assembly. Both were duly established in July 1999.

Area:	244 104 sq km (94 249 sq miles)
England:	130 423 sq km (50 356 sq miles)
Wales:	20 766 sq km (8018 sq miles)
Scotland:	77 167 sq km (29 794 sq miles)
N.Ireland:	14 120 sq km (5452 sq miles)
Population:	59.45 million; English, Scottish, Irish, Welsh, West Indians, Bangladeshis, Indians, Pakistanis
Languages:	English, Scottish and Irish forms of Gaelic, Welsh, many ethnic
Religions:	Anglican, Roman Catholic, Muslim, Presbyterian, Methodist, Sikh, Hindu, Jewish
Capital:	London
Administrative Districts:	
England:	35 counties, 6 metropolitan counties, Greater London, 50 unitary districts
Wales:	22 unitary districts
Scotland:	29 unitary districts, 3 island authority areas
N.Ireland:	26 districts
Date of Independence:	1801 (creation)
Neighbouring Country:	Ireland

English-speaking Name	Local Name	Former Names	Notes
UNITED KINGDOM			Officially the United Kingdom of Great Britain and Northern Ireland. The name was adopted when Great Britain and Ireland were united (1 Jan 1801). When the union with Ireland was terminated (1922) Ireland was qualified by adding 'Northern'.
Britain		Albion, Britannia	Albion is the first known name for Britain, possibly taken from the Latin *albus*, white, a reference to the white cliffs that sailors saw as they approached Dover, the shortest route across the English Channel from the continent; or it may be taken from the Indo-European word *alb*, mountain. The Greeks referred to the inhabitants of the island as Prittanoi, meaning 'figured people', that is those who decorated their bodies. In time this became Britons and they gave their name to the country. Britain includes all offshore islands except for the Isle of Man and the Channel Islands. The epithet 'Perfidious Albion' was bestowed on Great Britain by Napoleon in recognition of its self-centered determination to maintain the balance of power in Europe.
Great Britain			The name adopted when England, already incorporating Wales within its realm, was united with Scotland (1 May 1707). The title was used earlier, informally, to distinguish the larger Britain from the smaller Brittany in France to which refugee Britons fled to escape Anglo-Saxon invaders.
England			The 'Land of the Angles', a Germanic tribe which invaded (5th century). The name England was first mentioned by the Venerable Bead (c.730). The seven Anglo-Saxon kingdoms, known as the Heptarchy, were those of the East Angles (East Anglia), East Saxons (Essex), Middle Saxons (Middlesex), West Saxons (Wessex), South Saxons (Sussex),

		Kent, Mercia (the border folk) and Northumbria (the land north of the River Humber).
Scotland	Caledonia	Named after the Scots, originally Celts who came (5th century) from northern Ireland (which was then called Scotia). Caledonia comes from the Caledonii, a tribe who lived in the far north. Joined in a personal union with England (1603), although separate kingdoms were maintained, when James VI of Scotland (1566–1625) became James I of England (r.1603–25).
Ulster	Ulaid	Means 'Place of the Ulaidh', the people who inhabited the area. The northernmost of the four traditional provinces of Ireland. Refusing to accept Home Rule, it became a separate political division of six counties of the United Kingdom under the Acts of 1920 and 1922. The name is commonly used as an alternative to Northern Ireland.
Wales	Cymru	A principality. The name is derived from *walh*, foreigner (plural, *walas*), which the Anglo-Saxons called the Britons here; they called themselves *Cymry*, compatriots. Roman conquest completed (78). Eastern border demarcated (8th century) by earthworks built by Offa, King of Mercia (757–96), and known as Offa's Dyke. Conquered (1301) by Edward I. Incorporated into the English realm under the Acts of 1536 and 1543.
Aberdeen	Devena	Means the 'Mouth of the River Don' from the Celtic *aber*, mouth, although the Roman name means the 'Town of two waters' since the city lies astride the Rivers Dee and Don.
Antrim	Aontroim	One of the original six counties of Ulster and a town meaning 'One house' from the Irish *aon*, one, and *treabh*, house. The name remained even though the area became populated.

Armagh	Ard Mhacha	One of the original six counties of Ulster and a city meaning the 'Height of the (mythical) goddess Macha', a reference to the fortress around which the city developed. Seat of the Kings of Ulster (c.400 BC-333 AD).
Barnsley		Means 'Beorn's woodland clearing'.
Bath	Aquae Calidae, Aquae Sulis, Akemanchester	Named originally after the complex of Roman baths and sacred hot springs and then dedicated to the local pagan goddess Sulis. The Anglo-Saxon Akemanchester enjoyed local popularity as 'aching man's place', a reasonable description of the place to which those suffering from rheumatism went. Gave its name to Bath (invalid) chairs which were invented (c.1750) by James Heath of Bath, Bath buns and Bath Oliver biscuits.
Bedford		Means 'Beda's ford'.
Belfast	Béal Feirste	Means 'Mouth or crossing of the sandbank' where the River Farset flows into the River Lagan, a point where it could be crossed at low tide. Capital of Northern Ireland (1920).
Benbecula		An island in the Outer Hebrides meaning 'Mountain of the fords'.
Birmingham	Beormingaham	Means a 'Settlement of Beornmund's (or Beorma's) people' from *ham*, settlement or homestead.
Blackburn		Means 'Dark stream'.
Blackpool	Pool	Named after a stretch of dark, peaty water about a kilometre from the sea.

Place		Old forms	Meaning
Bolton		Bothilton	Derived from the Old English *bothl*, building or dwelling place, and *tun*, here its original meaning of enclosure; thus a place where people lived as opposed to worked.
Bradford			Means 'Broad ford', a reference to a wide crossing place over a tributary of the River Aire.
Brighton		Brighthelmstone	Means 'Beorhthelm's settlement'.
Bristol			Means the 'Place with the bridge' from *stow*, place, and *brycy*, bridge, a reference to a stone bridge built at the place where the Lower Avon and the Frome used to join.
Buckinghamshire		Buccingahamm	A county named after the 'Land in a river bend belonging to Bucca's people' which spread from the settlement of Buckingham to the status of a county.
Bury St Edmunds		Beodriceswyrth, St Edmundsbury	Originally meant 'Beodric's enclosure'. Then renamed after St Edmund (841–70), a martyr and King of the East Angles, who is buried here. Bury comes from the Old English *burh*, fort or town.
Caerphilly	Caerffili		Means 'Ffili's fort' from the Welsh *caer*, fort. Gave its name to the famous cheese.
Cambridge		Grantacaestir, Grantanbrycg	Originally indicated a Roman camp, *ceaster* from the Latin *castra*[1], on the River Granta. Later it became 'Bridge over the Granta'.
Canterbury		Durovernum, Cant-wara-burg	The original Roman name meant 'Walled town by the alder marsh'. The present name is a corruption of Cant-wara-burg meaning the 'Town (fort) of the men of Kent' (Cantware), from *ware*, dwellers. The word canter comes from pilgrims

		going at a 'Canterbury pace' to visit St Thomas Becket's shrine.
Cardiff	Caerdydd	Means 'Fort on the River Taff' from *caer*, built by the Romans (c.75), and Dydd, the Welsh name for the Taff. Capital of Wales (1955).
Carlisle	Luguvalium	Means 'The fort of Luguvalos', the Celtic god Lugus. 'Lisle' is a corruption of the Roman name and was added to the Celtic *cair*, 'fortified town'.
Cheshire		A county with a name shortened from Chestershire, Chester being its capital.
Chester	Deva, Castra Legionum, Legacaester	Means simply a Roman 'Fort', with no geographical distinction although Deva is associated with the River Dee, itself meaning 'The goddess', on which it lies. Garrison of the 20th Roman Legion, hence its Roman names which meant 'Camp of the legions'.
Chichester	Noviomagus Regnensium	Means the 'Roman fort belonging to Cissa', a son of Aelle, the first King of the South Saxons. The garrison of the 2nd Roman Legion.
Colchester	Colonia Victricensis, Camuloctunum, Colneceaste	The original name mean the 'Colony of the Victorious' after the defeat of Queen Boudicca (61) who razed it. The second Roman name, Fortress of Camulus, is taken from the Celtic war god, Camulos. The Saxon name means 'Camp on the River Colne'.
Cornwall	Cornubia, Cornwalas	A county, the name for which is derived from its shape (in Latin *cornu*, horn) which became a tribal name, Cornovii, the 'People who live on a promontory'. *Walh*,

Place	Old name	Meaning
		foreigner, was added by the Anglo-Saxons because the language spoken by the natives was different from their own.
Coventry		Means 'Cofa's tree'.
Cumbria		A county formed (1974) from the counties of Cumberland, Westmoreland and parts of Lancashire. It is derived from *cymry*, compatriots – land of the Welsh – which the locals used to describe themselves. Westmoreland means 'Land of the Westmoringas' – people who lived west of the moor.
Darlington	Dearnington	Means the 'Settlement of Deornoth's people'.
Derby	Northworthy, Deoraby	Means 'Village where the deer were seen' and thus 'Deer village' from the Old Norse *djur*, deer, and *by*, village. The English name of Northworthy, 'North enclosure', was changed by the Danes.
Devizes		From the Old French *devise*, boundary, to indicate that the castle was built (12th century) on the boundary between two hundreds.
Devon	Defnum	A county named after the Dummonii, the 'Deep ones'.
Doncaster	Danum	Indicates a Roman fort on the River Don which itself means 'Rapidly flowing river'.
Dorset	Thornsaeta	A county meaning 'Settlers of Dorn' from Old English *saete*, dwellers.
Dover	Dubris, Dofras	Derived from a Celtic word *dubro*, the waters.

Down		One of the six counties of Ulster meaning 'Fort' from the Gaelic *dun*.
Dumbarton	Alcluith, Dumbretain	Originally meant the 'Hill by the Clyde' and then the 'Fort of the Britons'; it was once the capital of the Strathclyde Britons.
Dumfries		Derived from *dun* and *preas*, copse, to give a fortified position in woodland.
Dundee	Dunde	May mean 'Daig's fort' from the Gaelic *dun*, fort, and Daig, a personal name, or possibly 'Wonderful hill' from *dun teagh*.
Durham	Dunholm	Derived from the Old English *dun*, hill, and Old Norse *holmr*, island. Durham is on a hill overlooking the River Wear and island here really means firm ground surrounded by marsh. The Normans later exchanged the *n* for an *r*.
Dyfed		A county in Wales created (1974) from the former counties of Cardiganshire, Carmarthenshire and Pembrokeshire. Named after the Demetae.
East Anglia		A region and one of the seven Anglo-Saxon kingdoms of England. The first large-scale settlement named after the East Angles who had come from Schleswig in southern Denmark. Mainly comprised the northern and southern folk (Norfolk, Suffolk).
Edinburgh	Duneideann, Edwinesburh	May mean 'Fort on a ridge' from *burh*, fort. A popular misconception is that it means 'Edwin's fort', Edwin being King of Northumbria (613–32, r.616–32), that is, after the fort was built (c.500). The city was developed round the

Place	Name	Description
		castle built by King Malcolm III (c.1031–93) on a rocky ridge. Capital of Scotland (1437).
Ely	Elge	Means 'eel district' on account of the great number of eels caught in the Fens.
Exeter	Isca Dumnoniorum	Originally named by the Romans after the Dumnonii tribe. A garrison for Emperor Vespasian's (9–79, r.69–79) legion on the River Exe, the ancient name for which was Isca. Thus Exeter is a corruption of 'Execeaster'. Withstood two sieges by the Danes until taken (1003) and surrendered to William I the Conqueror (1027–87, r.1066–87) after an 18-day siege (1068).
Falkirk		Means 'Speckled church', that is, a church built with mottled stone from *faw*, multi-coloured. Edward I (1209–1307, r.1272–1307) defeated the Scots at the Battle of Falkirk (1298). At the Battle of Falkirk (1746) Charles Edward Stuart (1720–88), the Young Pretender to the British throne, achieved one last victory over the government's forces before being crushed three months later and forced to flee.
Fermanagh	Fear Manach	A district of N.Ireland. Means 'District of the Monaig' who may have been monks.
Gateshead	Ad Caprae Caput	Means a 'Headland or hill grazed by (wild) goats' from the Old English *gat*, goat, and *heafod*, headland.
Glasgow	Glas Ghu	Means 'Green hollow' or 'glen'.
Gloucester	Nervana, Glevum	Founded (96) by the Roman Emperor Marcus Nerva (c.32–98, r.96–98). The first syllable of the present name is

Place	Former name	Description
		derived from Glevum which itself is derived from the Celtic root word for 'bright'. Thus it meant a 'splendid camp'.
Grimsby		Means the 'Village of a person whose name is not known', the Old Norse *grims* meaning a masked person, and *by*, village.
Hampshire	Hantescire, Hamtonshire	Named after Hammtun, the old name for Southampton, with shire, an administrative division, added to indicate a county. The first, Domesday Book, name is the origin of the present abbreviated form of Hants.
Hereford		Means an 'Army ford' from the Old English *here*, army, possibly to indicate that it was wide enough for marching soldiers to cross without breaking ranks.
Hull	Kingston-Upon-Hull, Wyke	Named after the River Hull. Kingston refers to Edward I who acquired the port in exchange for land elsewhere and renamed it (1293).
Inverness		Derived from the Gaelic *inbhir*, mouth (of the river) and its name, here the Ness.
Ipswich	Gipeswic	The name may be derived from a personal name, Gip, and the Old English *wic*, trading place.
Kenilworth		Means 'Cynehild's settlement' with *worth* meaning an enclosed settlement.
Kent	Cantium	A county, named after the Cantii who probably took their name from the Celtic *cant*, edge or rim, to refer to land on the water's edge.

Kidderminster	Stour-in-Usmere	Means 'Cydela's monastery' after land was given (731) by Aethelbald, King of Mercia (r.716–57), for the purpose of building a monastery.
Kilmarnock		Means the 'Church of St Ernan' from the Gaelic *cill*, church, and a corrupted version of the saint's name, *mo Ernanoc*, 'my little Ernan'.
King's Lynn	Lynn Episcopi	Belonged to the Bishop of Norwich. Renamed (1537) by King Henry VIII (1491–1547, r.1509–47) to indicate that he had acquired a manor here from the bishop. Lynn means 'Pool', a reference to the mouth of the River Ouse on which the town lies.
Kingston Upon Thames	Cyningestun	A corruption of the original name which meant 'royal estate' or 'royal palace' since seven Anglo-Saxon kings (in Old English *cyning*, king) were crowned here.
Lancaster		Means '(Roman) Fort on the (River) Lune'.
Leicester	Ratae Coritanorum, Ligera Ceaster	The first part of the name may refer to the Leire, a tributary of the River Soar on which the city stands. Thus the name probably meant the 'Roman fort by the dwellers on the Leire'.
Lichfield	Letocetum	Originally meant 'Grey wood' to which the Celtic *feld*, open field, was added.
Lincoln	Lindum Colonia, Lincolia	Means 'The place by the pool', a reference to the marshes of the River Witham. As a place for retired legionaries Colonia was added (c.90). The present name is a combination of the first syllables of each word. A Roman garrison for the 9th Legion and later the 2nd Legion. Site of the Battle of Lincoln in which King Stephen (c.1097–1154,

r.1135–54) was defected by rebellious barons (1141).

Lisburn	Lios na Gcearrbhach	Means 'Fort of the gamblers'. Renamed (17th century) perhaps as a result of the arrival of French Huguenots.
Liverpool		Means a 'Pool full of weeds' or 'muddy pool' after a tidal creek, now disappeared, known as the Pool.
London	Lundenwic, Londinium	The meaning is still uncertain. A Celtic name, possibly formed from a personal name, Londinos. Or it may refer to a marshy place, possibly on account of the state of the banks of the River Thames. It may be derived from the Old Welsh *llong dyn*, ship hill. The Roman Londinium was the Latinized version of the Celtic Lundenwic which was the centre of a large settlement just upstream of the Roman city. Sacked by the Iceni (61). Population decimated in the Great Plague (1665) and the Great Fire (1666) laid waste to four-fifths of the City. The Treaty of London (1839) established the international status of Belgium and followed on from a British and French guarantee (1831) of Belgium's neutrality. This triggered Britain's participation in the First World War when the Germans began their attack on France by a flanking manoeuvre through Belgium. The Treaty of London (May 1913) ended the First Balkan War. Capital since c.1145.
Londonderry	Doire, Derry	Derry is derived from the Irish word *doire*, oak wood. Granted to the city of London (1613) for colonization, hence the present name. Resisted a 105-day siege (1689) by James II (1633–1701, r.1685–88), King of both England and Scotland (as James VII).
Luton	Lygetun	Means 'Farm on the (River) Lea'.

Manchester	Mamucium	Founded (c.80) as a Roman fort, hence 'chester'. Corrupted from Mamucium, 'Place on the Breastlike Hill'. The centre of the cotton industry, it was sometimes known as 'cottonopolis'.
Merthyr Tydfil		Means the 'Grave of the Martyr Tudful', a Welsh Christian princess who was killed and buried here (5th century).
Middlesbrough		Means 'the Middlemost town'.
Middlesex		A county named after the Middle Saxons, that is, those between the East Saxons (Essex) and the West Saxons (Wessex).
Milton Keynes		Designated a new town (1967), it takes the name of the original village meaning 'Middle town' and Lucas de Kaynes who held the manor (1221).
Newcastle	Newcastle Upon Tyne	Named after the 'new' Norman castle that was built (1080) on the site of a Roman fort at the east end of Hadrian's Wall by Robert Curthose (c.1054–1134), the eldest son of William I the Conqueror.
Norfolk		A county named after the 'Northern folk', the northern East Anglian tribes.
Northampton	Hamtun, Hampton	Originally 'Home farm' with 'North' added later to distinguish it from Southampton. Under the Treaty of Northampton (1328) Scottish independence was recognized by England. As a result of the capture of Henry VI (1421–71, r.1422–61 and 1470–71) at the Battle of Northampton (1460), during the Wars of the Roses, it was agreed that Henry should remain king, but that he should be succeeded, not by his own son,

Northumberland	but by Richard. Duke of York. He was killed (1460) before Henry was deposed.
	A county. As Northumbria it was one of the seven Anglo-Saxon kingdoms and meant the 'Land north of the River Humber'. At that time it extended as far north as the River Forth in Scotland. After the land north of the River Tweed was ceded to Scotland (1018) the name referred to land only in England.
Norwich	Means 'Northern port', a reference to a port on the River Wensum; north in comparison to the port of Ipswich. Sacked by the Danes (1004).
Nottingham	Means the 'Village of Snot's people', the S being dropped by the Normans.
Oxford	Means a 'Ford for oxen'. The Royalist headquarters (1642–45) during the Civil War; after a siege, surrendered to Parliamentary forces (1646). OXFAM, the Oxford Committee for Famine Relief, was founded here (1942).
Paisley	May be said to mean 'Church' from the Latin *basilica*, having grown from a village based on a Cluniac abbey founded here (1163). Gave its name to the patterned shawls made here.
Peterborough	After destruction of the old monastery by the Danes (9th century) the settlement was re-developed and simply called *burh*, the town. The cathedral, started (1118), was dedicated to St Peter whose name was also transferred to the town.
Perth	After its first church was dedicated to St John the Baptist Perth was also known as St John's Town. Perth is probably Celtic and means 'bush'. Capital of Scotland (until c. 1452).

Northwic

Snotengham

Oxnaford

Medeshampstead

St John's Town

Plymouth	Sudtone, Sutton	Means a 'Place at the mouth of the River Plym'. Sutton meant 'South farm'.
Pontefract		Means 'Broken bridge' from the Latin *Pons fractus*.
Rugby	Rockbury	Originally meant 'Hroca's fort'. *Bury*, or *burh*, was then changed by the Danes to *by*, village. The game of rugby football originated (1823) at Rugby School when one of the players picked up the ball and ran with it during a game of football.
St Albans	Verulamium	Originally a town built (1st century BC) on the west bank of the River Ver, it was renamed after St Alban, a Roman who had converted to Christianity and who was martyred here (c.303). Sacked by Boudicca, Queen of the Iceni (61). Site of two battles during the Wars of the Roses: in the first (1455) Henry VI was captured by Richard, Duke of York; the second (1461) resulted in a Lancastrian victory.
Salisbury	Sorviodunum, Searobryg, Sarisberia, New Sarum	The origins of the city lie 2 km north at Old Sarum, the Latin name for Salisbury, an Iron Age fort subsequently developed by the Romans, the Saxons and then the Normans. This was abandoned (13th century) and a cathedral was started on the new site. The meaning of *Sorvio* is unknown; *dunum* means 'fort'. This gave way to *bryg* and *burh*, all with the same meaning. In line with Norman custom, the *r* was changed to *l*.
Sheffield	Escafeld	Means an 'Open area (field) by the River Sheaf'.
Shrewsbury	Scrobbesbyrig	Means 'Fortified place in scrubland' from a word related to *scrybb*, shrub. Henry IV (1367–1413, r.1399–1413) routed Henry Percy, Earl of Northumberland, at the Battle of Shrewsbury (1403).

Name	Old form	Description
Shropshire	Scrobbesbyrigscir, Salop(escire)	A county whose name is based on that of Shrewsbury. Salop was used officially (1974–80) and is derived from the Norman inability to pronounce the *Scr* of Scrobbesbyrig; Scrob became Salop, the *Sc* becoming *Sa* and the *r* being changed to *i* (as Sarisberie became Salisbury).
Solihull		May mean 'Muddy hill' from *sylig*, boggy, and *hyll*.
Somerset	Sumortunsaete	Means 'Dwellers at Somerton' from *saete*, dwellers, with Somerton meaning 'Summer farm'.
Southampton	Clausentum, Hammtun, Hampton	Southern Hampton (as opposed to Northampton) meaning a 'Settlement on the promontory' or 'by the water'.
Stafford		Means 'Ford by a landing place' from *staep*, landing place, on the bank of the River Sow.
Stoke-on-Trent		Means 'Settlement on the (River) Trent'.
Suffolk		A county named after the 'Southern folk' of the East Anglian tribes. Gives its name to the Suffolk Punch, the smallest breed of draft horse which originated in Suffolk, and a breed of hornless sheep.
Sunderland		Means 'Separate land', that is, private land separated from the main estate.
Surrey		A county meaning 'Southern district' to describe those Saxons living just south of the Thames (whereas the Saxons living north of the river were the Middle Saxons).
Sussex		Now two counties, East Sussex and West Sussex, both named after the South Saxons.

Place	Alternative	Description
Swansea	Abertawe	Means 'Sveinn's place by the sea' from Sveinn, a Viking commander, and the Old Norse *sóer*, sea. Abertawe means the 'Mouth of the (River) Tawe'.
Swindon		Means a 'Small hill, *don*, where swine were to be found'.
Telford		Named (1968) as a new town after Thomas Telford, a Scottish civil engineer.
Warwick		Means the 'Settlement by a dam' on the River Avon from Old English *wering*, weir, and *wic*.
Wiltshire		Means the 'Shire belonging to the people of Wilton', that is, the settlers on the River Wylye.
Winchester	Venta Belgarum	Venta, corrupted to Win, may mean 'Loved place' and Belgarum 'of the Belgae tribe'. It then became a Roman fort. Capital of England (871-c.1145) until burned.
Wolverhampton	Heatun	Means 'Wulfrun's high farm', Wulfrun being the lady of the manor. The land here was higher than the surrounding countryside.
Worcester		Means the 'Roman fort of the Wigoran tribe'. Oliver Cromwell (1599–1658), the leader of the Parliamentary forces and later Lord Protector of England (1653–58), defeated Charles II (1630–85, r.1660–85) at the Battle of Worcester (1651) which brought an end to the Civil War.
York	Eboracum, Evorog, Eoforwic, Eorvik, Jorvik	Temporary capital of the Roman Empire when Emperor Vespasian was present. Garrison for the 9th Legion after its move from Lincoln (71) and the 6th Legion. The name then became corrupted to Evorog which, to the Anglo-

Saxons, sounded like *eofor*, wild boar. They added *wic*, trading place or farm. Captured by the Danes (867), they changed *wic* to *vik* and Eofor became Jor; the name finally being corrupted to York. York races include the famous Ebor Handicap.

Channel Islands

So named because of their location in the English Channel. Not part of the United Kingdom but dependencies of the British crown (since 1066). The three biggest islands are Jersey, Guernsey and Alderney. Their names are of Scandinavian origin with the Old Norse suffix *ey*, island. It is also claimed that Jersey is derived from 'Caesar's island'. Gave its name to the pullover and a breed of dairy cattle; Guernsey also has a breed of cattle.

Isle of Man   Ellan Vannan   Mona

A self-governing dominion of the British crown and not a part of the United Kingdom. The name could be derived from a Gaelic word for mountain or from *menagh*, middle (i.e. between Britain and Ireland). According to a legend, it is named after a wizard, Mannanan, who hid the Land of Mann in cloud whenever it was threatened by invaders.

1. Towns ending in -caster, -cester or -chester indicate that they were the site of a Roman camp or town; from the Latin *castra* which gave the Old English *ceaster*.

YUGOSLAVIA

The history of the lands and peoples of what was Yugoslavia until 1991 has been shaped largely by stronger foreign powers which have dominated the Balkans, often while en route from Europe to the East or from Asia to Europe. By the beginning of the second century the Romans had more or less overcome the resistance of the native peoples of the Balkan Peninsula and eclipsed Greek influence. In 395 the Roman Empire was split along a line which ran northwards along the River Drina up to the River Sava. In the sixth and seventh centuries various groups of Slavs, an Indo-European people originating from east-central Europe, began to strike south into the Balkans: the Slovenes settled in the north-west, the Croats further south and the Serbs to the east of the dividing line. The Serbs later adopted Eastern Orthodox Christianity and the Cyrillic alphabet.

Although Serb tribes were scattered over modern-day Bosnia-Hercegovina and Yugoslavia, the first Serbian tribal state of Zeta was formed during the late eighth century on the Adriatic coast. It was of little consequence and it was not until 1169 that the first true Serb state, known as Raška, was formed by Stephen I Nemanija who set out to unite the Serbs. By the time of his death in 1196 he had largely succeeded, incorporating Zeta and gaining independence from Byzantine domination. His son exploited the disintegration of the Byzantine and Bulgarian Empires, extending his realm southwards.

At the death of Uroš IV (Stephen the Great Dušan), Tsar of the Serbs, Greeks, Albanians and Bulgarians, in 1355, the Nemanjić dynasty was at the height of its power, ruling over modern Albania, much of Bosnia, Macedonia, northern Greece, Serbia and Zeta. The seat of the dynasty was Prizren in Kosovo.

Zeta now re-emerged as a distinct entity and in the 15th century became known as Montenegro. The Montenegrins are Serbs as well as Montenegrins in the same way as the English are British and English, and both Serbs and Montenegrins are also Yugoslavs. Serb power was not to last long after Dušan's death. After its decisive, if glorious, defeat at the Battle of Kosovo Polje in 1389, Serbia slowly succumbed to the onslaught of the Ottoman Turks. The conquest was finally completed with the fall of Smederevo in 1459.

As Ottoman rule spread Serb refugees fled westwards and northwards. The population of Kosovo, the Serb heartland, fell dramatically and to compensate the Turks resettled more non-Slav Muslim Albanians in the region. For the next 370 years the Serbs endured Turkish domination and brutal misrule, but they did not succumb to the extent that the Bulgarians did. Montenegro, on the other hand, resisted fiercely and prevented Turkish domination.

Modern Vojvodina, north of the River Danube, comprising elements of Baranja, Bačka and the Banat, was part of Hungary. As the Turks pushed northwards Serb refugees fled before them. The Hungarians encouraged them to settle in Vojvodina where they could play a part in the defence of Hungary. In 1526, however, the Hungarians were defeated at the Battle of Mohács and the king was killed. A new power, Austria, began to exercise influence in the region and, after the failure of the Ottoman siege of Vienna in 1683, gradually started to win back land from the Turks. By 1718 Vojvodina had been regained and the Turks had retreated south of the Danube. North Serbia, also regained that year at the Treaty of Passarowitz (now Požarevac), and Belgrade were lost again to the Turks at the 1739 Treaty of Belgrade. In 1799 the Turks gave up trying to subdue Montenegro and recognized its independence.

In 1804 the Serbs under Karageorge, Black George, rose against the Turkish Janissaries and defeated them. Karageorge was declared the supreme hereditary Serbian leader, thus founding the Karageorgević dynasty. The Turks returned, however, and by 1813 had restored Serbia to its previous status and caused Karageorge to flee. Another uprising began in 1815, this time led by Miloš Obrenović who founded a new dynasty. The Serbian principality finally gained full autonomy from the Ottoman government in 1830. In 1876 Serbia and Montenegro felt strong enough to declare war on Turkey and they were joined by Russia in 1877. The result was a defeat for the Turks. The Russian-imposed Treaty of San Stefano upset the balance of power in the Balkans and was superseded at the insistence of Austria-Hungary and Britain by the Congress of Berlin four months later in June/July 1878. The frontiers of the Balkan states were redrawn. Serbian independence and large territorial gains were acknowledged, but its ambitions to have an outlet to the Adriatic and a common border with Montenegro were thwarted. The *sandžak* of Novi Pazar between the two was occupied by Austria-Hungary and this prevented the existence of a Greater Serbia. Serbia itself became a kingdom, Prince Milan IV becoming king in 1882. The dynasty came to an end with his brutal murder in 1903 and the Karageorgević dynasty was restored. The size of Montenegro was doubled and its independence internationally recognized. In 1910 Nicholas I, who had become the ruler of Montenegro in 1860, proclaimed himself king.

Serb strength and self-confidence grew. Blocked by Austrian administration in Bosnia from 1908 and deprived of access to the sea, its aspirations turned towards Macedonia. The final outcome of the two Balkan Wars in 1912 and 1913 (q.v. Macedonia) resulted in Macedonia being liberated from the Turks and then divided between Bulgaria, Greece and Serbia. Serbia almost doubled in size, regaining not only the Macedonia it had possessed in the late 13th century, but also Kosovo, which represented roughly a third of the Albanian nation. To the Albanians this was tantamount to conquest by a foreign power, but their struggle against Serbianization was largely unsuccessful.

By sharing the *sandžak* of Novi Pazar Serbia acquired a common border with Montenegro. Besides these territorial acquisitions the victory over the Turks had other far-reaching implications: all the South Slavs in Austria-Hungary, Bosnia-Hercegovina, Dalmatia, Croatia, Vojvodina and Montenegro began to feel that their future lay with Serbia.

Austria-Hungary's declaration of war on Serbia in July 1914 following the assassination of the heir to the Habsburg throne began the First World War. The establishment of a unified South Slav state was thus delayed for another four years. In the meantime, in 1917, it was agreed that the Serbs, Croats and Slovenes would form a South Slav state under the Serbian monarchy after the war. Despite displaying great courage during the fighting, Serbian troops were unable to hold out against superior Austrian, Bulgarian and German forces and were overwhelmed. Nevertheless, in 1918, Serbia emerged on the winning side. In anticipation of the Treaty of Versailles (1919), the Kingdom of the Serbs, Croats and Slovenes was proclaimed on 1 December 1918. Besides Serbia, Croatia and

Slovenia, the kingdom included the previously independent Kingdom of Montenegro, Vojvodina and other small districts acquired from Hungary, some additional parts of Macedonia taken from Bulgaria, Austrian Dalmatia, and Bosnia-Hercegovina. Ethnically, Yugoslavia became home to Slovenes, Croats, Serbs, Bosniaks (Bosnian Muslims), Albanians, Hungarians, Bulgarians and Macedonians, besides small groups of Romanians, Slovaks, Vlachs and gypsies.

Deep mistrust between, on the one hand, the Croats and Slovenes, and on the other, between Croats and Serbs soon grew into resentment and then a constitutional crisis. The fiercely nationalist Croats demanded a Croat state of their own. Rather than consent, King Alexander suspended the constitution, declared a royal dictatorship and changed the name of the country to Yugoslavia (the land of the 'South Slavs') in January 1929. At the same time he abolished the regional names and reorganized the country into nine *banovine* (governorships): the boundaries were drawn so that the Serbs were a majority in six, the Croats in two, the Slovenes in one and the Muslims and others in none. To reduce regional patriotism the new names were geographical with most based on large rivers. Much of Serbia was subsumed into the Dunavska (Danube) and Moravska (Morava) *banovine* while the Zetska *banovina* swallowed up Montenegro.

On 6 April 1941 German, Italian, Bulgarian and Hungarian troops invaded Yugoslavia. Ten days later the country capitulated. Yugoslavia was then dismembered, the Germans taking northern Slovenia and most of Serbia, including the Banat, while the Bulgarians took some of the eastern regions of Serbia and south-eastern Macedonia. The Italians obtained southern Slovenia, much of the Adriatic coastline and Montenegro. Albania, a vassal of Italy, got most of Kosovo and western Macedonia. Hungary took back its pre-1919 territories of Baranja and Bačka. The *Ustaša*, a fascist terrorist movement, soon established the independent puppet state of Croatia which included Bosnia-Hercegovina. All these lands were returned to Yugoslavia during the latter half of 1944 and the first half of 1945 when Tito's Partisans, together with Russian troops, liberated the country.

The 1946 federal constitution of the new Yugoslavia abolished the monarchy and recognized Serbia and Montenegro as two of the six nominally autonomous republics; Macedonia, hitherto known as South Serbia, was another. Within Serbia were the two autonomous provinces of Kosovo with a majority non-Slav Albanian population and Vojvodina with a 20 per cent Hungarian minority. Only Slovenia of the six republics did not contain substantial ethnic minorities. When he died in 1980 having been in charge of Yugoslav affairs since 1944, President Tito left a dismal economic legacy and political paralysis. His death also encouraged some Kosovar Albanians to demand republican status. As a result of Serb repression unrest broke out and continued throughout the 1980s; at the same time, the Serbs, about ten per cent of the population of Kosovo, began to feel increasingly beleaguered.

In 1989 fear of secession by the Albanian majority in Kosovo caused Serbia to strip the province of its autonomy and dissolve its government; Vojvodina suffered the same fate. Elsewhere nationalism began to wax as fears grew of Serbian irredentism until, on 25 June 1991, Slovenia and Croatia declared their independence from Yugoslavia; they were followed by Macedonia (8 Sep 1991) and Bosnia-Hercegovina (3 Mar 1992). On 27 April 1992 President Milošević of Serbia declared the creation of a new state: the

Federal Republic of Yugoslavia, comprising Serbia and Montenegro. The latter enjoys a considerable degree of independence, but in the aftermath of the Kosovo crisis in 1999, it contemplated holding a referendum on formal independence. For many reasons Serbia would oppose any moves towards independence, not least because Montenegro has strategic value in providing it with its only outlet to the Adriatic.

Although ruling Kosovo like a colonial backwater, the Serbs have always regarded it as the heartland of Serbia despite the huge Albanian majority. The Albanians, on the other hand, consider it to be the centre of Albanian nationalism, their national movement having been founded in Prizren in 1878. While the Kosovar Albanians aspired to independence, Milošević's long-term strategy was to reverse the ethnic balance. He began to resettle refugee Serbs from other parts of the former Yugoslavia in the province. Passive Albanian resistance broke into active hostilities in 1998. In March 1999 peace negotiations at Rambouillet in France broke down, largely because Milošević probably thought that they would lead to the *de facto* loss of Kosovo and its subsequent independence. Instead, he sent police and troops into the province to unleash a campaign of terror and ethnic cleansing. This was done with great brutality, leaving NATO little option but to carry out its threat of air attack.

The 78-day air onslaught on Yugoslavia forced the withdrawal of Serbian security forces from Kosovo. As hundreds of thousands of Albanian refugees returned to their homes Serbs began to flee the province, thus increasing the Albanian majority to around 90 per cent. A NATO-led force, KFOR, which included a Russian contingent, was deployed to bring about peace and stability, and protect both Albanians and the remaining Serbs from revenge attacks. Although Belgrade retains sovereignty, it is academic; *de facto* control resides with the UN Mission in Kosovo (UNMIK) making the province a UN protectorate.

As a result of the Yugoslav wars in the past decade Milošević now controls considerably less territory but more Serbs, as up to a million fled from Bosnia, Croatia and Kosovo. Nevertheless, Hungarians remain in Vojvodina, Muslims in the *Sandžak* and Albanians in southern Serbia.

Area:	102 173 sq km (39 449 sq miles)
Population:	10.5 million; Serbs, Albanians, Montenegrins, Hungarians, Muslims, Gypsies, Croats
Languages:	Serbian, Albanian, Hungarian
Religions:	Serbian Orthodox, Muslim, Roman Catholic
Capital:	Belgrade
Administrative Districts:	2 republics (Serbia and Montenegro); Serbia has 2 provinces (Kosovo and Vojvodina)
Date of Independence:	27 April 1992
Neighbouring Countries:	Hungary, Romania, Bulgaria, Former Yugoslav Republic of Macedonia, Albania, Bosnia-Hercegovina, Croatia

Former Yugoslavia 1999

English-speaking name	Local Name	Former Names	Notes
YUGOSLAVIA	Jugoslavija	Kingdom of the Serbs, Croats and Slovenes	Federal Republic of Yugoslavia (Savezna Republika Jugoslavija) since April 1992. Socialist Federal Republic of Yugoslavia (Apr 1963), Federal People's Republic of Yugoslavia (Nov 1945), Kingdom of Yugoslavia (Jan 1929), Comprises Serbia and Montenegro. Yugoslavia means 'Southern Slavia' or the 'Land of the South Slavs'. Dardania was a name given to modern Kosovo, south Serbia and Macedonia (1st millennium BC).
Serbia	Srbija	Raška	Constituent republic. Named after the Serbs.
Aleksandrovac		Kožetin	Named after King Alexander Obrenović (1876–1903), the last of the Obrenović dynasty (1815–1903).
Arandjelovac		Vrbica	Named after a church dedicated to the Holy Archangel, *Sveti Arhandjel*, built on the orders of Prince Miloš (1859), who decreed the name change.
Bela Crkva		Weisskirchen, Fehertomplon	Means 'White Church'. Called Weisskirchen when it was the headquarters of an Austrian regiment serving in the 'Military Frontier'. The 'War of Liberation' against the German invaders first began here (7 Jul 1941).
Bela Palanka		Remesiana, Mokro, Izvor, Bunar-Baši, Su-Hazor, Musa-paša Palanka	Means 'White town'. Became Izvor, spring, (14th century). The next two Turkish names meant 'Great spring' and 'Source of water'. It was then renamed Musa Pasha's Town (17th century); variations of this name were Mustafa-pašina Palanka (18th century) and Ak-Palanka (19th century) from which the town's present name is derived.
Belgrade	Beograd	Singidunum, Beli Grad,	Means 'White Fortress' from *beo*, white, and *grad*, city; in Turkish, 'House of the Holy Wars'[1]. Originally named after

	Dar ul Jihad, Weissenburg	the Singi. Later, according to legend, named after the white bluff at the confluence of the Rivers Danube and Sava. Under Serbian rule for the first time (1284). Capital of Serbia (1404). Seized by the Turks (1521). Changed hands between the Turks and Austrians several times (17th and 18th centuries). At the Treaty of Belgrade (1739) the Austrians lost all the territory gained at the Treaty of Passarowitz (1718) except the Banat of Temesvár. Liberated by the Serbs (1806). Capital again (1839). Finally independent of the Turks (1867). Captured and lost by the Austrians (Dec 1914), taken again (Oct 1915) and liberated (Nov 1918). Became capital of the Kingdom of the Serbs, Croats and Slovenes (1918) and of Yugoslavia (1929). Captured by the Germans (Apr 1941) and liberated (Oct 1944).
Ćuprija	Horeum, Ravno	Means 'Bridge' in old Turkish. Acquired its present name (mid-17th century) when the Grand Vizier Mehmed-Pasha Ćuprilić built a bridge, *ćuprija*, across the River Morava on the site of the former Roman bridge.
Donji Milanovac		Named after Milan, the son of Prince Miloš Obrenović (1780–1860). Milan died within a month of becoming Prince of Serbia (1839). *Donji*, lower, was added (1859) to distinguish it from Gornji Milanovac.
Gornji Milanovac	Despotovica	First named after the river on which it stood. By decree (1859) of Prince Miloš changed to commemorate Miloš's brother Milan who was born here. *Gornji* means 'upper'.
Knjaževac	Gurgusovac	Original name thought to be after Grgur, the eldest son of Despot Djuradj Branković. Renamed (1859) on orders of Prince Miloš. Means 'Prince town' from *knez*, prince.
Kosovo	Dardania, Kosovo i Metojiha, Kosova[2]	A province, autonomous (1974–89). The western half is called Metojiha by the Serbs and was included in the

name (1946–71); it is derived from the Byzantine Greek *metochia*, land of the monasteries. Kosovo i (and) Metohija was sometimes shortened to Kosmet. Called Kosova by the Albanians. The Albanian version of Dardania (modern Kosovo, south Serbia and Macedonia) was Dardhë, pear.

Kosovo Polje Fushë e Kosovë

Means the 'Field of Blackbirds' from *kos*, blackbird, and *polje*, field. It was the site of the Battle of Kosovo (15 Jun 1389, new style 28 Jun) when the Turks decisively defeated the Serbs after which Serbia went into decline. It is some 4.5 km north-west of Priština. A second Battle of Kosovo (1448) ended in a Turkish victory over invading Hungarian and Wallachian forces which aimed to eject the Turks from the Balkans. Seven kilometres to the south-west of Priština is the town of Kosovo Polje. The first battle is also commemorated in the village of Lazarevo, named after the Serb leader, Lazar, who was killed during the battle.

Kosovska Mitrovica Mitrovicë Titova Mitrovica

The town in Kosovo Province named (1948–92) after President Tito (1892–1980), Partisan leader during the Second World War, Prime Minister (1945), President (1953).

Kragujevac Karajovja

Named after the vulture (in Serbo-Croat, *kraguj*), large numbers of which used to nest in the woods nearby. Capital of Serbia (1819–39).

Kraljevo Karanovac, Janok

Means 'King's town' from *kralj*, king. Out of respect the Slavs took the name of Charlemagne (742–814), the Holy Roman Emperor, as their word for 'king'; in Latin his name was *Carolus Magnus* and in German *Karl der Grosse*; the Russian for king is *korol*. Renamed (1882) to mark the proclamation of Serbia as a kingdom and the coronation of King Milan Obrenović (1854–89).

Kuršumlija	Ad Fines, Toplica, Bela Crkva	The Roman name gave way (1019) to Toplica, one of the two rivers on which the town lies, and then to Bela Crkva, White Church, after the appearance of the cathedral which dominates the town. The Turks coined the present name from *kuršum*, lead, after the lead roofs of the churches.
Lazarevac	Šopić	Named (1889) after Prince Lazar (?-1389).
Niš	Naissus	Derived from the River Nišava on which it stands. The Roman Emperor Claudius (214–270, r.268–70) defeated the Goths (269). A mass suicide took place when the Serbs blew up the powder magazine to prevent it falling into the hands of the besieging Turks (1809).
Novi Pazar	Yeni Pazar	Means 'New Bazaar' in both Serbian and the earlier Turkish. It lies in the *Sandžak* (of Novi Pazar), originally the name for a Turkish military district.
Novi Sad	Petrovaradinski Šanac, Neoplanta, Neustaz/Neue Stadt, Újvidék	The original name, referring to the massive Austrian fortress, meant 'Petrovaradin's trench'. Renamed (1748) to mean 'New Plantation' or 'Orchard', or more realistically, 'New Settlement', as the two most recent names, the last one Hungarian, indicate. The Austrians under Prince Eugène of Savoy routed the Turks here (1716).
Obilić	Gllaboder	Named after Miloš Kobilić, a Serb hero of the Battle of Kosovo (1389), who assassinated the Ottoman ruler, Sultan Murat I (1326?-89, r.1360–89).
Peć	Siperant, Ipek, Pejë	The name is derived from *pećine*, caves, nearby.
Požarevac	Margus, Viminacium,	The Treaty of Margus (c.435) between Attila, King of the Huns, and the Eastern Roman Empire forced the Romans

Prištma	Passarowitz	to double the tribute paid to Attila. The Treaty of Passarowitz (1718) was signed between Austria, the Ottoman Empire and Venice and resulted in the Ottoman loss of considerable territory in the Balkans.
Priština	Prishtinë	*Prišt* in Serbo-Croat means 'boil' and this may be a reference to the seething waters of the nearby River Gračanka. Came under control of the medieval Serbian state (late 12th century) and later became capital until the Turkish victory at the Battle of Kosovo Polje (1389). Capital of Kosovo.
Prizren	Përzeren Theranda, Prizdijana, Prizrenum	The present name is derived from the earlier Byzantine names which superseded the Roman name after the fall of the Roman Empire. The first Albanian nationalist movement, called the Albanian League or the League of Prizren, was formed here (1878).
Prokuplje	Hammeum, Komplos	Named after the martyred St Procopius (303).
Smederevo	Mons Aureus, Semenderia	Capital of Serbia (1427–1459) until captured by the Turks (1459), an event which marked their final victory over Serbia. Remained the seat of the Turkish administrator until he moved to Belgrade (1521).
Sremska Mitrovica	Sirmium Dimitrovica	Named after St Demetrius, martyred here. Sremska indicates that this Mitrovica is located in the Srem region. Birthplace of four Roman Emperors and one of the four capitals of the Roman Empire (Lower Pannonia).
Sremski Karlovci	Castrum Caror, Karloca, Karlowitz	Originally took its name from its earliest rulers, the de Caron family; now from another ruler, Duke Karl, and the region of Srem. At the Treaty of Karlowitz (1699) between Austria-Hungary, Poland, Venice and the Ottoman Empire,

the Turks ceded practically all Croatia and much of Hungary to the Habsburg Emperor, Podolia and Ukraine to Poland, and Morea to Venice.

Subotica	Sent Maria, Maria Tereziopolis Szabadka	Long part of Hungary, named after Maria Theresa (1717–1780, r.1740–80), Archduchess of Austria and Queen of Hungary.
Uroševac	Ferizaj / Ferizović	Named after King Stephen Uroš II (1282–1321), having been named after a Turkish governor (14th century).
Užice	Titovo Užice	The Serbian town named (1948–1992) after President Tito who set up his headquarters here (1941).
Vojvodina	Bácsko	A province, the name meaning 'Duchy' and named after the title of the ruler (in Serbo-Croat, Vojvoda, Duke). Used for the first time (1849) to encapsulate the districts of Bačka, Banat and a small part of Baranja when it was incorporated into Croatia-Slavonia. Became part of Hungary (1867) until inclusion into the Kingdom of the Serbs, Croats and Slovenes (1918).
Žiča		Means 'thread' or 'cord'. According to legend, Stephen II Nemanija (?-1228), tribal leader of Serbia, was led to the future site of the monastery by a golden thread.
Zrenjanin	Bečkerek, Petrovgrad	Named after Žarko Zrenjanin.
Montenegro	Crna Gora / Zeta	An independent state (1878–1918) when it passed to Yugoslavia. Now a constituent republic. Means 'Black Mountain' from the dark appearance of Mt Lovćen. Montenegro is the Venetian form of the Italian Monte Nero.

Name	Alternative names	Description
Bar	Antivaris, Antibarum	Built as a defence against the Avars, hence its name. Capital of Zeta (1042).
Cetinje	Cettigne	Named after the small River Cetina. Capital (1484–1918).
Crnojevića		Named after Ivan Crnojević, known as Ivan the Black, ruler of Zeta (15th century).
Danilovgrad		Means 'Danilo's town' and named after Danilo Nikola Petrović, *vladika* or prince-bishop of Montenegro (late 17th century).
Herceg-Novi	Novi, Sveti Stefan, Castelnuovo	Originally named (1382) 'New' by King Stephen I Tvrtko of Bosnia as he tried to build a new port on the north side of the Bay of Kotor. Then re-named St Stephen but this name was generally ignored and Novi still used. As a result of development by the Bosnian general, Duke (Herceg) Stephen Vukčić Kosača who established a factory (1448) and who took refuge in the port after the fall of Bosnia to the Turks (1463), the town came to be known as Herceg-Novi.
Ivangrad	Berane	Renamed (1949) as 'Ivan's town' after a leading Yugoslav revolutionary, Ivan Milutinović, who was killed (1944). Still also called Berane.
Kotor	Akourion/Dekatharon, Acrruvium/Catharum, Catharo/Cattaro	The present name stems from the Greek Dekatharon. At one time or another the town has been under Bosnian, English, French, Hungarian, Greek, Italian, Roman, Russian, Saracen, Tatar, Turkish and Venetian control.
Nikšić	Anagustum, Onogošt	Originally named after the Gothic leader Angasta. Present name taken from Nikša, founder of the Nikšić clan.

Podgorica	Birziminium, Ribnica, Titograd	Capital of Montenegro (1946). Named Ribnica (13th century) after the river; renamed Podgorica (1326) which means 'Under Mt Gorica'. The Montenegrin town named (1948–1992) after President Tito.
Tivat	Crni Plat	According to legend, the name may come from Queen Teuta (3rd century BC) who had a summer residence here; or Tivat (Teodo) is a corruption of the Greek word *theos*, god.
Ulcinj	Colchinium, Ulcinium, Elkinion, Dulcigno	Named after seafarers from Colchis on the Black Sea.

1. The Ottomans divided the world into two: *Dar ul Islam*, the House, or Realms, of Islam, and *Dar ul Harb*, the House of War. A continuous Holy War was to be waged against the non-Islamic world, the *Dar ul Harb*, not necessarily to destroy it, but to subdue it, so that *Dar ul Islam* would always expand.

2. Although the Albanians used their own place names in Kosovo, the conflict there in 1999 brought these names into greater prominence as the Serbs left the province in huge numbers. Thus the Albanian name is given under 'local names'.

SELECT BIBLIOGRAPHY

Bernard, Jack F. (1971). *Italy: an Historical Survey*. David & Charles, Newton Abbot.

Cameron, Kenneth (1996). *English Place Names*. B. T. Batsford Ltd, London.

Carr, William (1991). *The Origins of the Wars of German Unification*. Longman, London.

Chandler, David (1967). *The Campaigns of Napoleon*. Weidenfeld & Nicolson, London.

Clissold, Stephen *et al* (1966). *A Short History of Yugoslavia*. Cambridge University Press, Cambridge.

Crampton, R. J. (1997). *A Concise History of Bulgaria*. Cambridge University Press, Cambridge.

Crankshaw, Edward (1970). *The Fall of the House of Habsburg*. Sphere Books Ltd, London.

Cronin, Vincent (1973). *A Concise History of Italy*. American Heritage Publishing Co. Inc, London.

Cuddon, J. A. (1974). *The Companion Guide to Yugoslavia*. Collins, London.

Curran, Joseph M. (1980). *The Birth of the Irish Free State 1921–1923*. University of Alabama Press, Auburn.

Davies, Norman (1996). *Europe: A History*. Oxford University Press, Oxford.

Davies, Norman (1984). *Heart of Europe: A Short History of Poland*. Clarendon Press, Oxford.

Davis, R. H. C. (1970). *A History of Medieval Europe: From Constantine to Saint Louis*. Longman, Harlow.

Dawson, Christopher (1932). *The Making of Europe: 400–1000 AD: An Introduction to the History of European Unity*. Sheed & Ward, London.

Derry, T. K. (1973). *A History of Modern Norway 1814–1972*. Clarendon Press, Oxford.

Donnelly, Desmond (1965). *Struggle for the World: The Cold War from its Origins in 1917*. Collins, London.

Dudly Edwards, R. (1972). *A New History of Ireland*. Gill and Macmillan, Dublin.

Fejtö, François (1974). *A History of the People's Democracies: Eastern Europe since Stalin*. Penguin Books, Harmondsworth.

Foster, R. F. (Ed.) (1989). *The Oxford Illustrated History of Ireland*. Oxford University Press, Oxford.

Gallagher, Tom (1983). *Portugal: A 20th Century Interpretation*. Manchester University Press, Manchester.

Gippenreiter, Vadim and Komech, Alexei (1991). *Old Russian Cities*. Laurence King Ltd, London.

Gjerset, Knut (1925). *History of Iceland*. Macmillan Co., New York.

Gooch, John (1989). *The Unification of Italy*. Routledge, London.

Gunther, John. *Behind Europe's Curtain*. The Right Book Club, London.

Gunther, John (1936). *Inside Europe*. Hamish Hamilton, London.

Gunther, John (1962). *Inside Europe Today*. Hamish Hamilton, London.

Hallam, Elizabeth M. (1980). *Capetian France 987–1328*. Longman, London.

Halász, Zoltán (1980). *Hungary*. Corvina Kiadó, Budapest.

Hawgood, John (1955). *The Evolution of Germany*. Methuen & Co.Ltd, London.

Hiden, John and Salmon, Patrick (1991). *The Baltic Nations and Europe: Estonia, Latvia and Lithuania in the Twentieth Century*. Longman, London.

Hills, George (1970). *Spain*. Ernest Benn Ltd, London.

Home, Gordon (1960). *Cyprus, Then and Now*. J. M. Dent & Sons Ltd, London.

Jackson, J. Hampden (1941). *Estonia*. George Allen and Unwin Ltd, London.

James, Edward (1988). *The Franks*. Basil Blackwell Ltd, Oxford.

Jelavich, Barbara (1983). *History of the Balkans (Vol 2)*. Cambridge University Press, Cambridge.

Jones, W. Glyn (1970). *Denmark*. Ernest Benn Ltd, London.

Jørgensen, Bent (1995). *Stednavne Ordbog, 2nd edition*. Gyldendal, Denmark.

Kennedy, Paul (1989). *The Rise and Fall of the Great Powers*. Fontana Press, London.

Kissinger, Henry (1994). *Diplomacy*. Simon & Schuster, New York.

Kossman, E. H. (1978). *The Low Countries*. Oxford University Press, Oxford.

Krofta, Dr Kamil (1934). *A Short History of Czechoslovakia*. Robert McBride & Co., New York.

Kütt, Alexander and Vahter, Leonhard (1964). *Estonia*. The Assembly of Captive European Nations, New York.

Lalaguna, Juan (1990). *A Traveller's History of Spain*. The Windrush Press, Moreton-in-Marsh.

Lappo, G. M. (1997). *Geografiya Gorodov*. Gumarnitarniy Izdatelskii Tsentr, Moscow.

Latvia in 1939–1942. (1942). Press Bureau of the Latvian Delegation, Washington, D.C.

Lockhart, Sir Robert Bruce (1953). *What Happened to the Czechs?*. Batchworth Press, London.

Luck, J. Murray (1985). *History of Switzerland*. SPOSS Inc, Palo Alto.

MacKay, Angus (1977). *Spain in the Middle Ages: From Frontier to Empire, 1000–1500*. Macmillan Education Ltd, Basingstoke.

Malcolm, Noel (1994). *Bosnia: A Short History*. Macmillan Publishers Ltd, London.

Mallinson, Vernon (1969). *Belgium*. Ernest Benn Ltd, London.

Mamatey, Victor S. (1971). *Rise of the Habsburg Empire 1526–1815*. Holt, Reinhart and Winston Inc, New York.

Mann, Golo (1990). *The History of Germany since 1789*. Penguin Books, London.

Maurois, André (1956). *A History of France*. Jonathan Cape, London.

McKitterick, Rosamond (1983). *The Frankish Kingdoms under the Carolingians 751–987*. Longmans, London.

Newton, Gerald (1978). *The Netherlands: An Historical and Cultural Survey 1795–1977*. Ernest Benn Ltd, London.

Palmer, Alan (1970). *The Lands Between*. Weidenfeld & Nicolson, London.

Panteli, Stavros (1984). *A New History of Cyprus: From the Earliest Times to the Present Day*. East-West Publications, London.

Pollo, Stefanaq and Puto, Arben (1981). *The History of Albania from Its Origins to the Present Day*. Routledge & Kegan Paul Ltd, London.

Rauch, Georg von (1974). *The Baltic States: The Years of Independence 1917–40*. C. Hurst & Co., London

Reuter, Timothy (1991). *Germany in the Early Middle Ages 800–1056*. Longman, London.

Riasanovsky, Nicholas V. (1993). *A History of Russia*. Oxford University Press, Oxford.

Rupnik, Jacques (1988). *The Other Europe*. Weidenfeld and Nicolson, London.

Syrop, Konrad (1982). *Poland in Perspective*. Robert Hale, London.

Stewart, George (1975). *Names on the Globe*. Oxford University Press, Oxford.

Taagepera, Rein (1993). *Estonia's Return to Independence*. Westview Press, Oxford.

Vaitiekunas, Vytautas (1965). *Lithuania*. The Assembly of Captive European Nations, New York.

Ward, Philip (1983). *Albania: A Travel Guide*. Oleander Press, Cambridge.

Ward, Philip (1982). *Bulgaria: A Travel Guide*. Oleander Press, Cambridge.

Westwood, J. N. (1993). *Endurance and Endeavour: Russian History 1812–1992, 4th edition*. Oxford University Press, Oxford.

Woodhouse, C. M (1984). *Modern Greece: A Short History*. Faber and Faber Ltd, London.

Wuorinen, John H. (1965). *A History of Finland*. Columbia University Press, New York & London.

Zaprudnik, Jan (1993). *Belarus: At a Crossroads in History*. Westview Press, Oxford.

Zeman Zbynek (1990). *The Masaryks: The Making of Czechoslovakia*. I. B. Taurus & Co. Ltd, London.

REFERENCE BOOKS:

David Crystal (Ed.) (1998). *The Cambridge Biographical Encyclopedia, 2nd edition*. Cambridge University Press, Cambridge.

Channon, John (1995). *Historical Atlas of Russia*. Penguin Books Ltd, London.

Dalby, Andrew (1998). *Dictionary of Languages*. Bloomsbury Publishing plc, London.

Ekwall, Eilert (1960). *The Oxford Dictionary of English Placenames*. Clarendon Press, Oxford.

Encyclopaedia Britannica (1995). Encyclopaedia Britannica Inc, London.

Farmer, David Hugh (1978). *The Oxford Dictionary of Saints*. Clarendon Press, Oxford.

Gilbert, Martin (1993). *Atlas of Russian History*. J. M. Dent, London.

Great Soviet Encyclopaedia (1974). Soviet Encyclopaedia Publishing House, Moscow.

Hupchick, Dennis P. and Cox, Harold E. (1996). *A Concise Historical Atlas of Eastern Europe*. St Martin's Press, New York.

International Institute for Strategic Studies. *Strategic Survey*. Oxford University Press, Oxford.

Iordan, Iorgu (1963). *Toponimia Romaneasca*. Editura Academiei Republicii Populare Romane, Bucharest.

Mangulis, Visvaldis (1983). *Latvia in the Wars of the 20th Century*. Cognition Books, Princeton Junction, New Jersey.

Marcato, Carla *et al* (Eds.) (1994). *Dizionario dei Nomi Geografici Italiani*. UTET, Torino.

Munro, David (1995). *The Oxford Dictionary of the World*. Oxford University Press, Oxford.

National Geographic Atlas of the World (1996). National Geographic Society, Washington, D.C.

Nicolaisen, W. F. X. *et al* (1970). *The Names of Towns and Cities in Britain*. B. T. Batsford Ltd, London.

Riley-Smith, Jonathan (Ed.) (1991). *The Atlas of the Crusades*. Guild Publishing, London.

Regional Surveys of the World (1997). *East Europe and CIS, 2nd edition*. Europa Publications Ltd, London.

Room, Adrian (1992). *Brewer's Dictionary of Names: People & Places & Things*. Helicon Publishing Ltd, Oxford.

Speake, Graham (Ed.) (1994). *Penguin Dictionary of Ancient History*. Blackwell Publishers, London.

Steinberg, S. H. (1979). *Historical Tables 58 BC–AD 1978*. Macmillan Press Ltd, London.

Turner, Barry (Ed.) (1999). *The Statesman's Yearbook 2000 (136th edition)*. Macmillan Reference Ltd, London.

The Times Atlas of European History. (1994). Times Books, London.

Treasures of Yugoslavia (1980). Yugoslaviapublic, Beograd.

Voennyi Entsiklopedicheskii Slovar (1983). Voennoe Izdatelstvo, Moscow.

Xenopol, A. D. *Istoria Romanilor din Dacia Traiana* (Vols 1–3). Editura Cartea Romaneasca, Bucharest.

INDEX OF PRESENT, LOCAL AND FORMER PLACE NAMES

Note: Some of these names appear in the historical profiles as well as in the tables.

Friesland 109, 209
Friuli-Venezia 165
Funchal 238
Fünfkirchen 143
Furmanov 262
Fushë e Kosovë 373

Gablonz an der Neisse 64
Gades 304
Gadir 297, 304
Gaditana 304
Gagarin ix, 262
Gaillimh 154
Galicia 6, 220, 221, 297, 298, 306, 327, 328
Gallaecia 237, 306
Gallia Transalpina 105
Galway 154
Gand 23
Garestine 49
Gascony 92, 101
Gasteiz 311
Gatchina 262–3
Gateshead 352
Gaul 96
Gazimağusa 55
Gdańsk 221, 227, 231
Gdynia 227
Gelderland 209
Gelendzhik 263
Genabum 104
Genava 324
Geneva 322, 324
Genève 324
Genf 324
Genoa 160, 165
Genova 165
Gent 23
Genua 165
Georgi Traikov 38
Georgiu-Dezh 269
Germany 6, 7, 8, 14, 21, 35, 59, 61, 65, 67, 68,
  70, 78, 91, 94, 96, 102, 103, *109–26*, 139,
  158, 170, 175, 176, 183, 191, 203, 206, 216,
  219, 221, 222, 223, 242, 252, 253–4, 285,
  292, 299, 314, 322, 339
Gesoriacum 98
Gezlev 339
Gharnatah 306

Ghent 23
Gheorge Gheorghiu-Dej 246
Gibraltar xi, xiv, 52, 297, 299, 300
Giola 107
Gipeswic 353
Girgenti 173
Girne 55
Giulia 165
Gjirokastër 4
Glas Ghu 352
Glasgow 352
Gleiwitz 227
Glevum 352–3
Gliwice 227
Gllaboder 374
Gloucester 352–3
Gnesen 227
Gniezno 227, 229
Godhavn 75
Godthåb 74
Golbshtadt 263
Golyama Kutlovitsa 38
Gomel 13, 17
Gomy 17
Gorgonzola 166
Gorkiy 271
Gorna Dzhumaja 37
Gornji Milanovac 372
Gorno-Altaysk 263
Göteborg 317
Gothenburg 317
Gotland 72, 317, 331–2
Göttingen 121
Gottwaldov 66
Gotvald 339
Gradac 50, 295
Gran 142
Granada 298, 306
Grantacaestir 348
Grantanbrycg x, 348
Gratianopolis 101
Graubünden 321, 324
Graz 9
Great Britain 6, 128, 170, 343, 345
Greece 1, 2, 34, 35, 51, *127–36*, 161, 193,
  194, 195, 365, 366
Greenland 69, 74, 213, 214
Grelibre 101

Semenderia 375
Semenovka 259
Semigallia 179
Sena Gallica 170
Sena Julia 170
Senigallia 170
Sent Maria 376
Septem Fratres 305
Serbia 1, 25, 33, 34, 114, 129, 193, 194, 196, 197, 365, 366, 367, 368, 371, 372, 373
Serdica 40
Sereda 262
Sergiyev 276
Sergiyev Posad 276
Sergo 333
Serov 277
Setúbal 239
Sevastopol 337
Severnaya Osetiya 271
Severodvinsk 277
Sevilla 310
Seville 298, 303, 310
Shakhty 277
Shamilkala 277
Shantariya 239
Sharypovo 277
Shcheglovsk 265
Shcherbakov 275
Sheffield 358
Shkodër 1, 4
Shlisselburg 277
Shqipëria 3
Shrewsbury 358
Shropshire 359
Shumen 39
Šiauliai 187
Šibenik 48
Siberia 251, 260, 268, 278
Sibir 278
Sibiu 247, 248
Sicilia 172–3
Sicily 157, 158, 159, 160, 161, 172–3, 297, 298, 299
Siebenbürgen 248
Siena 170
Sighişoara 247, 248
Siglufjördur 147

Silesia 6, 58, 59, 109, 111, 120, 219, 221, 222, 229
Simbirsk 278
Simeonovgrad 39
Simferopol 337
Sinbirsk 278
Singidunum 371–2
Sinigaglia 170
Siperant 374
Siracusa 173
Sirmium 375
Sisak 49
Sisapo 302
Siscia 49
Sisimiut 75
Skampis 3
Skaptopara 37
Skopje 193, 195, 196–7
Skoplje 196–7
Skupi 196–7
Śląsk 229
Slavgorod 18, 337
Slavhorod 337
Slavkov x, 66
Slavonia 6, 43, 49, 137
Slavonski Brod ix, 49
Slavyansk 336
Slawharad 18
Sligeach 155
Sligo 155
Sliven 39
Slovakia ix, 8, 57, 59, 61, 138, 139, 140, 223, 285–9, 329
Slovenia 7, 8, 45, 49, 140, 161, 291–5, 367
Slovenija 294
Slovenská 288
Smederevo 365, 375
Smolensk 278
Smolyan 40
Snæland 147
Snotengaham 357
Sofia 35, 40
Sofiya 40
Soissons 106
Soledar 337
Soli 31
Solihull 359